Condensed-Matter and Materials Physics

Basic Research for Tomorrow's Technology

Committee on Condensed-Matter and Materials Physics
Board on Physics and Astronomy

Commission on Physical Sciences, Mathematics, and Applications

National Research Council

NATIONAL ACADEMY PRESS
Washington, D.C. 1999

This project was supported by the Department of Commerce under Contract No. 50SBNB5C8819, the Department of Energy under Contract No. DE-FG02-96-ER45613, and the National Science Foundation under Grant No. DMR-9632837. Any opinions, findings, conclusions, or recommendations expressed in this publication are those of the author(s) and do not necessarily reflect the views of the organizations or agencies that provided support for the project.

International Standard Book Number 0-309-06349-3
Library of Congress Catalog Card Number 99-62179

Front cover: A scanning-tunneling microscope image that shows the wave nature of electrons confined in a "quantum corral" of 48 individually positioned atoms. See page 233. (Courtesy of IBM Research.)

Additional copies of this report are available from National Academy Press, 2101 Constitution Avenue, N.W., Lockbox 285, Washington, D.C. 20055; (800) 624-6242 or (202) 334-3313 (in the Washington metropolitan area); Internet, http://www.nap.edu; and

Board on Physics and Astronomy, National Research Council, HA 562, 2101 Constitution Avenue, N.W., Washington, DC 20418

Printed in the United States of America

COMMITTEE ON CONDENSED-MATTER AND MATERIALS PHYSICS

Preface

In the spring of 1996, the National Research Council's Board on Physics and Astronomy established the Committee on Condensed-Matter and Materials Physics to prepare a scholarly assessment of the field as part of the new survey of physics, *Physics in a New Era,* that is now in progress. This assessment has five objectives.

1. Identify future opportunities and priorities in the field.
2. Articulate the fundamental scientific challenges in the field.
3. Assess related infrastructure, institutional, resource, and educational issues.
4. Provide evidence of the societal impact of the field.
5. Provide a forum for coordinated community-wide communications with federal agencies, policy makers, and the public.

The committee was composed of individuals whose backgrounds reflect the diversity of the field and its close connections with related branches of science, including chemistry, biology, and engineering. The field spans research environments from principal investigators carrying out benchtop studies in universities to large collaborations carrying out experiments at major national facilities. It also spans the forefronts of many-body theory, the behavior of complex materials and fluids, and the design of semiconductor devices and circuits. Condensed-matter and materials physics research is carried out in various institutional settings, including university, government, and industrial research laboratories.

In the course of the study, the committee held two workshops on research frontiers and policy issues. These workshops brought together leading researchers in the field as well as leading policy makers from government, industry, and universities. The committee met several times to plan its work, debate the issues, and formulate its report. An early output of the study was the report *The Physics of Materials: How Science Improves Our Lives*, a short, colorful, and easy-to-read pamphlet illustrating how research in the field affects our daily lives. The committee generated several progress reports and held public forums at materials-related meetings of the American Physical Society and the Materials Research Society. The committee also sought input from the general science and engineering communities. We are particularly grateful to our colleagues in biology, chemistry, and materials and electrical engineering for their support and help in carrying out this study.

The committee would like to thank Donald C. Shapero, Daniel F. Morgan, and Kevin D. Aylesworth from the Board on Physics and Astronomy for their efforts throughout the course of this study. Special thanks also to Arthur Bienenstock, who served on the committee until the fall of 1997, when he assumed responsibilities at the Office of Science and Technology Policy. The committee gratefully acknowledges the contributions of the following individuals who provided material or particular advice that influenced its study: David Abraham, Eric J. Amis, Bill Appleton, Meigan Aronson, David Aspnes, John Axe, Arthur P. Baddorf, Samuel Bader, A. Balazs, N. Balsara, Troy Barbee, F. Bates, Bertram Batlogg, Robert Behringer, Jerzy Bernholc, Arthur Bienenstock, Jörg Bilgram, Howard Birnbaum, Stephen G. Bishop, Steve Block, Lynn A. Boatner, Eberhardt Bodenschatz, Greg Boebinger, William Boettinger, Bill Brinkman, R. Bubeck, David Cannell, Federico Capasso, G. Slade Cargill, John Carruthers, Robert Cava, Robert Celotta, David Ceperley, Paul Chaikin, Albert Chang, S.S. (Leroy) Chang, Eric Chason, Daniel Chemla, Shiyi Chen, S. Cheng, B. Chmelka, Alfred Cho, John R. Clem, Daniel Colbert, Piers Coleman, George Crabtree, George Craford, Harold Craighead, Roman Czujko, Elbio Dagatto, Adriaan de Graaf, Satyen Deb, Patricia Dehmer, Cees Dekker, David DiVincenzo, Russ Donnelly, Robert Doremus, J. Douglas, Mildred S. Dresselhaus, Bob Dunlap, J. Dutcher, Bob Dynes, Robert Eisenstein, Chang-Beom Eom, Evan Evans, Ferydoon Family, Matthew P.A. Fisher, Zachary Fisk, Paul Fleury, Mike Fluss, Judy Franz, Jean Fréchet, Glenn Fredrickson, Hellmut Fritsche, William Gallagher, E. Giannelis, Allen M. Goldman, Jerry Gollub, Matt Grayson, P. Green, G. Grest, Peter Grüter, Richard Hake, Thomas Halsey, Donald Hamann, Christopher Hanna, Bill Harris, Beverly Hartline, Kristl Hathaway, Lance Haworth, Frances Hellman, George Hentschel, Jan Herbst, Pierre Hohenberg, Susan Houde-Walter, Evelyn Hu, Robert Hull, David Huse, Eric Isaacs, Nikos Jaeger, Adam B. Jaffe, Sungho Jin, David Johnson, James Jorgensen, Malvin H. Kalos, A. Karin, Marc Kastner, Efthimios Kaxiras, Jeffrey Koberstein, Carl C. Koch, Kei Koizumi, J. Kornfield, Mark Kryder, Max Lagally, David V. Lang, Robert Laudise, G. Leal, Manfred Leiser, Ross Lemons, Joseph Levitzky, Peter Levy, David Litster, T. Lodge, Gabrielle Long, Steven Louie, Michael

Lowenberg, Tom Lubensky, C. Macosko, Richard Martin, Denis McWhan, Jim Meindl, Jim Merz, Burkhard Militzer, Andrew Millis, S. Milner, David Moncton, Jagadeesh Moodera, Donald Murphy, M. Muthukumar, Sidney Nagel, Al Narath, David Nelson, Jeff Nelson, Robert J. Nemanich, Robert Newnham, K.L. Ngai, William Oosterhuis, Stuart Parkin, Seevak Parpia, Michelle Parrinello, Kumar Patel, Eva Pebay-Peyroula, Stephen J. Pennycook, V. Percec, Pierre Petroff, Tom Picraux, Gary Prinz, Itamar Procaccia, Peter Pusey, R. Ramesh, R. Register, James Rice, Kevin Robbie, Mark Robbins, Jack Rowe, Michael Rowe, John M. Rowell, M. Rubinstein, Jack Rush, Robert Schafrik, Hans Scheel, Sheldon Schultz, Lyle Schwartz, Pabitra Sen, James Sethna, Don Shaw, K. Shull, Jerry Simmons, John Slonczewski, James Speck, Gene Stanley, Galen Stucky, Harold Swinney, Bruce Taggart, Andrew Taylor, Philip Taylor, Zlatko Tešanovi, Iran Thomas, Carl V. Thompson, David Tirrell, Matt Tirrell, Robert Trew, Ruud Tromp, Jeffrey Tsao, Dan Tsui, David Turnbull, Paul Umbanhowar, Priya Vashishta, Stephan von Molnar, Jim Voytuk, James Warren, John Weaver, Eicke Weber, Tom Weber, David Weitz, Steven White, Hollis Wickman, John Wilkins, Ellen D. Williams, Stan Williams, T. Witten, Horst Wittmann, Victor Yakhot, Sidney Yip, Andrew Zangwill, Richard Zare, Z. Zhang, and Thomas Zipperian. The committee also thanks Janet Overton, who edited the final production draft of the report.

The committee's work was supported by grants from the U.S. Department of Commerce, the U.S. Department of Energy, and the National Science Foundation. The committee thanks them for their support.

Acknowledgment of Reviewers

This report has been reviewed by individuals chosen for their diverse perspectives and technical expertise, in accordance with procedures approved by the National Research Council's (NRC's) Report Review Committee. The purpose of this independent review is to provide candid and critical comments that will assist the authors and the NRC in making the published report as sound as possible and to ensure that the report meets institutional standards for objectivity, evidence, and responsiveness to the study charge. The contents of the review comments and the draft manuscript remain confidential to protect the integrity of the deliberative process. We wish to thank the following individuals for their participation in the review of this report:

Phillip W. Anderson, Princeton University,
Steven Chu, Stanford University,
Esther Conwell, University of Rochester,
Robert Dynes, University of California, San Diego,
Val Fitch, Princeton University,
Paul Fleury, University of New Mexico,
Jerry P. Gollub, Haverford College,
David Moncton, Argonne National Laboratory,
Thomas Russell, University of Massachusetts, Amherst, and
Thomas Theis, IBM T.J. Watson Research Center.

Although the individuals listed above have provided many constructive comments and suggestions, the responsibility for the final content of this report rests solely with the authoring committee and the NRC.

Contents

Condensed-Matter and Materials Physics

Executive Summary

Condensed-matter and materials physics plays a central role in many of the scientific and technological advances that have changed our lives so dramatically in the last 50 years. Condensed-matter and materials physics gave birth to the transistor, the integrated circuit, the laser, and low-loss optical fibers so important to the modern computer and communications industries. The years ahead promise equally dramatic advances, making this an era of great scientific excitement for research in the field. Communicating this excitement and ensuring further progress are the main goals of this report.

In the decade since the last major assessment of the field, important results and discoveries have come rapidly and from unexpected directions. These results and discoveries have made possible advances that range from new experimental tools for atomic-scale manipulation and visualization, to the creation of new synthetic materials (such as buckyballs and high-temperature superconductors), to new physical phenomena such as giant magnetoresistance and the fractional quantum Hall effect. An enormous increase in computing power has yielded qualitative changes in visualization and simulation of complex phenomena in large-scale many-atom systems. Progress in synthesis, visualization, manipulation, and computation will continue to have an impact on many areas of research spanning different length scales from atomic to macroscopic. Strong impact may also be expected in "soft" condensed-matter physics, particularly where it interfaces with biology and chemistry.

The priorities of society are shifting from military security to economic well-being and health. Changing societal priorities, in turn, create shifting demands on condensed-matter and materials physics. Among these demands are an improved public understanding of science, better education of scientists and engineers for today's employment marketplace, and new contributions to the nation's industrial competitiveness.

There are four key challenges facing condensed-matter and materials physics:

- The intellectual vitality of the field must be nurtured, particularly by facilitating the research of individual investigators and small teams in areas that cross disciplinary boundaries.
- A state-of-the-art facilities infrastructure is essential for competitive research; such an infrastructure requires the creation of laboratory-scale microcharacterization facilities at universities and large-scale facilities at national laboratories.
- Efforts must be enhanced in research universities to improve integration of condensed-matter and materials physics education and research, particularly at the boundaries of disciplines, and to prepare flexible and adaptable physicists for the future.
- New modes of cooperation among universities, colleges, government laboratories, and industry need to be developed that will ensure the connection between the field and the needs of society and to ensure preservation of the fertile innovative climate of major industrial laboratories that have played a dominant role in condensed-matter and materials physics research.

In this report the committee makes a number of recommendations for steps to be taken to meet these challenges. They are outlined here and discussed more extensively in the Overview and in further detail in each of the chapters.

For the overall research effort to address the full range of problems facing the field, a hierarchy of approaches is necessary. The core of the research effort in condensed-matter and materials physics is in the work of individual investigators and small research groups. Some of the most innovative and creative developments originate in this mode of research. At the next levels, larger groups, centers, and entire laboratories cooperate on significant problems, aided by progressively more-complex instrumentation and facilities. Theoretical work and benchtop experiments are usually done by individual investigators. Small-scale centers located in universities and government laboratories play an essential role in a number of areas including microcharacterization, processing, synthesis, and state-of-the-art instrumentation development. The highest level in the hierarchy is exemplified by major facilities, including synchrotron light sources, centers for neutron-scattering research, and laboratories for high magnetic field studies. These major facilities address a broad range of problems. An area of particularly rapid growth is found in the use of these major facilities, particularly synchrotron light sources, in understanding soft condensed matter and biological systems. A key facilities problem is the critical gap in U.S. capabilities in the area of neutron sources.

The different modes of research—benchtop experiments, larger collaborations, and so on—are evolving steadily. The work carried out in these varied modes is complex and diverse and continuously expanding to encompass an increasing number of disciplines. The committee has paid special attention to describing the forefronts of condensed-matter and materials physics research in conjunction with a small number of research themes. These themes are discussed in some detail in the Overview and reappear in each of the chapters. Throughout this study the themes of new experimental and computational capabilities, the ability to address problems of increasing complexity, and the importance of relationships with other fields are interwoven with discussion of subdisciplines of condensed-matter and materials physics. One of the subdisciplines that has captured the imagination of theorists and experimenters alike is the structure and properties of materials at reduced dimensionality—for example, in planar structures. Developing large-scale integrated circuits depends on understanding the behavior of semiconductors in such configurations, so the potential for impact is apparent.

A number of actions are required to maintain and enhance productivity in the field of condensed-matter and materials physics. These actions involve each level of the hierarchy of research modalities *and the interactions among the various levels and the various performers*. The principal recommendations of the committee are summarized here:

- The National Science Foundation, the U.S. Department of Energy, the U.S. Department of Defense, and other agencies that support condensed-matter and materials physics research should continue to nurture the core research at the heart of the field. The research areas described in the Overview provide a guide to the scientific arenas at the forefront of this work.
- The agencies that support and direct research should plan for increased investment in modernizing the condensed-matter and materials physics research infrastructure at universities and government laboratories.
- The National Science Foundation should increase its investment in state-of-the-art instrumentation and fabrication capabilities, including centers for instrumentation R&D, nanofabrication, and materials synthesis and processing at universities. The Department of Energy should strengthen its support for such programs at national laboratories and universities.
- The insufficiency of neutron sources in the United States should be addressed in the short term by upgrading existing neutron-scattering facilities and in the long term by the construction of the Spallation Neutron Source.
- Support for operations and upgrades at synchrotron facilities, including research and development on fourth-generation light sources, should be strengthened.
- The broad use of synchrotron and neutron facilities across scientific disciplines and sectors should be considered when establishing agency budgets.

- Federal agencies should provide incentives for formation of partnerships among universities and government and industrial laboratories that carry out research in condensed-matter and materials physics.
- Universities should endeavor to enhance their students' understanding of the role of knowledge integration and transfer as well as knowledge creation.

Overview

INTRODUCTION

Condensed-matter and materials physics is the branch of physics that studies the properties of the large collections of atoms that compose both natural and synthetic materials. The roots of condensed-matter and materials physics lie in the discoveries of quantum mechanics in the early part of the twentieth century. Because it deals with properties of matter at ordinary chemical and thermal energy scales, condensed-matter and materials physics is the subfield of physics that has the largest number of direct practical applications. It is also an intellectually vital field that is currently producing many advances in fundamental physics.

Fifty years ago the transistor emerged from this area of physics. High-temperature superconductivity was discovered by condensed-matter physicists, as were the fascinating low-temperature states of superfluid helium. Scientists in this field have long-standing interests in electronic and optical properties of solids and all aspects of magnetism and magnetic materials. They investigate the properties of glasses, polymeric materials, and granular materials as well as composites, in which diverse constituents are combined to produce entirely new substances with novel properties.

Condensed-matter and materials physics has played a key role in the technological advances that have changed our lives so dramatically in the last 50 years. Driven by discoveries in condensed-matter and materials physics, these advances have brought us the integrated circuit, magnetic resonance imaging (MRI), low-loss optical fibers, solid-state lasers, light-emitting diodes, magnetic recording disks, and high-performance composite materials. These in turn have led to the spectacular growth of modern computer and telecommunications industries and, consequently, to the information revolution.

For many years after the invention of the transistor, the major intellectual challenge facing researchers in condensed-matter and materials physics was to understand the physical properties of nearly perfect single crystals of elements, simple compounds, and alloys. Most of these materials occur in some form in nature. On a basis of increased knowledge and powerful new synthesis techniques, today's condensed-matter and materials physics is directed toward creating entirely new classes of materials—so-called "artificially structured" materials—that do not exist in nature and whose sizes reach all the way down to the atomic domain. At the same time, a growing number of researchers are using new theoretical and experimental tools to extend our understanding to much more complex forms of matter—high-temperature superconductors, multicomponent magnetic materials, semicrystalline polymers, and glasses. These tools are, in turn, giving greater insight into more complex phenomena like the fracture of solids and the continuous transition from liquid to glass in the process of cooling. Ever in view in current condensed-matter and materials physics are research opportunities presented by dramatic progress in the biological sciences. Condensed-matter and materials physicists are working with biological scientists to develop a new field of "physical biology" in which physics-based techniques and approaches are applied to the study of biological materials and processes.

Indeed, condensed-matter and materials physics is distinguished by its extraordinary interdependence with other science and engineering fields. It is a multifaceted and diverse interdisciplinary field, strongly linked to other science and engineering disciplines that both benefit from and contribute to its successes. Important examples of this collaboration include fullerenes (physics and chemistry), macromolecules (physics, chemistry, and biology), structural alloys (physics and materials engineering), and silicon technology (physics and electrical engineering). Condensed-matter and materials physics also has strong interrelationships to other branches of physics. Prominent examples include Bose-Einstein condensation (with atomic physics) and the fractional quantum Hall effect (elementary-particle physics). Its practitioners include those who discover and develop new materials, those who seek to understand such materials at a fundamental level through experiments and theoretical analysis, and those who apply the materials and understanding to create new devices and technologies. This work is done in universities, in industry, and in government laboratories. Advances in basic research inspire new ideas for applications, and applications-driven technological advances provide tools that enable new fundamental investigations. Technological advances provide new tools such as synchrotrons, neutron sources, electron microscopes, computers, and scanning-probe microscopes. These new tools are leading to new advances in the fundamental understanding of materials and to a wide-ranging impact on other fields—biology, chemistry, environmental sciences, and engineering.

A NEW ERA

The world of condensed-matter and materials physics is entering a new era. Extraordinary advances in instrumentation are providing access to the world of atoms and molecules on an unprecedented scale. Powerful new experimental tools, from national synchrotron and neutron facilities to bench-scale atomic-probe microscopes, are opening up new windows for visualizing and manipulating materials on the atomic scale. Applications range from nanofabrication of electronic devices to probing the secrets of superconductivity and protein folding. These changes are far-reaching. Many research areas, previously inaccessible, are yielding to new and unanticipated advances in atomic-scale synthesis, characterization, and visualization.

Advances in computational power have made it possible, for the first time, to simulate the behavior of complex materials systems and large assemblies of atoms. As a result, numerical simulation is approaching parity with laboratory experiments and analytic theory in many areas of condensed-matter and materials physics research. Based on benchmarks provided by experimentation, and enlightened by a proper consideration of theory, the new computational tools provide synergy to accelerate the understanding of ever more-complex systems. Again, this is a qualitative change—with each new generation of computational power, opportunities are emerging that could only be imagined a few years earlier.

The combined power of the new experimental tools and computational advances are having an enormous impact on condensed-matter and materials physics, particularly in those areas where the ability to span length scales from the atomic to macroscopic is of fundamental importance, that is, where the properties of atoms and molecules—especially quantum phenomena—become relevant to large-scale phenomena. This new capability to span length scales is bringing the world of atoms and molecules closer to the world of our experience, from the mysteries of quantum mechanics, to the mechanical properties of materials, to the self-assembly of biological systems. Many of these problems, which underlie technological innovation and revolution, could not have been addressed on a fundamental basis even a few years ago.

The developments described in this report present a condensed-matter and materials physics profoundly different than it has been at any other time in history. The ability to control and manipulate atoms, to observe and simulate collective phenomena, to treat complex materials systems, and to span length scales from atoms to our everyday experience, provides opportunities that were not even imagined a decade ago. These developments underlie current progress in condensed-matter and materials physics and provide tremendous optimism for the future vitality of the field. They also underlie a new unity in science. Advances in condensed-matter and materials physics increasingly interface with and relate to nearly all areas of science and engineering, including atomic and molecular physics, particle physics, materials science, chemistry, biology, and computational sciences. The next decade will bring extraordinary benefits from this unity, especially as the new capabilities in condensed-matter and materials physics bridge the gap between physics and biology, revealing the molecular-physics basis of biological phenomena.

THE SCIENCE OF MODERN TECHNOLOGY

The information age owes its rapid growth to technological advances that depend on progress in science. Scientific understanding of fundamental phenomena has been closely tied to the development of materials with special properties as carriers and controllers of electrical current, light waves, and magnetic fields. Equally important is understanding, on both fundamental and quantitative levels, the processes that enable cost-competitive manufacturing of devices and systems based on these materials for the rapidly changing electronics and telecommunications industries.

Silicon is the foundation of today's integrated circuit technology. The science and technology of this material have built on each other, leading to progressively smaller, faster, and more-complex devices. In turn, semiconductor fabrication technology has enabled the construction of exotic quantum devices that are essentially man-made two-dimensional atoms. A range of new insights into the behavior of large collections of electrons have come from studying electron behavior in these devices. As these circuits continue to develop, doubling their speed and power every 18 months in accordance with the empirical Moore's Law, they march farther into the quantum domain where new physical phenomena arise that must be understood and controlled to ensure continuing progress.

Compound semiconductors such as gallium arsenide, gallium nitride, and others are essential to the field of telecommunications. They have characteristics that enable the production of electronic and optoelectronic devices with exceptional performance characteristics. They also underlie the solid-state laser technology that converts digital electronic signals into optical signals that travel great distances on fiber-optic cable. The development of lasers at the right wavelengths is essential to minimize losses and to enable the amplification of the signal by purely optical (nonelectronic) means. A particularly exciting recent development is the gallium nitride laser, which produces blue light suitable for next-generation optical storage systems. Another important development is vertical-cavity surface-emitting lasers, which will bring the mass-production advantages of integrated circuits to diode lasers. Their geometry will enable chip-to-chip optical interconnects in future generations of integrated circuits.

Fiber-optic cable, an extraordinary product of optical materials research, enables the transmission of information by modulation of light waves traveling in a glass fiber. Because light, the carrier of the information, has a frequency of at least 10^{14} Hz (much higher than ordinary radio waves), the rate of information transmission achievable with fiber optics is extremely high. The mechanisms that scatter light in these fibers had to be understood in great detail before it was possible to develop today's low-loss fibers that enable a signal to travel 800 km (500 miles) before it is attenuated. Modern optical fibers are so perfect that the dominant loss mechanism is the scattering of light by density fluctuations frozen in when the fiber is cooled to become a glass. Fiber-optic amplifiers have been developed that greatly extend the reach of these systems. A great challenge is to develop an inexpensive interface that will bring fiber information links to the home user. Optics enables data storage in compact disks, in the new high-capacity digital video disks (DVDs), and high-density storage systems now in development. To continue this progress, materials scientists must develop commercially manufacturable solid-state lasers that emit blue light. Optics is also essential to display, printing, and copying technologies. Optics is playing an increasingly important role in the telecommunications revolution. Further research into the properties of materials for fiber-optic amplifiers, fast optical switches, and many other optical technologies will be necessary for future advances in information technologies.

Magnetism has presented physics with some of its most challenging theoretical problems and also with some of its most important applications. Driven by the need for progressively more data-storage capacity, the science of magnetism has yielded new technology for the devices that read and write data on computer disk drives. Using the recently discovered phenomenon of giant magnetoresistance, technologists have found ways to fabricate read/write heads that have allowed development of these devices to keep pace with rapid improvements in integrated circuitry. Progress in magnetic materials has also yielded a new class of small motors and new transformer core materials for power distribution. And magnetoelectronic devices have moved from the laboratory to applications with amazing speed. Among the sensors based on these new technologies are superconducting quantum interference devices (SQUIDs) that enable the detection of minuscule magnetic fields emitted by the human brain and heart, and magnetic force microscopes that can image magnetic properties with nearly atomic-scale resolution.

As integrated circuitry becomes ever smaller and closer to the realm of quantum physics, new ways of constructing the logic functions that are the building blocks of these circuits may be discovered. It may even become possible to develop a new form of logic circuitry that exploits the properties of the strange world of quantum mechanics. Until then, a number of challenges face the manufacturers of information systems. The tiny aluminum (and recently, copper) layers that connect the logic devices and their insulating glassy sheaths may have to be replaced with something better in order to continue increasing speed. Optical interconnections may play a role. Will some new way of producing digital switches, the building blocks of computer logic, emerge? Can optical technology be developed so that the essential functions of the communications network (such as switching, now carried out electronically) can all be carried out with optical devices? Will the key information technologies be reducible to the atomic-size scale? Can the complex physics and chemistry of current and future materials be mastered to enable us to meet this grand challenge?

NEW MATERIALS AND STRUCTURES

The effort to develop new materials and new configurations of matter, driven largely by the potential for innovation that such novelties make possible, has led to the creation of structures that have turned out to be microphysics laboratories on the atomic scale. Some of these new materials and structures have provided the environment for completely unexpected phenomena to emerge. A few examples include high-temperature superconductors, organic superconductors, buckyballs and related structures, and giant magnetoresistance materials.

High-temperature superconductors are superconducting above the boiling point of liquid nitrogen. Before the discovery of superconductivity in this class of materials, the effort to increase the superconducting transition temperature (T_c) of materials focused on painstaking efforts to combine various metals into superconducting alloys. This effort seemed to hit a brick wall at about 23 degrees above absolute zero—too low for widespread applications outside of research. Unlike conventional superconductors, the new high-temperature materials are not metals but ceramics, which one would expect to be insulators. The fact that they conduct electricity at all is quite surprising. These materials have a complicated perovskite crystal structure (like many naturally occurring minerals) with planes of copper and oxygen atoms. At the right temperatures, these planes can function as superconductors. These ceramics are much more complex than most of the materials previously studied in condensed-matter physics. There are many possible permutations and combinations of constituents, making these materials difficult to prepare and characterize. Because of the short range of superconducting correlations, the materials are extremely sensitive to defects, which adds to the difficulty. Notwithstanding the complexity, high-T_c superconducting films have already found application in the SQUID devices described above.

The discovery of high-T_c superconducting materials has had a profound influence on researchers in condensed-matter physics. The classic paradigm in the field was to seek systems that exhibit some special simplification as an aid to understanding the physics of solids, even the simplest of which are fairly complicated. But the emergence of startling properties in materials with more complex composition and structure has convinced researchers that admitting higher complexity to the field of study can also offer the possibility of new and unexpected phenomena and insights in all areas of condensed-matter and materials physics. One effect of the study of high-T_c materials was a renewed interest in complex perovskites. That, in turn, led to the discovery that the electrical resistance of lanthanum manganate can be extremely sensitive to the presence of magnetic fields. This effect, known as "colossal magnetoresistance," is of great interest because of its potential application to read data from a new generation of ultra-high-density magnetic storage disks.

Until recently, carbon was thought to exist in only two crystal structures—diamond and graphite. The discovery of new crystalline structures, generically called "buckminsterfullerenes," was a great surprise. Variations of these structures can have amazing properties, including forms with a tensile strength 100 times that of steel.

The name "buckminsterfullerene" was chosen because the structure of the first one discovered resembles that of a geodesic dome. Their properties depend primarily on this special shape. The structure can be imagined by starting with a two-dimensional hexagonal lattice of carbon atoms found in graphite. If pentagons are substituted for some of the hexagons, the surface develops positive curvature and can be made to close on itself, forming a soccer-ball structure (a "buckyball") or various other possible shapes, including tubes. These tubes can vary from metallic to semiconducting, depending on their geometry. By exposing fullerene molecules (C_{60}) to alkali or alkaline-earth metal vapors, organic superconductors can be prepared.

These examples have all been startling breakthroughs. But amazing outcomes can also result from the steady, evolutionary development of the properties of materials as some property is refined past previous technological barriers. For example, steady improvements in the purity of semiconductor materials used in high-frequency applications such as cellular phones eventually led to the fabrication of quantum-dot structures. These structures, fabricated on the quantum-size domain, have energy states similar to those of atoms but with optical and electronic properties that can be tailored for a wide variety of applications.

Several themes and challenges are apparent—the role of molecular geometry and reduced dimensionality, the synthesis and processing and understanding of more complex materials, tailoring the composition and structure of materials on very small scales, and incorporation of new materials and structures in existing technologies. Progress in these areas holds the promise of further startling breakthroughs, yielding materials with unexpected and useful properties and extending the understanding of condensed-matter and materials physics.

NOVEL QUANTUM PHENOMENA

Perhaps the most important lesson learned from studying the physics of systems that contain many particles is that when the number of particles in the system is large enough, entirely new phenomena can appear. These new behaviors of the whole system may not have any obvious relationship to the properties of the individual particles, but rather may arise from collective or cooperative behavior of all the particles. Such phenomena are often referred to as "emergent phenomena" because they emerge as the complexity of a system grows with the addition of more particles.

In some materials, at sufficiently low temperatures, small motions of the crystal lattice create interactions among the electrons that cause them to pair in such a way that they form an electron liquid that no longer experiences friction when it flows. Thus, because the flow of electrical current experiences no resistance in such materials, they are known as "superconductors." Superconductivity is a good example of a quantum emergent phenomenon. Its existence was unanticipated when only the properties of individual atoms were considered.

A rich variety of other new effects emerge when large ensembles of atoms are brought together. The length scale of an atom is about 1 Å, and typical quantum energy levels of the electrons in atoms are in the range of 1 to 10 electron volts (eV). As atoms are assembled in a solid, and more collective effects emerge, length scales become larger and energy scales become smaller. It is often convenient to view such systems not in terms of their building blocks, but rather in terms of "elementary excitations" that may have properties very different from those of the electrons and atoms that compose the system. As it becomes possible to make materials and have more and more control over their structure, impurities, and imperfections, collective excitations at smaller and smaller energy scales have been observed. These excitations sometimes have bizarre properties, such as carrying a charge that is a fraction of the electron charge.

Because these emergent phenomena present new and unexpected properties, it can be said that one of the new frontiers of condensed-matter physics is at low energies and at length scales large compared to atoms. By contrast, in elementary-particle physics, the frontier is at increasingly higher energies and shorter length scales.

A particularly fascinating emergent phenomenon occurs in helium at very low temperatures. Helium has an integer spin, so it is not bound by the Pauli exclusion principle. As a result, many helium atoms can all occupy the same quantum state, and this effect actually occurs at very low temperatures. In a manner somewhat analogous to the loss of electrical resistivity in superconductivity, liquid helium at very low temperatures loses all viscosity. It can flow without resistance through very small orifices and climb up the walls of its containers. This phenomenon is called "superfluidity." The general phenomenon whereby many integer-spin particles occupy the lowest energy state is called "Bose-Einstein condensation," after Bose and Einstein, who first described the statistical behavior of ensembles of quantum-mechanical particles with integer spin.

Physicists were surprised to learn that helium-3 atoms can also exhibit superfluidity. Helium-3 is a half-integral spin atom, so it must obey the Pauli exclusion principle, and it should not be able to undergo Bose-Einstein condensation. But two helium-3 atoms can pair up to make a particle with integer spin, and that composite particle can then undergo condensation. This phenomenon is similar to the way electrons in a superconducting solid form pairs. Helium-3 forms a superfluid only at very low temperatures because the particles of which it is formed are, themselves, only weakly bound together. Creating the super-low-temperature environment necessary to achieve superfluidity in helium-3 was an amazing experimental tour de force. The discovery of superfluidity enabled by these experimental techniques was awarded the 1996 Nobel Prize for physics. (Table O.1 lists Nobel Prizes awarded for research related to condensed-matter and materials physics.) The mechanism involved may have analogies to those responsible for high-T_c superconductivity.

An even more extreme form of Bose-Einstein condensation was recently achieved by the atomic physics community when alkali metal (such as sodium) vapor was held in an atomic trap and cooled to temperatures within a few microdegrees of absolute zero. The gas is very dilute and the atoms are far apart, but the atoms behave coherently and can be directed in a laser-like beam. This incredibly difficult experiment is at the outermost reaches of the low-energy, large-distance quantum frontier.

These achievements have built bridges among the condensed-matter, atomic-physics, and quantum-optics communities because condensation phenomena are important in all of these areas. For example, optical lasers depend on the ability to place a large ensemble of atoms in the same excited energy state. The atoms then decay in a collective, coherent transition to a lower energy state, releasing coherent light in the process.

An issue at the forefront of condensed-matter theory concerns quantum spin chains and ladders. Quantum spins are typically associated with magnetic dipoles. Magnets are simply solids in which the atomic spins in the crystal lattice and their associated magnetic dipoles are all pointing in the same direction. An important consequence of the discovery of high-T_c superconductors has been progress in learning how to synthesize compounds that have spins arranged in unusual configurations. Among these configurations are two-dimensional planes, one-dimensional chains, and other more complex structures such as ladders. The synthesis of these and other new families of organic and inorganic compounds has reinvigorated the study of quantum magnetism. The role of magnetism in high-T_c superconductors is among the outstanding questions in this area.

Probably the most remarkable collective phenomenon discovered in the latter half of the twentieth century is the quantum Hall effect. The ordinary Hall effect arises when electric current passes through a semiconductor film in the presence of a magnetic field perpendicular to the plane of the film. The current-bearing electrons moving in a magnetic field experience a force perpendicular to both the magnetic field and the direction of motion. As a result, electrons are pushed to one side of the film, which creates a transverse electric field and a voltage across the film. The more current is passed through the film, the greater the voltage. The ratio of the applied current to this voltage has the units of a conductance (or inverse electrical resistance) and is called the "Hall conductance."

If this experiment is carried out at high magnetic fields and low temperatures, quantum mechanics comes into play and the Hall conductance becomes

TABLE O.1 Nobel Prizes Awarded for Research Related to Condensed-Matter and Materials Physics Since 1986

Year	Field	Citation	Laureates
1986	Physics	For design of the first electron microscope (Ruska) and the scanning-tunneling microscope (Binnig and Rohrer)	Ernst Ruska, Gerd Binnig, and Heinrich Rohrer
1987	Physics	For discovery of superconductivity in ceramic materials	Johannes Georg Bednorz and Karl A. Müller
1991	Physics	For discovery of methods for studying order phenomena in complex forms of matter, particularly liquid crystals and polymers	Pierre-Gilles de Gennes
1994	Physics	For development of neutron-scattering techniques for studies of condensed matter	Clifford G. Shull and Bertram N. Brockhouse
1996	Chemistry	For the discovery of fullerenes	Harold Kroto, Robert Curl Jr., and Richard E. Smalley
1996	Physics	For the discovery of superfluidity in helium-3	David M. Lee, Douglas D. Osheroff, and Robert C. Richardson
1997	Physics	For development of methods to cool and trap atoms with laser light	Steven Chu, Claude Cohen-Tannoudji, and William D. Phillips
1998	Chemistry	For development of the density-functional theory (Kohn) and computational methods in quantum chemistry (Pople)	Walter Kohn and John A. Pople
1998	Physics	For discovery of a new form of quantum fluid with fractionally charged excitations	Robert B. Laughlin, Horst L. Störmer, and Daniel C. Tsui

quantized in precise integer multiples of the fundamental constant e^2/h where e is the charge of the electron and h is Planck's constant. It is remarkable that this universal result is completely independent of all microscopic details of the sample such as the density and types of impurities, the precise value of the magnetic field, etc. Among the consequences of this extraordinary phenomenon is that it is possible to make a high-precision measurement of the fine-structure constant (which expresses the strength of electromagnetic forces) and also to realize a highly reproducible standard of electrical resistance. This phenomenon is used in standards laboratories throughout the world to maintain the unit of electrical resistance (the ohm).

An even more surprising phenomenon occurs with samples of very high purity at very low temperatures and in very high magnetic fields. In these conditions it is possible to observe a Hall effect in which the conductivity is a *fraction* of e^2/h—for example, $(1/3)e^2/h$. Physicists were quite startled by this observation. It turns out to result from the formation of quasiparticles whose effective charge is one-third (or various other rational fractions) of the electron's charge. These quasiparticles are a collective mode of a quantum fluid. The low-energy excitations of this weird fluid consist of vortices bound to a fraction of an electron charge. These objects have been recently observed by direct measurement of their charge and by tunneling experiments in which an electron added to the system breaks up into three excitations, each with one-third of the electron's charge. This discovery, which earned the 1998 Nobel Prize in Physics (see Table O.1), offers a whole universe of intriguing possibilities for experimental and theoretical exploration of new collective modes.

NONEQUILIBRIUM PHYSICS

Nonequilibrium physics is the study of systems that are out of balance with their surroundings. They may be changing their states as they are heated or cooled, deforming as a result of external stresses, or generating complex or even chaotic patterns in response to forces imposed on them. Examples include water flowing under pressure through a pipe, a solid breaking under stress, or a snowflake forming in the atmosphere.

Understanding nonequilibrium phenomena is of great practical importance in such diverse areas as optimizing manufacturing technologies, designing energy-efficient transportation, processing structural materials, or mitigating the damage caused by earthquakes. At the same time, the theory of nonequilibrium phenomena contains some of the most challenging and fundamental problems in physics. A central theme in this field is that the physics of ordinary materials and processes is a rich source of inspiration for basic research.

Because nonequilibrium physics touches on such a wide range of different areas of science and technology, it is an important channel through which physics makes contact with other disciplines. For example, its concepts help explain galaxy formation, climate change, and—perhaps most important—some of the physical mechanisms that underlie biological phenomena. Indeed, the most characteristic property of living systems may be their ability to maintain their extraordinarily complex nonequilibrium states for extended periods of time. Because of the variety of the different disciplines involved, each with its own culture and approach to research, continued progress in nonequilibrium physics will depend in large measure on the ability of institutions and scientists to bridge the culture gaps that separate their different communities.

Much progress has been made in the last decade. For example, we now have a detailed understanding of how complex patterns emerge in many apparently simple hydrodynamic, metallurgical, and chemical situations. Pattern-forming systems are intrinsically unstable against small perturbations and often exhibit chaotic motion; it is now possible to understand how this happens in some cases. The extreme sensitivity to small perturbations shown by many of these systems makes it difficult to predict or control their behavior.

This field is very large and only a few topics have been selected in this report to illustrate the major issues. These include pattern formation and turbulence in fluid flow, processing and performance of structural materials, and some topics in solid mechanics, specifically, friction, fracture, granular materials, and polymers and adhesives. The chapter on nonequilibrium physics also includes brief discussions of nonequilibrium phenomena in the quantum and biological domains, and yet briefer remarks about nonequilibrium complexity and limits of predictability. To convey some of the flavor of the field, the committee mentions just a few of those topics in the paragraphs that follow.

A beautiful example of pattern formation in fluid dynamics is provided by Rayleigh-Benard convection. If a fluid is heated very gently from below, the heat diffuses slowly up through the liquid and is dissipated at the upper surface without any flowing motion of the fluid. But if the heat is turned up, the fluid starts to convey the thermal energy upward by convection. That is, the lower, hotter layers of fluid begin to rise in plumes to the top. The hot plumes hit the top of the fluid, cool, and then sink back. If the fluid is spread out (like gravy in a sauce pan, for example), convection cells may appear and arrange themselves over the surface of the liquid in a regular pattern of squares or hexagons. This phenomenon is commonly seen in the kitchen. Theoretically, the behavior of such a fluid is completely described by a set of well-known equations (named after Navier and Stokes) derived directly from Newton's laws of motion. Indeed, this theory works quite well in practice until the driving force—in this case, the heat—is turned up so high that the fluid motion becomes turbulent. Although there is no reason to believe that the Navier-Stokes equations lose their validity with turbulent flows, solving the equations in the turbulent domain is a much harder theoretical problem. It is important in many areas of both basic and applied research. Here, too, modern experimental techniques, computer simulations, and new analytic approaches are enabling progress.

One classic nonequilibrium problem in materials physics is fracture in solids. How does one characterize a material that is prone to fracture? We usually describe such a solid as brittle, as opposed to ductile; but what does that distinction mean at the microscopic level? What is the role of crystal structure, and how do amorphous or glassy materials differ from crystals in their failure mechanisms? Some progress in addressing these issues has been achieved through computer simulation and modeling, especially in engineering applications where the primary interest has been in predicting failure criteria. It is much harder, however, to predict what happens after failure begins. How do fractures propagate, and how can propagation be controlled or modified? Progress in answering these questions would have far-reaching consequences.

The science of friction is another example of a classic part of materials research that is becoming amenable to physical understanding on a microscopic basis. Friction is related to fracture dynamics because, in many ways, two solid surfaces sliding past each other look like a propagating shear crack. Friction is a rich class of phenomena with a variety of underlying mechanisms. When two imperfect surfaces slide past each other, friction is produced by the cohesion and decohesion of atomic-scale contact points that are strongly coupled to the deformation modes of the material. Novel techniques, including atomic-scale probe microscopies, are giving new insight into the dynamic details of these processes.

Although progress in these areas of long-standing interest is very promising, it is likely that the next major frontier in nonequilibrium physics will be in the area of biological materials and phenomena. Optical tweezers and other physics-based probes are enabling molecular-scale observation of biological processes. Forces between cellular membranes can be measured, and the physical mechanisms whereby proteins are formed and transported within a cell can be observed. The cell provides a rich new universe of complex nonequilibrium phenomena for study by physicists. The rewards to society of detailed physical understanding of fundamental life processes could be enormous.

COMPLEX FLUIDS AND MACROMOLECULAR AND BIOLOGICAL SYSTEMS

Big molecules have bumpy or sticky places that enable them to assemble themselves into regular or functional arrangements that are extremely sensitive to varied environments and conditions of formation. In artificial preparations these assemblies can form "complex fluids" whose morphology is easily modulated by changing temperature, dilution, or electrical currents.

Digital watches and portable computer monitors depend on this modulation in their liquid-crystal displays. The structure, and hence the optical properties, of the liquid crystal can be altered by small voltages applied across the fluid. Flashlight batteries now come with liquid-crystal testers that respond to the heat generated by a resistor. An inexpensive fever thermometer can be made from a plastic strip that contains liquid crystals; it is accurate enough to indicate a normal reading or to signal the need for a more accurate measurement by a digital thermometer (which will itself use a liquid-crystal display). Because these liquid crystals are so sensitive, their use consumes very little power.

Complex fluids occur in bewildering variety. To describe and control the different liquid phases, theorists are developing new concepts of molecular organization. With progressively better understanding of the rules of formation, these fluids can be used to construct new kinds of materials with unusual topologies. Extremely light and strong aerogels, used in insulation for example, have a huge surface area trapped in a relatively small volume.

Polysaccharides are a class of materials that form complex fluids that are important to industry. In this class are the xanthan, guar gums, and carrageenan essential to modern food preparation and stabilization. Xanthan, used in salad dressing, is so stable and controllable that it can even be used in an oil field to stimulate petroleum recovery. Such materials have a wide range of viscous and elastic properties. A better understanding of the relationship between their structure and properties should lead to many new materials and applications.

Polyolefins, another class of complex fluids, are produced in large quantities—about as much annually by weight as steel. These molecules, formed into long-chain polymers, have many uses such as less-expensive alternatives to nylon. With the aid of new catalysts, it became possible during the past decade to put monomers like ethylene onto the chain in a highly controlled way to create synthetic materials with a range of properties such as hardness and ductility. The techniques of physics have great potential to characterize these materials and to elucidate the relationship between the process by which they were prepared and their final structure and properties.

Living systems create macromolecules with an enviable degree of specificity and startling properties. A strand of spider silk, for example, can have a greater tensile strength than a steel wire of the same diameter. If the physics governing the synthesis, processing, and related properties of such materials could be understood, it might open up the possibility of new classes of supermaterials. There is an opportunity for physicists to learn from biologists in this and other areas.

There are many other areas where useful insights have come from a physics-based approach to objects or to phenomena that traditionally have been in the domain of biology. Among these are protein-protein interactions, molecular motors, protein folding, photosynthesis, and nerve action. Modern molecular biology would not be the same without the physics-based facilities that have played key roles in these developments. By determining the structure of very large and complicated biological molecules at the atomic scale of resolution, physics is having a significant impact on biology.

Most structure determination is done via x-ray diffraction from crystallized material. Intense tunable x-ray sources allow these determinations on small samples. The premier facility where this is done is the Advanced Photon Source

(APS) at the Department of Energy's Argonne National Laboratory. Because of the brightness of this source and the time structure of its output, it is even possible to do time-resolved studies of changes in the configuration of biological molecules and aggregate structures. Biologists are the fastest growing class of users at the Advanced Photon Source. These physics-based facilities may turn out to be a great venue for creative interactions between physicists and biologists.

Because of its sensitivity to atomic motions and light (low-Z) atoms, neutron scattering is an important tool for studying large biological molecules. Neutrons are particularly sensitive to the vibrational modes that provide clues to the dynamical behavior of large molecules. Biological molecules are made mostly of carbon, hydrogen, nitrogen, and oxygen; neutrons are particularly good probes of such light elements, providing information that is complementary to that obtained with x-ray sources. Although neutron sources in the United States are very active, the dearth of high-intensity sources here is an obstacle to progress in this area; this obstacle will be removed when the Spallation Neutron Source comes online in 2005.

Physics has come to define the language for the study of protein folding, in which a linear chain of complex biological molecules arranges itself into the compact tangles that proteins adopt in living systems. Like structure determination, observation of the dynamics of folding depends on the availability of intense, tunable x-ray sources.

There are reasons to be optimistic about the ability of physicists to provide understanding of very complex materials and to engage in productive collaborations with biologists and chemists in the study of increasingly complex molecules and systems. Considering the potential for outcomes that will benefit humanity, it makes sense to look for mechanisms to promote this collaboration.

A number of the general recommendations in this report address the need for collaboration. Applying these recommendations to actual efforts to encourage collaboration between physicists and life scientists is a special challenge. Among the efforts that may work particularly well to bring biologists and physicists together are interdisciplinary workshops, summer schools with laboratories for scientists at all career stages (including nascent Ph.D.s), new cross-disciplinary courses (biologists need to learn some physics, physicists need to learn some chemistry and biology, and so on), inclusion of new biology-related material in introductory physics courses, and new texts that mix both physics and biology frameworks. The National Institutes of Health should continue to take the lead in this area, working with the National Science Foundation and the Department of Energy.

NEW TOOLS FOR RESEARCH: FROM THE BENCHTOP TO THE NATIONAL LABORATORY

The arsenal of research tools for condensed-matter and materials physics includes a wide variety of experimental equipment. This equipment enables the creation of extreme environments in which to explore the behavior of matter and the synthesis of materials with extraordinary properties. It also provides new "eyes" to observe and new "hands" to manipulate at the atomic scale. Examples of such equipment range from benchtop-scale atomic-force microscopes used by individual investigators, to storage rings the size of a small town that generate the x-rays used by collaborative research groups with many members.

These experimental tools enable new insights into systems of recognized importance and the exploration of completely new regimes. One example of scientific progress that depended on modern instruments of research involved unraveling the properties of the high-T_c superconductors. Developing today's understanding of these materials depended on the use of experimental tools:

- Neutron diffraction was used to determine atomic coordinates.
- Synchrotron radiation was used to determine the electronic structure.
- Electron microscopy made it possible to determine the microstructure.
- Neutron scattering was used to determine magnetic order.
- High magnetic fields and high pressures were used to gain understanding of charge transport.

Many of these studies were performed at large- or medium-scale facilities.

Although large facilities are critical to condensed-matter and materials physics, another theme that pervades this report is the importance of atomic-scale observation and manipulation. Two of the most important tools for this purpose are the scanning-tunneling microscope and the transmission electron microscope. The equipment is on the small-to-medium scale.

Scanning-tunneling microscopes work by placing a probe that is sharp, on an atomic scale, so close to the sample that the quantum wave function of the electron allows it to jump the gap. By scanning this probe over the sample, using sophisticated positioning technology, the surface can be mapped atom by atom. The resolution is far better than anything that can be achieved with light waves, because the wavelength of light is thousands of times too large to visualize atoms. Various kinds of scanning-probe microscopes are now commonly available, including instruments that can examine chemical reactivity, magnetism, optical absorption, mechanical response, and other properties. A particularly promising development is the imaging of molecules, including rather large ones that play a role in biological processes.

Scanning-tunneling microscopes can go beyond measuring structures to actually creating them by positioning individual atoms. In principle, it is possible to create any structure to test our understanding of the physics of devices at this scale. One of the challenges in this area is to learn to control the stability of atomic-scale structures. In general, individual atoms placed on a surface will not stay put unless the temperature is extremely low. Another challenge is that, if this technique is ever to lead to practical devices, it would have to be much faster than it is now.

Electron microscopes, which use beams of electrons to probe the sample, are able to penetrate below the surface. They have much higher resolution than optical microscopes because of the shortness of the electron wavelength. Instruments with 1-Å resolution have been demonstrated. Development of electron microscopes has occurred primarily in Europe and Japan. These instruments show great promise for reconstructing three-dimensional structures of biological interest as well as for studying the properties of amorphous and disordered materials.

We can be sure that improved brightness, spectroscopy, and resolution in electron microscopes will allow more precise determination of structure and composition in ultra-small volumes of materials, even when these are embedded below the surface. This will have a continuing major impact on the microstructural study of all materials, for example, identifying interfacial structures, solving important problems of support effects on small clusters, and understanding the structural basis of adhesion and fracture in materials. These new capabilities will require a significant reinvestment in infrastructure and increased investment in instrumental research and development.

An equally important theme is that, to a growing degree, significant advances in a number of sciences depend on large national facilities. The United States has been particularly strong in the development of synchrotron radiation sources. These sources depend on the fact that when an electron is accelerated, it gives off light. The most sophisticated (third-generation) synchrotron radiation source in the United States is the Advanced Photon Source (APS) at Argonne National Laboratory. The APS uses devices called "undulators" that wiggle an electron beam by passing it through an array of powerful magnets to generate tunable beams of very high intensity x-rays. The power and controllability of the x-rays from the APS have made possible a new generation of experiments that have resolved structures with unprecedented precision.

The technology now being developed for a fourth generation of light sources may be eight orders of magnitude more powerful than even the APS and will have pulse lengths less than a picosecond. These devices, called "free electron lasers," use undulators configured so that the radiation given off bathes the electron beam and stimulates further emission of radiation in a process closely analogous to the operation of a laser. If past experience is a guide, the greater intensity and coherence that these devices will one day offer will lead to new classes of experiments and new insights into the structure of materials.

Synchrotron radiation is being used to conduct research in a number of areas. Inelastic x-ray scattering has provided unique information about the dynamics of fluids and glasses. Photoemission experiments have provided much information about the electronic structure of solids, which is essential to understanding the physical details of the operation of semiconductor devices and integrated circuits. X-ray studies of disordered systems have given insight into inorganic glasses and biological structures. Information can be obtained about chemical states and environments of individual atoms in the structure of disordered systems. Structural changes associated with the functioning of metalloproteins in processes like nitrogen fixation, photosynthesis, and respiration can be observed.

The most rapidly growing and arguably the most important use of synchrotron radiation is in protein crystallography, in which the structure of biologically important proteins is determined. Crystallography of big molecules using synchrotrons is the primary source of insight into three-dimensional biological structures. Biologists are just scratching the surface of the possibilities. Approximately two-thirds of all new structures published in *Science* and *Nature* in 1995 used synchrotron radiation, and the number is growing. Progress in this area in the next 10 years should be breathtaking, both in its intellectual scope and its real-world implications. Pharmaceutical companies are pursuing a "drugs-by-design" approach to relating the structure of complex molecules to their function and activity in biological systems; this approach has the potential to revolutionize the industry.

Another area in which the use of synchrotron radiation has been of benefit involves providing the means for researchers in the life sciences to understand the workings of a variety of biological structures:

- Scientists now understand how muscle contraction works at the molecular level.
- Scientists now know how living cells mobilize energy.
- Scientists understand how nitrogen-fixing bacteria work.
- Scientists have learned the structure of ribozyme, a catalytic form of RNA.
- Scientists have determined plant and animal viral structures.

One of the key factors that enabled these results was that both the physics community and the sponsoring agencies (primarily the Department of Energy) adopted a philosophy that synchrotron facilities are national resources that should be designed and implemented to serve all the branches of science. That has meant investing resources in making the machines and experimental facilities reliable, predictable, and easily used by researchers unfamiliar with accelerator facilities. It has meant providing the human infrastructure necessary to support such users. It has meant husbanding the special institutional arrangements necessary to make such users successful. Nonphysicist users are now in the majority at most synchrotron facilities.

Neutron scattering is a technique particularly sensitive to spin states and low-atomic-weight atoms. It is therefore particularly well suited for the study of magnetism, high-T_c materials, polymers, and biological materials. The major research facilities for neutron scattering in the United States have been the Department of Energy's high-flux reactors at Brookhaven National Laboratory and at Oak Ridge National Laboratory, the Department of Commerce's reactor at the

National Institute of Standards and Technology, and accelerator-based pulsed sources at Argonne National Laboratory and Los Alamos National Laboratory. These facilities have not kept pace with developments in Europe, and in the last decade, leadership decisively passed to the Institute Laue-Langeven reactor facility in France and the ISIS pulsed facility in the United Kingdom. Among the recent successes of neutron scattering has been the demonstration that the high-T_c superconductors have a common feature—nearly square planar arrays of copper and oxygen atoms. Neutron scattering also was used to image vortex lattices in high-T_c superconductors. These studies gave evidence for two of the most important new ideas about superconductivity. The first is that solids can become superconducting by mechanisms analogous to those responsible for superfluidity of helium-3. The second is that vortices in superconductors can have intricate phase diagrams. Structural information about the perovskite manganites that exhibit "colossal magnetoresistance" has also come from neutron scattering.

So far we have discussed strategies for research that involves using powerful tools to probe the structure of matter, to visualize what is happening at the atomic and molecular levels. Another strategy that has proved effective in flushing out new insights is placing matter in extreme conditions of temperature, pressure, or magnetic field. A recurring theme of this study has been exploration of what might be called the "low-energy frontier." One aspect of this frontier is extremely low temperatures. In this regime, very gentle, low-energy effects such as superconductivity and superfluidity begin to emerge; but these effects are washed out by the chaotic thermal motion of the atoms and electrons at higher temperatures. An experimental tour de force of the last decade was the discovery of collective nuclear spin order at nanokelvin temperatures in elemental copper.

The rapidly growing capability of computers for experimental control, visualization, and numerical simulation is having a significant impact on condensed-matter and materials physics. Computers used to be viewed not as part of the instrumentation armamentarium but as something outside it. But that may be changing. Modern computers and their associated visualization capabilities are now so powerful, they are beginning to provide "virtual laboratories" that offer new ways to acquire physical insight by exploring the effects of varying physical parameters. Real physical experiments will always be necessary, but computation may provide a virtual experimental space that will greatly speed up decisions about what kinds of experiments to do and where to look for new phenomena.

A major event of the decade for individual-researcher laboratories has been the development of scanning-probe microscopes as routine, off-the-shelf analytical tools. The technologies involved hold the promise of control and manipulation of surface materials at the atomic scale as well as study of large molecules in fine detail.

On the intermediate scale of infrastructure, there have been three important developments: greater access to electron microscopes and related equipment, the exploitation of university-based microfabrication centers, and reinvigoration of the U.S. high-field magnet research through the establishment of the National High-Field Magnet Laboratory.

The establishment of third-generation synchrotron light sources at Lawrence Berkeley National Laboratory and at Argonne National Laboratory and the decision to construct the new Spallation Neutron Source (SNS) have dominated the decade as far as large national facilities are concerned. It is particularly crucial to move forward with construction of the SNS to make the United States competitive in neutron-scattering studies. Once the SNS is commissioned at Oak Ridge National Laboratory, U.S. researchers can begin the process of recapturing the lead in the use of neutrons to study structure and spins in superconductors, polymers, and other materials.

We can confidently predict a rapid growth in knowledge and understanding of biological materials and living organisms resulting from the exploitation of the Advanced Photon Source by the biology community. That progress will be accelerated once the SNS becomes operational. The extensive deployment of various microscopies in many laboratories can be expected to implement great strides in understanding surface physics.

The rapid pace of development of research at various scales has depended on accelerator science and research on the physics of smaller instrumentation. This work has had to be parasitic on various enterprises, despite the fact that advances in scientific equipment propel science forward in great leaps. How much better could we exploit the leverage that instrumentation research has if we were to recognize it as an important enterprise worthy of *planned* investment! The institutional frameworks for such investments clearly depend on scale, but there are natural environments at each level. Expertise in tomorrow's beam physics, for example, partly resides at the major centers of high-energy physics research, but development for low-energy applications will likely occur elsewhere. The materials research science and engineering centers are one set of obvious potential homes for intermediate-scale instrumentation development.

FINDINGS AND RECOMMENDATIONS

Condensed-matter and materials physics is entering an era of great excitement and anticipation as powerful new experimental and computational capabilities are brought to bear on some of the most fundamental scientific and technical challenges of our time. Underlying these challenges is the knowledge that drives the information revolution, modern materials technology, and biotechnology enabled by understanding of the molecular basis of life. We have seen astounding developments over the past decade such as buckyballs and carbon nanotubes, high-temperature superconductivity, giant and colossal magnetoresistance, and large-scale quantum phenomena. The next decade, enriched by powerful new research tools, promises to be even more extraordinary.

New capabilities for condensed-matter and materials physics research include spectacular advances in the atomic-scale characterization and manipulation of materials, computer simulations of large interacting systems, and the ability to relate properties and phenomena from molecular- to macroscopic-length scales. These new capabilities are uniting the worlds of atomic-scale behavior and macroscopic phenomena in ways that provide avenues for understanding and designing materials and processes from the atoms up. In turn, this new understanding holds the promise of breakthroughs at a time when the limits of incremental progress are being tested in materials-based technologies ranging from integrated circuits and magnetic storage devices to the synthesis of advanced polymers to the performance of materials under extreme conditions. Perhaps the greatest impact will be felt at the interface between biology and physics, where the convergence of condensed-matter and materials physics and molecular biology is expected to drive important advances in the fundamental understanding of biological processes.

Condensed-matter and materials physics faces critical challenges in realizing this future. Investments in facilities and research infrastructure are essential to provide a world-class research environment and to enable breakthrough opportunities. Partnerships across disciplines and among universities, government laboratories, and industry are essential to leverage resources and strengthen interdisciplinary research and connections to technology. Finally, special attention must be given to condensed-matter and materials physics education to ensure the availability of intellectual capital to sustain the vitality of the field and its contributions to society.

Research Infrastructure

The United States has a strong foundation of research groups and small-scale centers located in universities and government laboratories. Centers play an essential role in a number of areas including microcharacterization, processing, synthesis, and state-of-the-art instrumentation development. Research groups and centers are a crucial reservoir of expertise. They also play an important institutional role by providing a meeting ground for research and development personnel in industry and students and researchers in universities and government laboratories. Centers bring together the problem-definition capabilities of industry with the educational role of the universities and the research missions of government laboratories. As a result, leading-edge research capabilities are applied to important areas of microcharacterization, processing, synthesis, and instrumentation. Centers also bring a long-term commitment to applying intellectual excellence to research problems and to developing expertise in the next generation of researchers in these essential areas of study.

The role that small-scale centers now play has been fostered also in major industrial research laboratories as well as by the research strategy of the Department of Defense. But the burden now falls much more heavily on research groups and small centers in universities and government laboratories. Therefore, it is appropriate to strengthen this part of the nation's research infrastructure. Additional support, long-term commitments, and oversight structures that involve all the interested parties (universities, government laboratories, and industry) are necessary to accomplish this fortification.

Recommendations for Upgrading the Infrastructure

- The National Science Foundation, the U.S. Department of Energy, the U.S. Department of Defense, and other agencies that support condensed-matter and materials physics research should continue to nurture the core research at the heart of the field. The research areas described in the Overview provide a guide to the scientific arenas at the forefront of this work.
- The agencies that support and direct research should plan for increased investment in modernizing the condensed-matter and materials physics research infrastructure at universities and government laboratories.
- The National Science Foundation should increase its investment in state-of-the-art instrumentation and fabrication capabilities, including centers for instrumentation R&D, nanofabrication, and materials synthesis and processing at universities. The Department of Energy should increase its support for such programs at national laboratories and universities.

Major Facilities

The emergence of national synchrotron and neutron facilities has revolutionized our understanding of the atomic-scale structure and dynamics of materials. The nation is fortunate to have world-class facilities for synchrotron research. However, the situation is strikingly different for neutrons, where we find ourselves with fewer facilities than those judged inadequate by national review committees more than a decade ago. Many of the advances in structural biology, polymers, magnetic materials, and superconductivity depend on access to state-of-the-art neutron-scattering facilities. Without a new neutron source, the nation cannot be competitive in these and other areas of enormous scientific and technological significance. This is an urgent and immediate need, and the committee strongly recommends construction of the Spallation Neutron Source (SNS). Upgrades at existing neutron-scattering facilities are also essential to sustaining neutron-scattering research in the United States during SNS construction as well as to strengthen the field and provide broad access to the user community.

Over the past decade there has been an explosion in the use of synchrotron facilities. A great success of these facilities has been the rapid growth in their use across the broad spectrum of science. At national synchrotron facilities biologists are attacking the structure of biological molecules, chemists are improving drug designs, and environmental scientists are following the migration of environmental pollutants, all alongside materials researchers. In fact, life scientists are the fastest growing user community at national synchrotron facilities and currently occupy more than 25 percent of available beam time. This large influx of users from other scientific disciplines is remarkably productive and is creating a healthy scientific melting pot, but it is also straining the ability of the facilities to respond. Modest investments at existing synchrotron facilities can greatly expand current capabilities and help alleviate this problem. On the horizon are fourth-generation light sources that offer enormous gains in intensity and coherence over existing sources. The committee recommends upgrades at existing synchrotron facilities and research and development to explore possible options for fourth-generation sources.

Recommendations for Major Materials Research Facilities

- The insufficiency of neutron sources in the United States should be addressed in the short term by upgrading existing neutron-scattering facilities and in the long term by the construction of the Spallation Neutron Source.
- Support for operations and upgrades at synchrotron facilities, including research and development on fourth-generation light sources, should be strengthened.
- The broad use of synchrotron and neutron facilities across scientific disciplines and sectors should be considered when establishing agency budgets.

Partnerships

Condensed-matter and materials physics is becoming increasingly interrelated with other fields of science and technology, with important links to many disciplines including other branches of physics, chemistry, materials science, biology, and engineering. At the same time, the field has advanced to the point where it is often impractical and sometimes impossible to assemble in one place all of the intellectual resources and specialized equipment for a given research project. Continued progress in the field depends on establishing effective partnerships across disciplines and among universities, government laboratories, and industry. These partnerships enable cross-disciplinary research, leverage resources, and provide awareness of technological drivers and potential applications. The extraordinary scientific and technological success of the major industrial laboratories over the past half-century resulted from their ability to integrate long-term fundamental research, cross-disciplinary teams involving experimentalists and theorists, materials synthesis and processing, and a strategic intent. Virtual elements of this fertile ground exist in potential partnerships among universities, government laboratories, and industry. Federal R&D agencies should encourage partnerships that recreate this environment in appropriate subfields of condensed-matter and materials physics.

Recommendations for Research and Development Partnerships

• Federal agencies should provide incentives for formation of partnerships among universities and government and industrial laboratories that carry out research in condensed-matter and materials physics.

These research and development partnerships should be encouraged in order to:

— Optimize the use of infrastructure and facilities,
— Enable cross-disciplinary research,
— Improve university and government laboratory appreciation of industry priorities and needs,
— Share the risks and returns of long-term research, and
— Assemble teams that can recreate the fertile research environment of the large industrial research laboratories of the past half-century.

These partnerships should be fostered by:

— Making resources available through special programs that stimulate partnerships,
— Developing effective protocols for resolving intellectual property issues in cooperative research,
— Encouraging university and government laboratory internships and sabbaticals in industry, and
— Requiring partners to have a stake in the partnership (e.g., for universities and government laboratories, the partnership should add value to core missions).

Education

Intellectual capital is probably the single most important investment for science and technology. Intellectual capital in condensed-matter and materials physics occupies a special place in the national economy, underpinning many of the technological advances that drive economic growth. The U.S. system of graduate education, research universities, government and industrial laboratories, and national facilities for condensed-matter and materials physics is a major reason for rapid progress in research and technological applications. Maintaining this progress requires continued commitment to strengthening these institutions. In addition, condensed-matter and materials physics must play a crucial role in engaging undergraduates in research and improving their understanding of science and technology. Making investments to develop the human capital essential for leadership in condensed-matter and materials physics and related technologies will pay rich dividends to the nation. Successful accomplishment of these objectives will also help the larger field of physics to adjust to a new role in which economic security becomes the dominant justification for national investments in research.

Recommendations for Education

- Universities should endeavor to enhance their students' understanding of the role of knowledge integration and transfer as well as knowledge creation.
- Universities should develop ways to bring the excitement and creativity of research and discovery into education at an earlier stage.
- University departments should consider new professional degree programs that link undergraduate physics education with, for example, engineering-oriented disciplines.
- Universities should foster joint academic appointments across departments to break down disciplinary barriers. Campuses should experiment with the creation of "virtual departments" to aid intellectual restructuring to better achieve their research and education missions in changing times.
- Universities need to expand existing programs that enable undergraduates to have research experiences in faculty laboratories or summer internships in industry or at national laboratories.
- Universities should recognize the importance of knowledge integration and transfer in addition to knowledge creation.
- Applied physics departments and programs should link to industrial liaison programs, which generally are strong in colleges of engineering.
- Agencies, particularly the National Science Foundation, should provide incentives for action in these areas.

RESEARCH THEMES

Throughout this study the themes of new experimental and computational capabilities, the ability to address problems of increasing complexity, and the importance of relationships with other fields pervade the subdisciplines of condensed-matter and materials physics (see Box O.1). These themes provide a sense of vitality and optimism for the future of condensed-matter and materials physics. Maintaining scientific excellence, a long-term perspective, and a world-class environment for research are essential. Investing in facilities, encouraging partnerships, integrating research and education, and encouraging discovery are critical elements. But where is the field headed? Although it is often dangerous to predict the future in science, the committee identified 10 areas that span and underpin the subdiscipline-specific scientific priorities of condensed-matter and materials physics as described in the body of this report. These areas, listed in Box O.1, encompass the committee's view of the high-level strategic priorities that have emerged from the internal dynamics of the field and that are likely to

BOX O.1 Strategic Scientific Themes in Condensed-Matter and Materials Physics

- The quantum mechanics of large, interacting systems
- The structure and properties of materials at reduced dimensionality
- Materials with increasing complexity in composition, structure, and function
- Nonequilibrium processes and the relationship between molecular and mesoscopic properties
- Soft condensed matter and the physics of large molecules, including biological structures
- Controlling electrons and photons in solids on the atomic scale
- Understanding magnetism and superconductivity
- Properties of materials under extreme conditions
- Materials synthesis, processing, and nanofabrication
- Moving from empiricism toward predictability in the simulation of materials properties and processes

characterize condensed-matter and materials physics research over the next decade.

Condensed-matter and materials physics lies at the heart of revolutionary advances in broad areas of science and technology. The next decade promises exciting new discoveries and technology impacts as powerful new capabilities in synchrotron and neutron research, atomic-scale visualization, nanofabrication, computing, and many other areas probe the secrets of materials and materials-related phenomena. This is a new era. Vast new arenas, ranging from subtle quantum phenomena, to macromolecular science, to the realm of complex materials, are increasingly accessible to fundamental study. It is a time of exceptional opportunity to perform pioneering research at the technological frontier—a frontier enabled by advances in condensed-matter and materials physics.

1
Electronic, Optical, and Magnetic Materials and Phenomena: The Science of Modern Technology

Important and unexpected discoveries have been made in all areas of condensed-matter and materials physics in the decade since the Brinkman report.[1] Although these scientific discoveries are impressive, perhaps equally impressive are technological advances during the same decade, advances made possible by our ever-increasing understanding of the basic physics of materials along with our increasing ability to tailor cost-effectively the composition and structure of materials. Today's technological revolution would be impossible without the continuing increase in our scientific understanding of materials, phenomena, and the processing and synthesis required for high-volume, low-cost manufacturing. The technological impact of such advances is perhaps best illustrated in the areas of condensed-matter and materials physics discussed in this chapter, which will examine selected examples of electronic, magnetic, and optical materials and phenomena that are key to the convergence of computing, communication, and consumer electronics.

Technology based on electronic, optical, and magnetic materials is driving the information age through revolutions in computing and communications. With the miniaturization made possible by the invention of the transistor and the integrated circuit, enormous computing and communication capabilities are becoming readily available worldwide. These technological capabilities, which enabled the information age, are fundamentally changing how we live, interact, and transact business. Semiconductors provide an excellent demonstration of the strong

[1]National Research Council [W.F. Brinkman, study chair], *Physics Through the 1990s*, National Academy Press, Washington, D.C. (1986).

interplay between and interdependence of science and technology. Perhaps in no other area are advances in technology more closely linked to advances in understanding.

This chapter is not intended to be comprehensive; rather, it seeks to illustrate the pivotal role of condensed-matter and materials research in providing the understanding required to develop enabling technologies. At the same time, the development of these new technologies has greatly expanded the tools and capabilities available to scientists and engineers in all areas of research and development, ranging from basic research in physics and materials to other areas of physics and to such diverse fields as medicine and biotechnology. The examples discussed also make evident the importance of long-term, sustained research in realizing the benefits to society of improved scientific understanding of materials (see Figure 1.1).

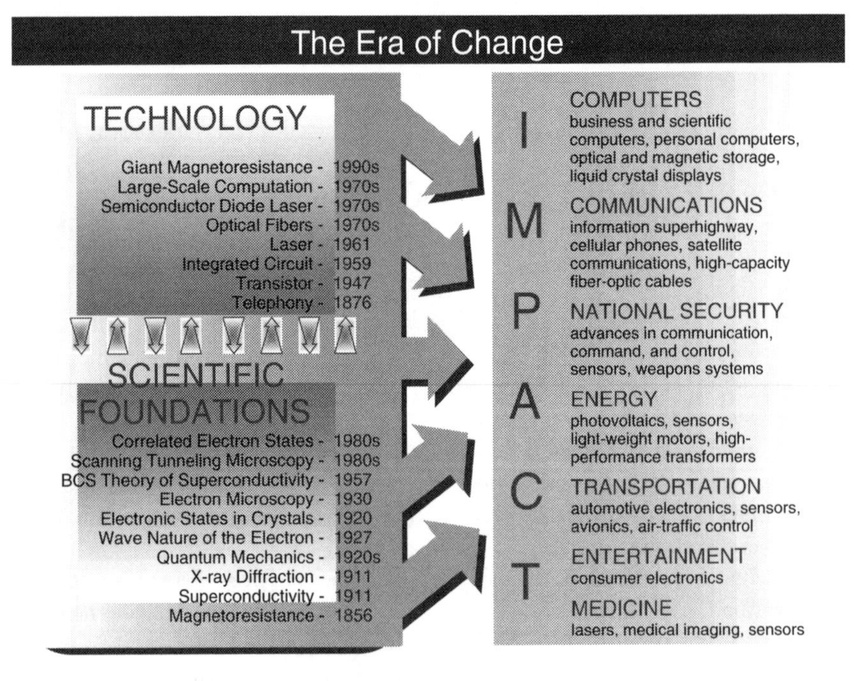

FIGURE 1.1 Incorporation of major scientific and technological advances into new products can take decades and often follows unpredictable paths. Illustrated here are some selected technologies supported by the foundations of electronic, photonic, and magnetic phenomena and materials. These technologies have enabled breakthroughs in virtually every sector of the economy. The two-way interplay between foundations and technology is a major driving force in this field. The most recent fundamental advances and technological discoveries have yet to realize their potential.

Although technological advances today are most often associated with the information age or communications and the computing revolution, impressive advances continue to be made across a broad spectrum of technologies and scientific disciplines (see Box 1.1). For example, progress in condensed-matter and materials physics has led to advances in biology, medicine, and biotechnology. New tissue diagnostics based on diffusing light probes use understanding borrowed directly from the physics of carrier transport in mesoscopic random materials. The development of new optical microscopies, such as two-photon confocal, optical coherence, and near-field optical microscopy, together with the widespread use of optical tweezers, have started a revolution in the observation and manipulation of submicrometer-sized objects in cell biology, in new forms of spectroscopic endoscopy, and in gene sequencing techniques. The emergence of high-power solid-state lasers and solid-state detectors and the widespread use of fiber optics make new optical approaches for diagnostics, dentistry, and surgery increasingly easy. A new form of magnetic resonance imaging enabled by semiconductor laser pumping of spin-polarized xenon gas has allowed the three-dimensional mapping of lung function. The generation of femtosecond pulses of light by the use of new solid-state lasers has begun another revolution in our understanding of the subpicosecond dynamics of biological molecules on the important frontiers of molecular signal processing and protein folding. Although not covered in detail here, such advances in the use of optics in medicine and biology are discussed in detail in another National Research Council report.[2] In addition, semiconductor and other solid-state lasers or enhanced solid-state detector arrays, offshoots from condensed-matter physics, are enabling major advances in the fields of atomic and molecular physics, physical chemistry, high-energy physics, and astrophysics. New optical materials and phenomena are also responsible for a number of advances in the technologies associated with printing, copying, video and data display, and lighting.

In the realm of magnetic materials, the loss of cobalt in the 1980s because of political unrest in Zaire prompted an intense research effort to find cobalt-free bulk magnetic materials. This led to major advances in creating magnetic structures from neodymium and iron, which had superior properties and lower cost compared with cobalt alloys for electric motors and similar applications requiring magnets with high permanent magnetization. These new magnets, which are achieved through complex alloys and even more complex processing sequences, are vastly expanding the industrial use of bulk magnetic materials.

Advances in magnetic materials and their applications are not limited to bulk materials with high permanent magnetization and magnetic materials used in information storage. Improvements in soft bulk magnetic materials play an important role in transformers used in the electric power distribution industry. In-

---

[2]National Research Council [C.V. Shank, study chair], *Harnessing Light: Optical Science and Engineering for the 21st Century*, National Academy Press, Washington, D.C. (1998).

BOX 1.1 The Science of Information-Age Technology

The predominant semiconductor technology today is the silicon-based integrated circuit. The silicon integrated circuit is the engine that drives the information revolution. For the past 30 years, this technology has been dominated by Moore's Law: that the density of transistors on a silicon integrated circuit doubles about every 18 months.* Moore's Law articulates the increased functionality-per-unit cost that is the origin of the information revolution. Today's computing and communications capability would not be possible without the phenomenal 25 to 30 percent per year exponential growth in capability per unit cost since the introduction of the integrated circuit in about 1960. That sustained rate of progress has resulted in high-density memories with 64 million bits on a chip and complex, high-performance logic chips with more than 9 million transistors on a chip. This trend is projected to continue for the next several years (see Figures 1.1.1 and 1.1.2).

If the silicon integrated circuit is the engine that powers the computing and communications revolution, optical fibers are the highways for the information age. Although fiber optics is a relatively recent entrant into the high-technology arena, its impact is enormous and growing. Fiber is now the preferred technology for transmission of information over long distances. There are already approximately 30 million km of fiber installed in the United States and an estimated 100 million km worldwide. In part because of the faster than exponential growth of connections to the Internet, optical fiber is being installed worldwide at an accelerated rate of

PROCESSING POWER ON SINGLE CHIP

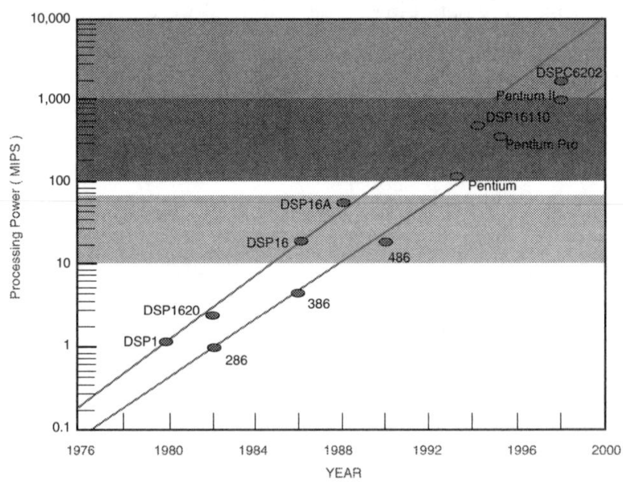

FIGURE 1.1.1 Computing power versus time in microprocessors. (Courtesy of Intel.)

*Moore's Law, first articulated by Gordon Moore of Intel Corporation, is not a statement of physics. It is a statement that the industry will perform the R&D necessary and supply capital investment at the rate required to achieve this doubling rate.

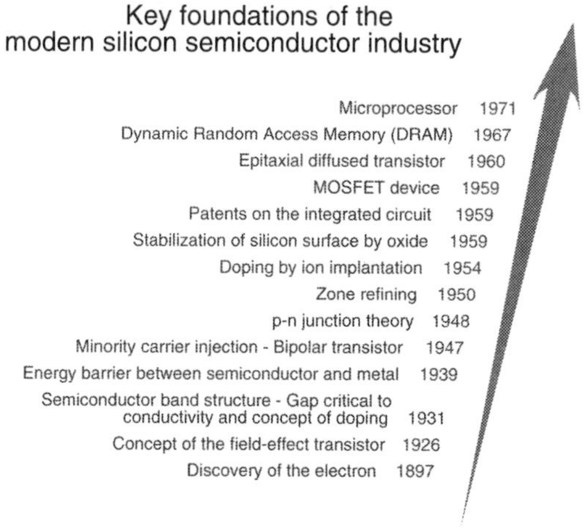

FIGURE 1.1.2 History of semiconductor technology.

more than 20 million km per year—more than 2,000 km per hour, or around Mach 2. In addition, the rate of information transmission down a single fiber is increasing exponentially by a factor of 100 every decade. Transmission at 2.5 terabits per second has been demonstrated in the research laboratory, and the time lag between laboratory demonstration and commercial system deployment is about 5 years. The analog of Moore's Law for fiber transmission capacity, which serves as a technology roadmap for lightwave systems, is shown in Figure 1.1.3. Figure 1.1.4 summarizes the history of optical communications technology.

Compound semiconductor diode lasers provide the laser photons that transport information along the optical information highways. Semiconductor diode lasers are also at the heart of optical storage and compact disc technology.

In addition to their use in very high-performance microelectronics applications, compound semiconductors have proven to be an extremely fertile field for advancing our understanding of fundamental physical phenomena. Exploiting decades of basic research, we are now beginning to be able to understand and control all aspects of compound semiconductor structures, from mechanical through electronic to optical, and to grow devices and structures with atomic layer control, in a few specific materials systems. This capability allows the manufacture of high-performance, high-reliability, compound semiconductor diode lasers that can be modulated at gigahertz frequencies to send information over the fiber-optic networks. High-speed semiconductor-based detectors receive and decode this information. These same materials provide the billions of light-emitting diodes sold annually for displays, free-space or short-range high-speed communication, and numerous other applications. In addition, very high-speed, low-power compound semiconductor electronics play a major role in wireless communication, especially for portable units and satellite systems.

Another key enabler of the information revolution is low-cost, low-power, high-density information storage that keeps pace with the exponential growth of com-

BOX 1.1 Continued

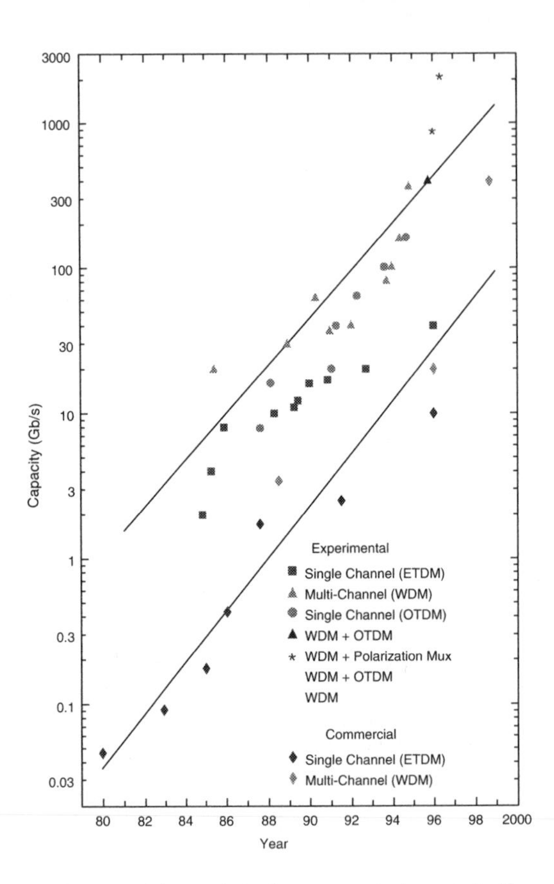

FIGURE 1.1.3 Exponential growth in data transmission rate in fibers. (Courtesy of Bell Laboratories, Lucent Technologies.)

puting and communication capability. Both magnetic and optical storage are in wide use. Recently, the highest performance magnetic storage/readout devices have begun to rely on giant magnetoresistance (GMR), a phenomenon that was discovered by building on more than a century of research in magnetic materials. Although Lord Kelvin discovered magnetoresistance in 1856, it was not until the early 1990s that commercial products using this technology were introduced (see Figure 1.1.5). In the past decade, our understanding of condensed matter and materials converged with advances in our ability to deposit materials with atomic-level control to produce the GMR heads that were introduced in workstations in late 1997. It is hoped that with additional research and development, spin valve and colossal magnetoresistance (CMR) technology may be understood and applied to workstations of the future. This increased understanding, provided in part

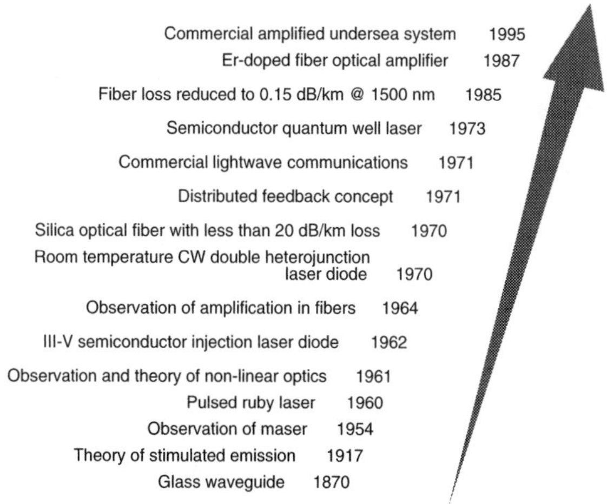

Commercial amplified undersea system 1995
Er-doped fiber optical amplifier 1987
Fiber loss reduced to 0.15 dB/km @ 1500 nm 1985
Semiconductor quantum well laser 1973
Commercial lightwave communications 1971
Distributed feedback concept 1971
Silica optical fiber with less than 20 dB/km loss 1970
Room temperature CW double heterojunction laser diode 1970
Observation of amplification in fibers 1964
III-V semiconductor injection laser diode 1962
Observation and theory of non-linear optics 1961
Pulsed ruby laser 1960
Observation of maser 1954
Theory of stimulated emission 1917
Glass waveguide 1870

FIGURE 1.1.4 History of optical communications technology.

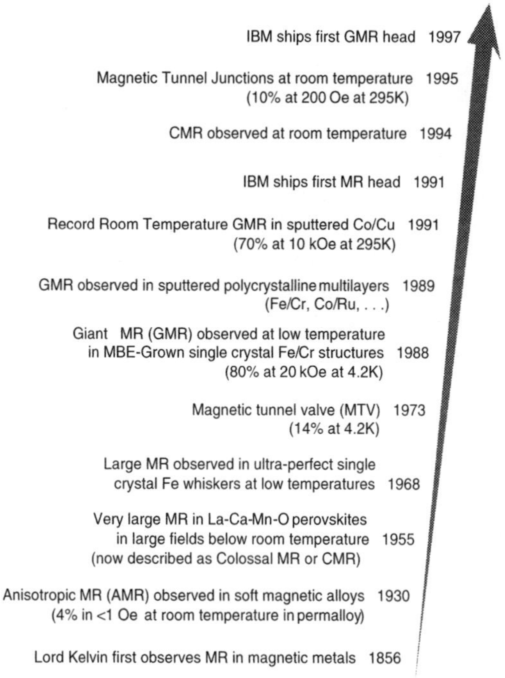

IBM ships first GMR head 1997
Magnetic Tunnel Junctions at room temperature 1995 (10% at 200 Oe at 295K)
CMR observed at room temperature 1994
IBM ships first MR head 1991
Record Room Temperature GMR in sputtered Co/Cu 1991 (70% at 10 kOe at 295K)
GMR observed in sputtered polycrystalline multilayers 1989 (Fe/Cr, Co/Ru, . . .)
Giant MR (GMR) observed at low temperature in MBE-Grown single crystal Fe/Cr structures 1988 (80% at 20 kOe at 4.2K)
Magnetic tunnel valve (MTV) 1973 (14% at 4.2K)
Large MR observed in ultra-perfect single crystal Fe whiskers at low temperatures 1968
Very large MR in La-Ca-Mn-O perovskites in large fields below room temperature 1955 (now described as Colossal MR or CMR)
Anisotropic MR (AMR) observed in soft magnetic alloys 1930 (4% in <1 Oe at room temperature in permalloy)
Lord Kelvin first observes MR in magnetic metals 1856

FIGURE 1.1.5 History of magnetoresistance. (Courtesy of IBM Research.)

BOX 1.1 Continued

by our increased computational ability arising from the increasing power of silicon integrated circuits, coupled with atomic-level control of materials, led to exponential growth in the storage density of magnetic materials analogous to Moore's Law for transistor density in silicon integrated circuits (see Figure 1.1.6).

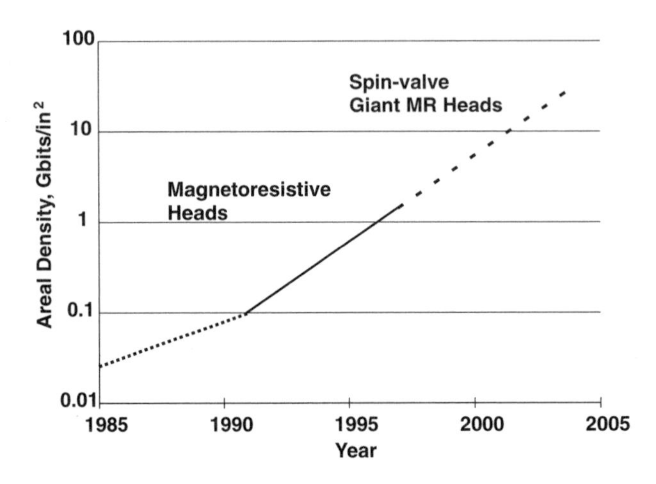

FIGURE 1.1.6 "Moore's Law" for magnetic storage: logarithm of storage density versus time. (Courtesy of IBM Research.)

creases in the magnetic permeability and decreases in the losses by incremental improvements in amorphous metglass leads to decreases in the losses suffered in transmission and distribution. Magnetoelectronics, an emerging area based on advances in the understanding of the properties and processing of magnetic materials, shows promise for future applications.

Despite the numerous recent discoveries and technological advances in the understanding and use of magnetic materials, our fundamental understanding of magnetism remains remarkably incomplete. Some of the basic questions and important challenges in magnetism facing the scientific community are discussed in this chapter.

ELECTRONIC MATERIALS AND PHENOMENA

Materials and Physics That Drive Today's Technology

Silicon Technology

As noted in the introduction, semiconductor technology is the key enabler of the information age. The science of materials as a specific discipline is a relatively modern development. The physics and materials science of semiconductors is an even more recent development. Metals and ceramics were commercially important materials when the transistor was demonstrated about 50 years ago. Despite the fact that the science of semiconductors is relatively new compared to that of metals and ceramics, the commercial importance of semiconductors is now comparable to that of metals and ceramics. Advances in semiconductor technology are driving the rapid growth of business sectors involved with computing, communications, consumer electronics, and software, and are enabling emerging fields such as biotechnology. Today's transistor performance and the incredible advances of integrated circuits in silicon technology are the result of more than 50 years of dedicated research in electronic materials. The understanding achieved from this focused research has enabled high-volume manufacturing of circuits with ever-increasing complexity and performance.

In addition to driving computing and communications, the steady decrease in cost-per-function has created literally hundreds of applications for silicon integrated circuits. Semiconductors are ubiquitous. Microprocessors are used almost everywhere today—from household appliances, to banking and smart cards, to automotive and aircraft control systems, automatic fuel injection engines, and cockpit instrumentation—and will be in even more applications tomorrow. The same sustained rate of progress that permitted the widespread application of semiconductors has created a global semiconductor industry with 1997 revenues of about $150 billion, supported by a materials and equipment infrastructure of about $60 billion. Semiconductor technology is also the heart of the $1 trillion global electronics industry and vital in many other areas of the approximately $33 trillion global economy.

The increasing functionality of integrated circuits, which comes as a by-product of scaling to smaller feature sizes, has been achieved by comparable increases in their complexity and that of the attendant manufacturing process. Today's leading-edge microprocessors are manufactured with minimum feature sizes of 250 nm and require six levels of metallization to connect the transistors and circuit components. A beneficial by-product of the steady decrease in feature size is higher speed devices and circuits. Based on technology projections that form the basis of the *National Technology Roadmap for Semiconductors*,[3] the semiconductor industry expects to manufacture integrated circuits with feature sizes of 180 nm in 1999 and 150 nm by 2001. If the scaling trend continues as indicated by Moore's Law, which the industry has followed since its inception, integrated circuits with minimum feature sizes of 50 nm will be manufactured in high volumes within 15 years (see Box 1.2). Continuing to advance this technology requires that the industry invest in expensive new manufacturing facilities and an ever-increasing scientific understanding and control of semiconductor

[3]Semiconductor Industry Association, *National Technology Roadmap for Semiconductors*, SEMATECH, Austin, Tex. (1997).

BOX 1.2 A Brief History of Ion Implantation

In the early 1940s, the basic machines that were later adapted for ion implantation in the semiconductor business were used at Oak Ridge National Laboratory for uranium isotope separation. This was a critical part of the Manhattan Project. Ion beams were first used as part of semiconductor-device processing at Bell Laboratories in 1952. Bell filed a comprehensive patent in 1954 covering the use of ion implantation for doping semiconductors, but it was not until 1966 that implantation was actually used to manufacture commercial semiconductor devices.

Hughes Research Laboratory used the technique to form junctions in the manufacturing of diodes. In 1970 Texas Instruments began using ion implantation in integrated circuit manufacturing to set threshold voltages. Concurrent with these developments in processing, several companies attempted to enter the implant-tool manufacturing business with only moderate success, most successful among them being Accelerators Incorporated. In 1971, however, a new company, Extrion, was formed to build commercial implanters specifically designed for integrated circuit manufacturing. Extrion soon became the primary supplier of implant tools. This led to the development of a whole new industry in America.

Today, ion implantation is used in several steps of the integrated circuit manufacturing process to control the concentration and depth distribution of dopants. Ion implantation tool manufacturing, an almost exclusively U.S. industry, has grown to a more than $1 billion per year business. Three U.S. companies (Applied Materials, Eaton, and Varian) supply virtually all the commercial ion-implant systems worldwide.

materials and manufacturing processes. Conversely, our rate of understanding has been greatly enhanced by the technology created by the rapid advances in semiconductor-related technologies.

Many daunting scientific and engineering problems must be overcome in order to continue at the Moore's Law rate of progress for the next 15 or even 10 years. For instance, the number of wires needed to connect the transistors grows as a power of the number transistors. As transistor dimensions are shrunk, computer chip manufacturers pack an ever-increasing number of them into their devices. The complexity of wiring the transistors in these devices may eventually reach the limits of known materials. Moreover, the cost of manufacturing increasingly layered and complex wiring structures may limit the performance of these systems. Even if solutions to the interconnect problem can be identified, continued scaling of silicon technology will ultimately encounter fundamental limits. For example, metal-oxide semiconductor transistors can be built today with gate lengths of 30 nm (only about 150 atoms long) that display high-quality device characteristics. Manufacturing complex circuits that rely on devices with these feature sizes will require several hundred processing steps with atomic-level control. However, the performance of complex integrated circuits with tens of millions of transistors may be degraded because of nonuniform operating characteristics. In time, continued decreases in device dimensions may result in the information being carried by an ever-decreasing number of charge carriers; ultimately, simple statistical fluctuations will limit the uniformity of device characteristics as the number of charges used to convey information decreases.

To delay this limitation as long as possible, research is under way on new materials with a high dielectric constant for memory applications or to limit leakage from tunneling currents. As understanding of synthesis and processing increase, ferroelectric materials are being introduced for nonvolatile memory applications. Even with these advances, as feature sizes continue to decrease, integrated circuits based on field-effect transistors will eventually encounter fundamental limits, such as interconnect delays caused by the ever-increasing number of interconnects, heat generation, or quantum limits of transistors that are too small to confine the electrons in the channels. Today's approach to the design and manufacture of integrated circuits will no longer be extendible to smaller feature sizes and higher densities. The fundamental limits of the current technology and our addiction to exponentially increasing computational power offer exciting scientific and engineering challenges in the search for the materials and device structures of the future.

Compound Semiconductor Technology

Compound semiconductors, which consist of more than one element, offer intrinsically higher speed and lower noise compared with silicon. These advantages have been exploited to develop very high frequency electronic devices and circuits for microwave and wireless communication applications.

The worldwide electronics market for compound semiconductors is estimated to be growing at about 40 percent per year and is expected to be about $1 billion in 2000. In addition to the high-speed microwave applications for which they have long been the materials of choice, discrete components are widely used in the low-noise receivers of telephone handsets. Compound semiconductors such as GaAs, AlGaAs, InGaAs, SiGe, GaN, and GaAlN are key to the development of next-generation wireless telephones, which will use higher frequency microwaves to transmit more information and allow more channels. GaN transistors, for example, have a high breakdown voltage and great robustness, although extensive research and development is required before the material can be understood and fabricated in a well-controlled fashion. Advancing the limits of semiconductor materials technology is essential for increasing the speed of transistors and advancing our ability to modulate light-emitting diodes and semiconductor lasers for high-speed optical information transmission. Because compound semiconductors are composed of more than one element, they offer a vastly increased range of materials from which to create structures with desired electronic properties.

The technology of modern compound semiconductor device fabrication is predicated on the ability to produce extended planar layers of uniform composition and thickness. Band-structure engineering through complex heterostructures formed from combinations of compound semiconductors greatly increases the performance, potential applications, and research opportunities. Heterostructure devices such as heterojunction bipolar transistors, field-effect transistors, semiconductor lasers, or light-emitting diodes require the presence of epitaxial layers with different compositions and well-controlled thicknesses in the same device.

This promise can be realized with manufacturing techniques such as molecular beam epitaxy or metal-organic chemical vapor deposition, developed in the 1960s by materials physicists. These techniques allow atomic layer control in the growth of one material on another in single atomic layers to produce materials not found in nature. The use of novel forms of microscopy for fabrication and testing, and the development of comprehensive modeling techniques that take into account all of the materials physics and carrier dynamics of the structures, will determine our ability to design and build such structures on the atomic scale with feature sizes comparable to the quantum de Broglie wavelength of the electrons.

The ability to span the broad spectral region from ultraviolet to long-wavelength infrared was enhanced by the invention of strained-layer systems. Very high crystal quality layers can be grown for systems with different equilibrium atomic spacing, provided the thickness of the layer is less than the critical layer thickness at which detrimental dislocations nucleate and grow. Superlattices of such systems are called "strained-layer superlattices," whereas single layers sandwiched between two layers with identical lattice constant are called either "strained-quantum well" or "pseudomorphic structures." The recognition that alloys with different equilibrium lattice constants, such as $GaIn_{1-x}As_x$ on GaAs, could be grown epitaxially to create strained layers opened up major opportunities for "band-gap engineering." These so-called pseudomorphic layers removed the constraint that epitaxial materials must have equilibrium atomic spacings that match the atomic lattice spacings of the substrate. Recently, dislocation-free growth of a different lattice constant material has been obtained on a thin layer of compliant substrate that has been wafer bonded at an angle with respect to the bulk substrate lattice. Such strained thin layers offer an additional degree of control over the electronic band structure of the resulting artificially structured material.

The ability to create nearly ideal two-dimensional electron gases (2DEGs) through the growth of artificially structured materials with charge carriers confined to potential wells, and modulation doping, in which the dopant impurities that donate charge carriers are located far away from the potential wells that confine the charge carriers, allowed the development of very high performance, high electron-mobility transistors (HEMTs). Today's highest performance transistors, in terms of speed-power product and noise, are pseudomorphic HEMTs.

The ability to fabricate complex structures with atomic-level control permits research into fundamentally new structures for technology applications. One example of a concept that is beginning to receive attention is that of quantum-state logic, in which a device can be switched between multiple states, in contrast to a field-effect transistor that is either "on" or "off."

The creation of heterostructures and quantum wells also led to the development of new tools to study the fundamental physical phenomena associated with these artificially structured materials, greatly increasing our understanding of their electronic band structure and optical properties. One of the recent examples is the use of ballistic electron energy microscopy to measure the energy band offsets in buried heterostructures (see Box 1.3).

The Science Underlying Semiconductor Microelectronics Technology

As noted earlier, for the past 30 years silicon technology has been dominated by Moore's Law: the density of transistors on a silicon integrated circuit doubles about every 18 months. This increased functionality-per-unit cost is the basis of the information revolution. The same technology that allows us to shrink the size of devices has allowed us to learn new physics. The synergy of technological development and new physics has been remarkably successful in the past few years, and it is difficult to anticipate the many new directions that synergy will take in the next decade or two.

The technology that fuels Moore's Law rests on the ability to make high-quality silicon field-effect transistors. In these transistors a metal gate confines electrons near the interface between silicon and a SiO_2 gate oxide. As feature sizes decrease, quantum mechanical effects become observable. Combining the small feature size available with electron-beam lithography with very high quality artificially structured materials that can be grown in compound semiconductors led to the discovery of entirely new physical effects, such as the fractional quantum Hall effect (FQHE) discovered in two-dimensional electron systems. The elegant physics underlying the FQHE is discussed in Chapter 3.

To continue the increasing levels of integration beyond the limits mentioned above, new approaches and architectures are required. One of the alternative approaches that has received recent attention is quantum-dot (see Box 1.4) cellular automata (QCA). QCA is an approach that takes advantage of the quantum tunneling between dots to perform the operations that transistors perform. In today's digital integrated circuit architectures, transistors serve as current switches to charge and discharge capacitors to the required logic voltage levels. In QCA, logic states are encoded by the positions of individual electrons rather than by voltages. Such structures are scalable to molecular levels, and the performance of the device improves as the size decreases; artificially structured QCA cells studied to date operate only at low temperatures, but molecular-sized QCA cells would function at room temperature.

Electron-beam lithography was developed to make very small semiconductor devices. The short wavelength of electrons allows one to pattern structures with

BOX 1.3 Ballistic Electron Emission Microscopy

Ballistic electron emission microscopy (BEEM) is a variant of scanning-tunneling microscopy (STM) which probes, with nanometer lateral resolution, the subsurface electronic properties of materials. The technique was first developed in 1988 to investigate lateral variations of Schottky barriers formed at metal/semiconductor interfaces and has since been applied to a wide variety of materials systems. The study of semiconductor heterostructures and quantum structures buried beneath the Schottky barrier has shown particular promise because BEEM is able to access length scales smaller than those available to more traditional techniques.

In BEEM, an STM tip locally injects hot carriers through a thin metal layer and over the Schottky barrier without making direct contact. Carriers that are further transmitted through the heterostructure are collected by a third terminal usually located at the semiconductor substrate. As the tip is scanned or its bias is changed, changes in the collected BEEM current give spectroscopic information about the heterostructure, such as heights of band offsets or positions of quantized electronic states.

Figure 1.3.1 shows schematically a BEEM measurement of ~50 nm-diameter GaSb self-assembled quantum dots located ~7.5 nm below an Au/GaAs interface. Because the dots are close to the surface, their profiles can be seen in the STM topography, so that the tip can be located on and off the dot for comparison. Fig-

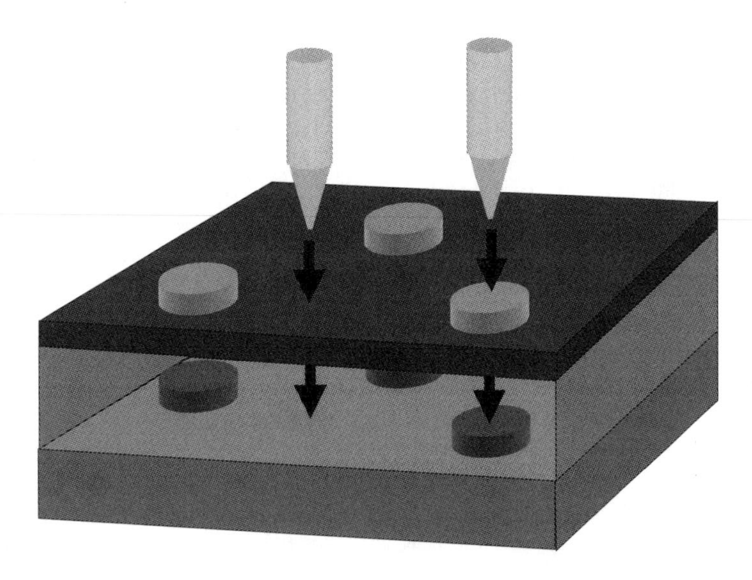

FIGURE 1.3.1 BEEM measurement of self-assembled quantum dots. [Reprinted from L.D. Bell and V. Narayanamurti, *Current Opinion in Solid State and Materials Science* **3**, 38 (1998).]

ure 1.3.2 shows concurrently scanned STM and BEEM images of a single dot. The BEEM image clearly shows a reduction in BEEM current through the dot, which shows the presence of a localized conduction band offset at the dot.

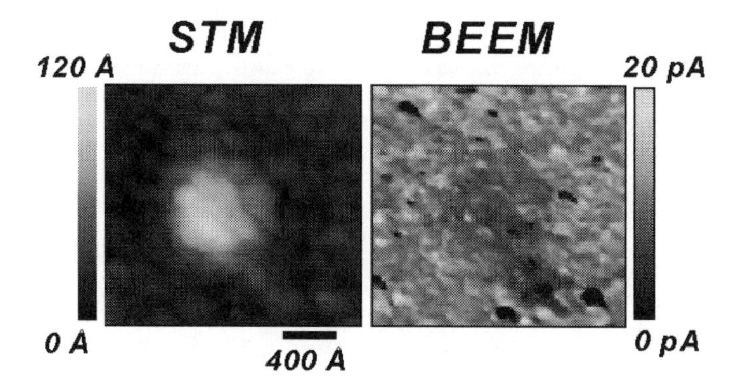

FIGURE 1.3.2 STM and BEEM images of a single dot. [Reprinted with permission from M.E. Rubin, H.R. Blank, M.A. Chin, H. Kroemer, and V. Narayanamurti, "Local conduction band offset of GaSb self-assembled quantum dots on GaAs," *Applied Physics Letters* **70**, 1590 (1997). Copyright © 1997 American Institute of Physics.]

dimensions that are smaller than 100 nm. Such structures have been used as gates on submicron GaAs/AlGaAs devices (see Box 1.5), eliminating the 2DEG under them. In this way, confinement both in the plane and perpendicular to the plane of the 2DEG can be achieved. The simplest structure of this kind is a narrow constriction in the 2DEG that exhibits a resistance quantized in units of h/e^2.

Electron-beam lithography can be used to make nanometer-sized metal wires and rings. This opened the field of mesoscopic physics: the study of systems that are larger than atoms but small enough that they are not bulk materials. In such mesoscopic systems, the wavelength of the carriers is comparable to the device dimensions and to the mean free path for phase breaking, and statistical averaging does not eliminate quantum mechanical phenomena. One dramatic phenomenon of this kind is universal conductance fluctuations.

Most mesoscopic effects for systems in one- or two-dimensional confinement are subtle. However, when electrons are confined in all three dimensions the results can be dramatic. Structures in which electrons confined to metals and semiconductors with tunnel junctions connecting the confined regions to the leads (essentially "artificial atoms") enhance the electron-electron correlations,

BOX 1.4 Transport in Quantum Dots

Quantum dots are semiconducting or metallic regions so small that the electrons are confined in all three dimensions. Like an atom, a quantum dot contains a finite number of charges and has discrete energy levels. The study of electron transport through these minuscule conducting regions has revealed a variety of fascinating phenomena including observable effects caused by individual electrons. Current versus voltage measurements for a quantum dot show discrete staircases where each successive plateau represents the addition of one electron to the quantum dot (see Figure 1.4.1).

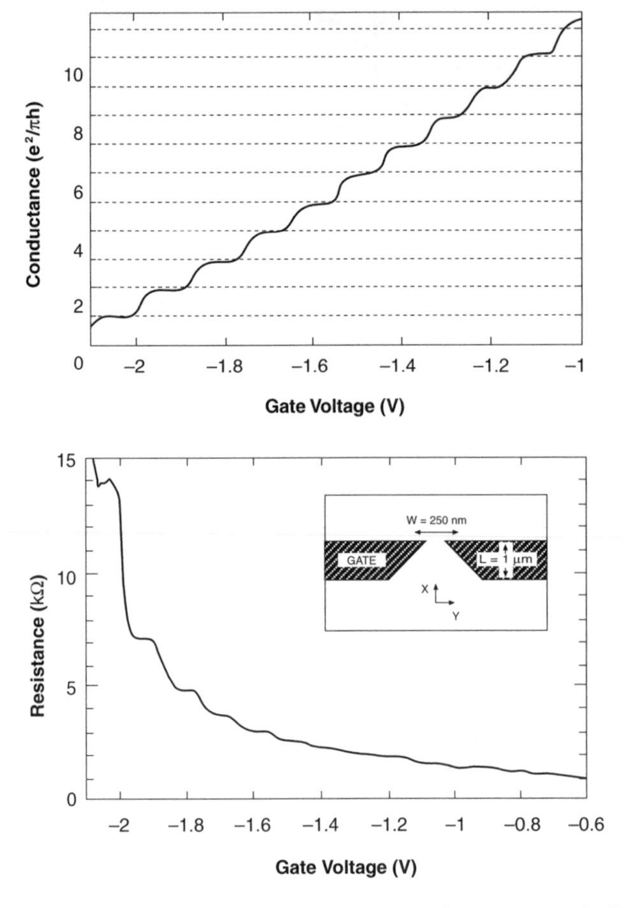

FIGURE 1.4.1 Current versus voltage measurements for a quantum dot illustrating the discrete electronic states. Each successive plateau represents the addition of one electron to the quantum dot. (Courtesy of Massachusetts Institute of Technology.)

Transmission through the dot can be measured by application of a bias voltage and measurement of the current (Figure 1.4.1, top). If, in addition, one steadily changes the potential of the dot by the application of a bias voltage to a gate electrode, a succession of resonance peaks are observed (Figure 1.4.1, bottom). The peaks occur when the energy levels in the quantum dot coincide with the Fermi level of the electrons in the leads.

Although the transmission probability (which has a shape independent of the number of electrons initially on the dot) is understood, these measurements did not provide information about the coherence of the tunneling electrons. Recent measurements of the phase of the electrons demonstrated that the tunneling was coherent. In the experiments, a semiconducting path around an insulating region penetrated by magnetic flux lines was constructed. This approach exploited the Aharonov-Bohm effect, in which an electron acquires a phase shift as it goes around a magnetic flux line. Interference was observed between the parts of the electron wave function that traveled on opposite sides of the flux line. Using a four-terminal configuration, the change in phase versus potential is measured as an electron enters the quantum dot and interacts with a quasi-bound state and leaves. The phase shift on both transmission and reflection were measured, yielding the unexpected result that the phase shifts are independent of the number of electrons initially on the quantum dot, a result that is not yet understood.

resulting in the quantization of both charge and energy. The energy levels of these artificial atoms can be measured by the voltage required to add an extra electron. These artificial atoms are large enough to display behavior that is not observed in natural atoms; for example, the superconducting energy gap in mesoscopic Al structures is quantized.

From the perspective of potential applications, single-electron transistors (SETs) (see Box 1.6) can be realized in systems with three-dimensional confinement provided by structures with two tunnel junctions and a gate. SETs turn on and off again every time an electron is added. This device not only functions as a transistor, but also provides insight into the physics of mesoscopic structures. Using the sharp peaks associated with the addition of an electron, the equilibrium ground-state energy of the droplet of electrons, as well as some low-lying excited states of the droplet, can be measured. Furthermore, application of a magnetic field reveals phase transitions between different states of the system. The magnetic field alters the balance between the confining potential, which favors a high electron density, and the Coulomb interaction, which favors a low electron density.

Based on recent successes in nanostructures, we can speculate about the kinds of nanostructures likely to yield new physics and technology. Three different physical effects in nanostructures that can be exploited for nanoelectronics are illustrated in Figure 1.2. In resonant tunneling, the probability for charge carriers to tunnel through barriers is greatly enhanced when the energy levels on

BOX 1.5 Double Electron Layer Tunneling Transistor (DELTT)

The double electron layer tunneling transistor (DELTT) is a recently developed quantum transistor that, unlike previous quantum transistors, does not require tight control over lateral dimensions and is thus easy to reproducibly fabricate in large numbers. It is based on the gate control of two-dimensional-two-dimensional (2D-2D) tunneling between the two electron layers in a double quantum well (QW) heterostructure, first investigated in 1990. Because the DELTT exhibits negative differential resistance, it is multifunctional, allowing the same circuit functions to be performed by fewer devices. It is also expected to be exceptionally fast, although such high speeds have yet to be demonstrated.

The source and drain terminals of the DELTT are formed by electrically contacting both QWs, and then locally depleting electrons (by gating or etching) from the QW one does not wish to contact (see Figure 1.5.1). A third gate terminal controls the tunneling between the two QWs. Close-proximity backgates—necessary for both good I-V peak-to-valley ratios and for small size and high speed—are achieved by an epoxy-bonding flip-chip technique that yields a total device thickness of less than one micron.

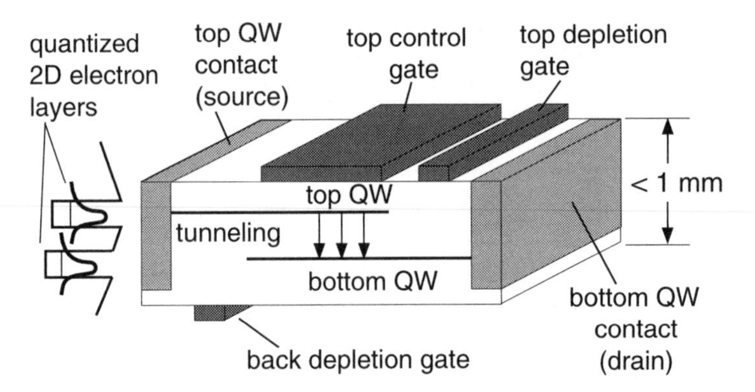

FIGURE 1.5.1 Schematic illustration of the DELTT. The energy band diagram of the double quantum well heterostructure is shown at left. [Reprinted with permission from J.A. Simmons, M.A. Blount, J.S. Moon, S.K. Lyo, W.E. Baca, J.R. Wendt, J.L. Reno, and M.J. Hafich, "Planar quantum transistor based on 2D-2D tunneling in double quantum well heterostructures," *Journal of Applied Physics* **84**, 5626 (Nov. 15, 1998). Copyright © 1998 American Institute of Physics.]

Current in the DELTT flows only if both energy and momentum can be conserved in a tunneling event. Because both layers are two-dimensional, this is equivalent to the QW subbands being aligned. This can be achieved by applying a source-drain bias, a control gate bias, or both. Figure 1.5.2 shows source-drain I-V characteristics at several control-gate voltages for an AlGaAs/GaAs DELTT. Both the height and position of the resonant current peak are clearly controlled by the gate. Similarly good behavior has been obtained at 77 K, and bistable memories and digital logic gates have been demonstrated. Although obstacles remain, the DELTT shows excellent promise as a practical, room-temperature quantum transistor.

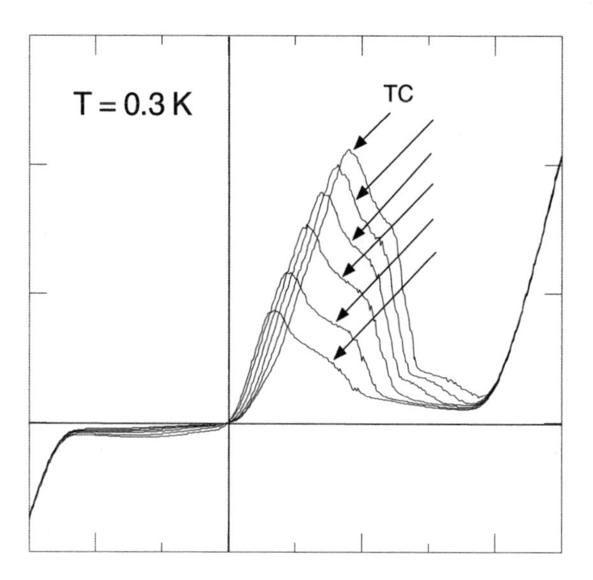

FIGURE 1.5.2 Source-drain current versus source-drain voltage for several values of top control gate voltage (VTC). Both the height and position of the resonant tunneling current peak is controllable by the gate. [Reprinted with permission from J.A. Simmons, M.A. Blount, J.S. Moon, S.K. Lyo, W.E. Baca, J.R. Wendt, J.L. Reno, and M.J. Hafich, "Planar quantum transistor based on 2D-2D tunneling in double quantum well heterostructures," *Journal of Applied Physics* **84**, 5626 (Nov. 15, 1998). Copyright © 1998 American Institute of Physics.]

BOX 1.6 Kondo Effect in an Artificial Atom

The analogy of a quantum dot to an artificial atom has been extended with the demonstration that a quantum dot interacts with nearby metallic leads in much the same way that a single magnetic impurity interacts with a surrounding metal—in the phenomenon known as the Kondo effect. Kondo behavior was found recently in a single-electron transistor, which consists of a semiconductor quantum dot sandwiched between two metallic leads. This miniature device turns on and off as individual electrons controlled by a nearby gate flow on and off the dot.

The theory of the Kondo effect was developed in the early 1960s to explain a long-standing puzzle about the resistance of some metals: Why does the resistance start to increase as the metal is cooled below a certain temperature? According to the picture that has emerged, the increased resistance comes from magnetic impurities whose local magnetic moments couple antiferromagnetically to those of the conduction electrons. The coupling becomes stronger and increasingly impedes the flow of current as the temperature is decreased.

The concept of the Kondo effect is intriguing because it involves the pairing of a localized electron with an electron in an extended state in the metal. Its manifestation in a quantum dot is no less compelling. Although interactions between electrons in quantum dots are known to be important, the Kondo phenomenon is a true many-body effect requiring a coherent state resulting from the coupling of the localized electrons in the dot and a continuum of electron states outside the dot.

Experimenters have tried to see a manifestation of the Kondo effect in quantum dots ever since its presence was predicted in the late 1980s, but succeeded only recently. Kondo behavior for a single spin had been observed in resonant tunneling through a charge trap created unintentionally in a point contact. A collaborative experiment involving the Massachusetts Institute of Technology (MIT) and the Weizmann Institute in Israel has attracted additional interest because it shows the Kondo effect in a way that will allow one to explore the phenomenon in a system with many tunable parameters.

Kondo-like effects in quantum dots are observable only under a very narrow set of conditions. To see the effects of coupling between the dot and the leads, one needs to make the rate for tunneling of electrons between the dot and the leads as high as possible. The higher this rate, the higher the temperature at which the Kondo effect survives. However, if one makes the rate too high, the electrons on the dot become completely delocalized. With a smaller dot, the electrons are more localized to begin with, and a higher rate is possible.

To make a semiconductor quantum dot, one starts with a two-dimensional electron gas of electrons confined in a plane at the boundary between two semiconducting materials. Additional semiconductor layers go on top of this boundary region. At the top of the structure, one lays down electrical gates; the electrical potentials created by these gates confine the electrons in the plane below the gates to a very small region. Typically the quantum dots lie 100 nm below the surface. The MIT-Weizmann team made a much smaller artificial atom by forming the two-dimensional electron gas closer to the surface.

The conductance of a single electron transistor displays a peak when the sum of the voltage (V_g) on one of the gates and of the voltage (V_{ds}) between the two leads on either side of the dot, each multiplied by the appropriate capacitance, is large enough to add an electron to the dot. A gray-scale plot of the conductance (see Figure 1.6.1) therefore consists of a series of bright diagonal bands, marking the positions of the peaks, whose slopes are determined by the relative capacitances. The highest peaks occur where the bands intersect on the $V_{ds} = 0$ axis. These maxima cluster in pairs along the $V_{ds} = 0$ axis, with the intra-pair peak separation smaller than the inter-pair separation. (One pair is shown in the figure.) The two peaks correspond to the addition of a pair of electrons to the same spatial state; one electron enters the state with spin up and the other with spin down. The next electron must go into the next spatial state. Thus, in the region between the paired peaks the artificial atom has an odd number of electrons.

The peak structure described so far is that expected for any artificial atom. One tip-off in the data to the presence of the Kondo effect is the non-zero conductance between the paired peaks, the bright, narrow vertical line along the $V_{ds} = 0$ axis. In this region the quantum dot has an unpaired electron, which is free to form a singlet with the electrons in the leads. This singlet state couples electrons from the

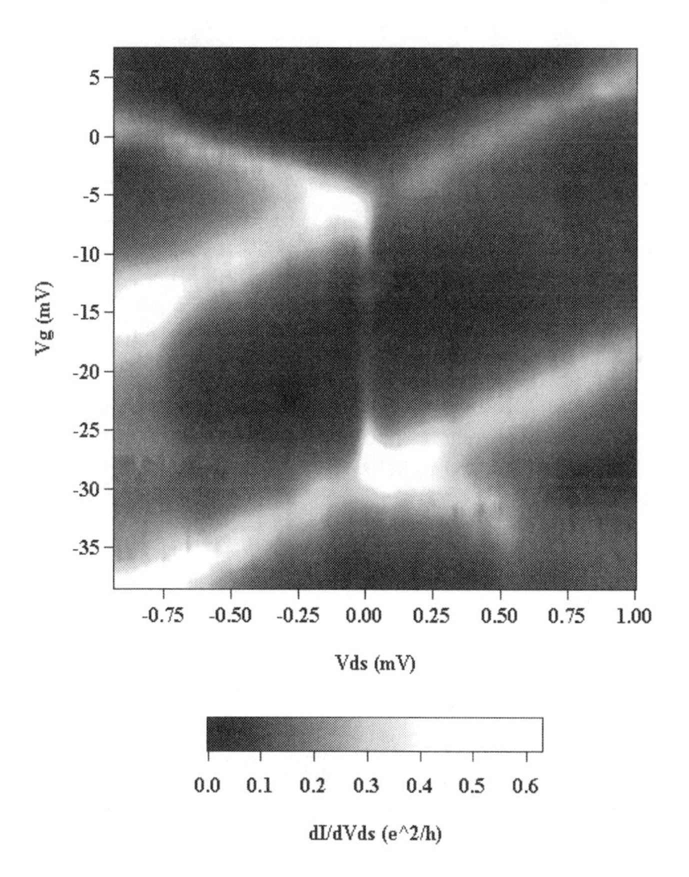

FIGURE 1.6.1 Evidence for the Kondo effect in a single electron transistor. (Courtesy of Massachusetts Institute of Technology and the Weizmann Institute.)

BOX 1.6 Continued

left lead to the unpaired electron on the dot and thence to the right lead, giving conductance in a region where none is ordinarily expected. As predicted by theory, this interpeak conductance increases as the temperature is decreased.

If the enhanced conductance that appeared between the two peaks were due to the Kondo effect, it would require a symmetric interaction of the unpaired electron on the quantum dot with electrons in both leads. But if one applies a voltage V_{ds} across those two leads, separating the Fermi energy levels of the two reservoirs, that interaction is no longer symmetric, and the conductance must fall. Another signature of the Kondo effect is the disappearance of the enhanced conductance as the voltage between the leads is increased, leading to the narrowness of the vertical line in the figure.

Finally, a magnetic field splits the unpaired electrons, causing the conductance peaks to split as well, by an amount equal to 2 $g\mu_B B$. This signature is also observed.

the two sides of the barrier are identical. Large changes in the tunneling current are realized with small changes in the bias voltage across such structures. In structures that confine electrons to regions with dimensions comparable to the electron wavelength, quantum interference effects can be used to switch electronic currents. A conceptual approach to a transistor based on quantum interference is shown in Figure 1.3a.

Quantum confinement structures can be created that serve as electron waveguides, conceptually similar to the waveguides encountered in optical structures. Nanostructure switches based on guided-wave coupling can be created in quantum confinement structures, illustrated in Figure 1.2b. In these switches, illustrated in Figure 1.3b, electrons in input channel 1 (IN1) can either exit through output channel 1 (OUT1) or be switched to output channel 2 (OUT2) depending on the gate bias voltage.

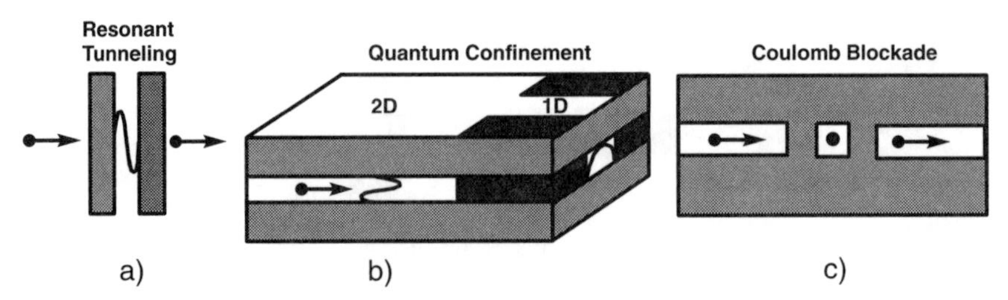

FIGURE 1.2 Illustration of physical effects realizable in nanostructures: (a) resonant tunneling; (b) quantum confinement; and (c) Coulomb blockade. (Courtesy of Stanford University.)

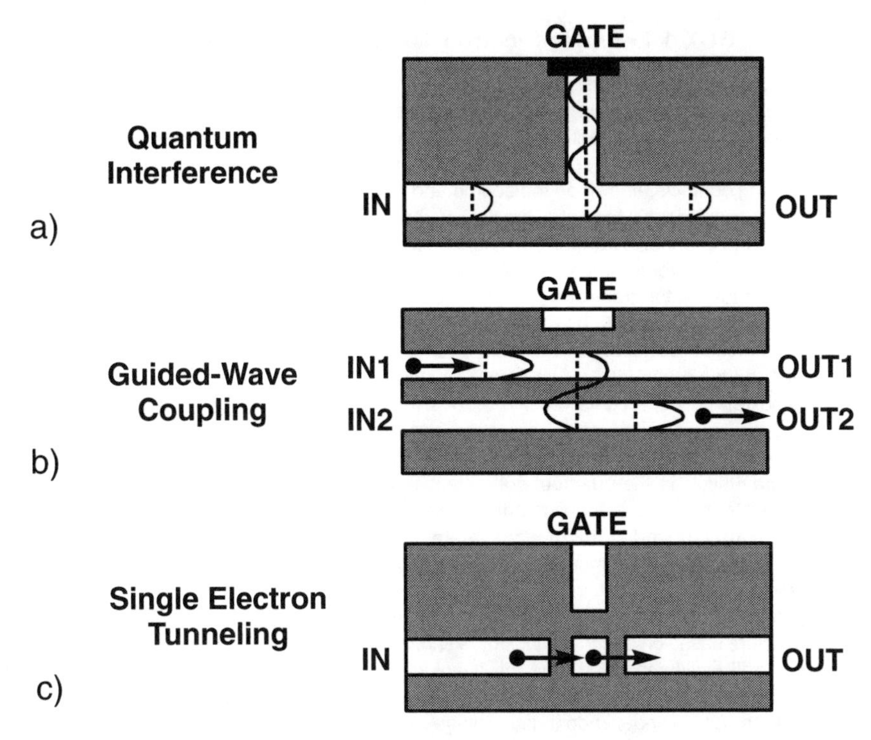

FIGURE 1.3 Schematic illustrations of nanoswitching concepts based on the physical effects illustrated in Figure 1.2. (Courtesy of Stanford University.)

In the Coulomb blockade structure illustrated in Figure 1.2c, adding an electron to a quantum dot creates a Coulomb field that repels the addition of another electron. A SET can be realized by placing the quantum dot between the input and output channels as illustrated schematically in Figure 1.3c. The transistor is switched between on and off by changing the voltage on the gate.

A number of novel circuits based on capacitively coupled arrays of artificial atoms have been proposed based on SETs; however, because the original SETs operated only at very low temperatures, the effort in this area was initially limited. The operating temperature scales with the energy required to add an electron to the artificial atom, which increases with decreasing size of the mesoscopic structure. Recently, nanometer-size SETs and single-electron memories (SEMs) have been demonstrated that have quantum dots sufficiently small to operate near room temperature (see Box 1.7), stimulating increased interest in these mesoscopic structures.

BOX 1.7 Single-Electron Transistors and Memories

Single-electron transistors (SETs) rely on quantum dots of small semiconducting or metallic regions in which electrons are confined in all three dimensions. In a conventional field effect transistor, electrons (or holes) travel through a semiconducting region, known as a channel. The channel connects two electrically conducting contacts, known as the source and drain. A gate electrode located above the channel is used to control the flow of charge in the channel. In complementary logic, the channel is normally in a nonconducting state. Application of a voltage to the gate electrode increases the conductivity of the channel, allowing charge to flow across it; the transistor turns from off to on.

SETs can be made by replacing the channel between the source and drain in conventional devices with quantum dots surrounded by an insulator. The key is that the quantum dots are so small that, when an electron is on the island, Coulomb repulsion prevents other electrons from crowding on. Changing the gate voltage on the quantum dot aligns the energy level of the dot and the electrode, allowing the electron to tunnel to the electrode. Because the separation of the energy levels in the quantum dot increases with decreasing size of the dot, very small structures, analogous to artificial atoms with dimensions on the order on 1 to 4 nm, are required for room-temperature operation of a SET.

Single-electron memories (SEMs) are made using a variation on the SET. The crucial difference is that, in a SEM, the quantum dots sit between the channel and the gate electrode rather than replacing the channel as they did in the SET. A SEM structure based on silicon technology is shown in Figure 1.7.1. The source and drain are sufficiently separated from the quantum dots to prevent spontaneous electron tunneling. In the uncharged, or 0 state, there are no electrons on the quantum dot and a given geometry-dependent gate voltage allows current to flow between the source and drain. In the 1 state, an electron is injected into the quantum dot. When an electron is on the quantum dot, a different voltage is required to turn on the transistor because of the Coulomb field caused by the presence of the electron. Using quantum dots formed from polycrystalline silicon and standard integrated-circuit processing technology, researchers at Hitachi have recently fabricated 128-Mbit single-electron dynamic random access memories (DRAM) that operate at room temperature.

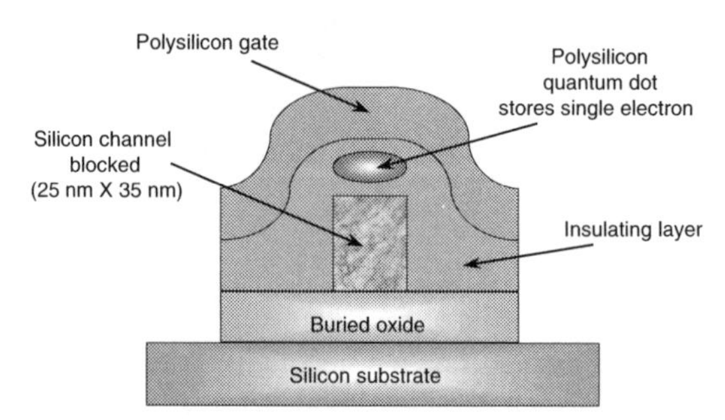

FIGURE 1.7.1 Single-electron memory structure based using a silicon quantum dot. (Courtesy of the University of Wisconsin.)

Challenges, Priorities, and Frontiers of Electronic Materials and Phenomena

Based on the progress in scanning microscopies, we can speculate that the successor to electron-beam patterning for fabrication of nanostructures could be atom-by-atom assembly. In recent years, scanning-tunneling microscopes have been used to arrange atoms on surfaces and to measure changes in the energies of the surface electrons by tunneling spectroscopies. Scanning probes have been used to construct single-atom switches in which the movement of an atom from one position to another opens or closes a circuit created by assembling rows of atoms on a surface.

The inexorable pressure to reduce the dimensions of semiconductor devices has introduced to condensed-matter physicists and materials scientists the concept of self-organizing structures in the nanometer-size regime. Self-assembly promises a broader frontier of nanostructures. A self-assembled array of quantum dots can be grown either by strained-layer nucleation of islands between quantum wells during growth or by assembly of nanometer-sized dots grown by solution chemistry into a macrocrystal. Both approaches are discussed in detail in Chapter 4. Such nanocrystals are a different approach to forming artificial atoms analogous to the artificial atoms discussed above. A major research area will be learning how to control self-assembled materials to create ordered one-, two-, and three-dimensional structures. These self-organized structures will be technologically useful to the degree that their size and nucleation density can be controlled.

The committee concentrates, in this section, on the electronic properties of quantum structures, but photonic lattices are beginning to emerge and are discussed in more detail in the following section. In photonic lattices, photons are confined in arrays of structures with dimensions comparable to the wavelength of light in the medium. Combining such lattices of quantum wells with feature sizes comparable to the wavelength of electrons will lead to coupled electron-photon systems with new and interesting electronic and optical properties.

Numerous outstanding scientific and technological problems have been identified in the research in electronic materials. Beginning with silicon integrated-circuit technology, major materials-related technical questions include, What interconnect technology will be used beyond copper and low *k* (i.e., beyond normal metals and dielectrics) for silicon integrated circuits? and How do we manufacture SETs and SEMs at reasonable operating temperature and cost?

Recognizing that one cannot continue to scale silicon integrated circuits to smaller feature sizes indefinitely raises the question, What is beyond silicon? Additional questions that need to be investigated to address this question are the following: How can self-assembled materials be controlled to create the desired one-, two-, and three-dimensional structures? How does one create hybrid structures that exploit the best properties of, e.g., organics or plastics and semiconductors, magnetic materials and semiconductors, or superconductors and semiconductors? How do we understand and exploit quantum state logic?

To continue to advance our fundamental knowledge, and to use this knowledge to continue to advance technology based on electronic materials, the committee offers the following recommendations:

- Perform the long-range research required to allow the semiconductor industry to follow Moore's Law and maintain its historical rate of productivity improvement.
- Accelerate research into materials, structures, and technologies that will go "beyond silicon," i.e., to discover what technology will be used after today's silicon integrated-circuit technology reaches its limits.
- Expand the research efforts in self-assembled materials to create structures that promise needed technologies.
- Continue to develop the processing and characterization tools required to create and evaluate ever more-complex, ever-smaller, artificially structured materials.

OPTICAL MATERIALS AND PHENOMENA

Materials and Physics That Drive Today's Technology

Optical Communications

As mentioned in the introduction, fueled by the explosion of Internet use and the globalization of voice and data communications, lightwave communication systems capacity and installation are growing exponentially with a growth rate of about a factor of 10 every 6 years. The current global market for lightwave systems is about $8 billion per year and is expected to grow to about $15 billion per year by 2000. Optical telecommunication was introduced into the market in 1980; today, not only is optical fiber the medium of choice for long-distance voice and data communications, but it is also rapidly growing to be a leading player in the local area network (LAN) market. Optical fiber is predicted to have revenues of about $20 billion in the year 2000 and dominate the analog cable television and fixed wireless loop markets.

The first undersea optical cable was installed in 1988, with a capacity of about 8,000 voice circuits per cable, at a cost of about $400 per circuit per year. More than 300,000 km of undersea lightwave cable had been installed by the end of 1996. Cable installed in 1996 cost less than $30 per year per voice channel and had a capacity of 120,000 voice channels per cable (5 Gb/s per fiber).

The first major terrestrial lightwave system installed in the United States linked Washington and New York with a capacity of 90 Mb/s per fiber in 1983. A similar system linked New York and Boston in 1984. More than 100,000 km of fiber had been installed in terrestrial systems, one-third of it in the United

States, by the end of 1996. The latest systems incorporate wavelength division multiplexing (WDM) that uses many separate wavelength channels per fiber, dispersion-shifted fiber, and optical amplifiers. Currently in deployment are 120-Gb/s-per-fiber systems using 48 channels with 2.5 Gb/s per channel. In the next 2 years 400-Gb/s, 80-channel systems will be introduced into the market.

These advances in technology were made possible by advances in our understanding of materials and growth techniques that reduced the transmission losses of silica-germania optical fiber from 400 dB/km in 1965 to about 0.15 dB/km by the early 1990s. Such losses allow a signal to travel 800 km (about the distance between Washington and Boston) before the signal intensity decreases to about 1/100 of its original value. These advances were accompanied by recent major advances in InP-based electrically modulated single-wavelength semiconductor diode lasers operating in the 1.3-μm and 1.5-μm wavelength regions, where the lowest loss in silica fiber occurs; in fast avalanche photodiode detectors; in erbium-doped fiber amplifiers and other fiber devices (see Box 1.8); and in high-power semiconductor diode lasers used to pump the fiber devices.

Current digital optical telecommunications networks typically use the NRZ (non-return-to-zero) format to transmit data in the linear amplitude regime. Future systems could use nonlinear effects in fiber with high-power lasers to exploit the properties of soliton transmission.

Solitons are wave packets that propagate without changing shape. They are solutions to the electromagnetic wave propagation equation in fiber waveguides that arise from the nonlinear effects of self-phase modulation and dispersion in the group velocity. Solitons are dispersion-free and exhibit a pulse shape that retains its waveform over long distances because the two nonlinear effects are exactly counterbalanced. They were first proposed as a means of data transmission in optical fiber in the 1970s and observed in the research laboratory in 1980. They offer extremely high bit-rate transmission (>100 GHz) at a single wavelength. Extensive research on the use of other nonlinear effects in both fibers and semiconductors and in artificially poled piezoelectric materials is under way to enable future ultrahigh-speed all-optical processing devices.

Local area networks (LANs), optical data links, and optical signal processing are emerging growth areas enabled by new technologies such as vertical-cavity surface-emitting lasers, smart pixels and microelectromechanical systems. All of these technologies were implemented as devices within the past decade. The emergence of low-loss graded index multimode plastic optical fibers in the past 5 years could lead to a low-cost medium to deliver high bandwidth communications over short links from a single-mode glass fiber backbone to the desktop. The advantage of extremely low-cost connectors and low-cost transceivers could outweigh the high cost of fluorinated polymer materials compared with the cost of glass fiber based LANs. The predicted annual market for optical data links is $1 billion by the year 2000 and $3.3 billion by 2005, with approximately half in computers and half in LANs.

BOX 1.8 Fiber Devices

The revolution in optical communications over the last decade began with the invention of the erbium (Er)-doped optical-fiber amplifier in the late 1980s. With the invention and implementation of a number of key optical-fiber devices, evolution of an all-optical network architecture has begun. Fiber gratings were first made in 1975. The technological revolution in fiber devices was enabled by the discovery in 1993 that when exposed to ultraviolet (UV) light, an index of refraction change as large as 0.01 occurs in the cores of silica fiber doped with hydrogen-loaded germania. This UV-induced irreversible chemical change permits stable fiber Bragg gratings to be easily written into the cores of standard optical fiber. These Bragg gratings serve as key building blocks for a large range of both active and passive fiber devices such as filters, amplifiers, fiber lasers, dispersion compensators, pump laser reflectors, demultiplexers, and equalizers.

The optical power in communications systems increased sharply with the introduction of the Er-doped fiber amplifier. This amplifier is basically a single-pass laser consisting of several meters of a spliced silica fiber doped with 1,000 ppm Er^{3+} in the core with an input coupler for the pump light. Optical amplification can be achieved through stimulated emission from the excited states of Er atoms in the glass if a population inversion is created with pump light from a semiconductor diode laser.

The optical properties of rare-earth impurities in a glass matrix were first studied in the 1960s. Rare-earths are ideally suited for use as an amplification and lasing medium; they have strong optical transitions in the infrared and their properties are nearly independent of the host material. Er, pumped with 1488-nm or 980-nm light, is ideal for an amplifier in the 1.55-μm communications window.

Optical amplification has many advantages over electronic regeneration: amplification occurs over a relatively wide (80-nm) gain curve, ideal for dense wavelength division multiplexing systems; amplification is transparent—i.e., independent of modulation format and bit rate—and in principle the gain is bidirectional. It also allows watts of optical power, which is important for higher data rates of wavelength division multiplexing, increased passive split architectures where one source is split into many channels, and extended repeater spacings. Systems with optical amplification are far simpler to upgrade to higher bit-rate systems after initial installation because all optical repeaters are independent of bit rate. Er-fiber amplifiers were first deployed in undersea communications systems in the mid-1990s.

Because silica fiber has more wavelength dispersion in the 1.55-μm region than at its minimum in dispersion at 1.31 μm, additional dispersion compensating fibers were installed at intervals in the system. New fiber designs have shifted the dispersion minimum to 1.55 μm. New installations use this fiber to minimize the effects of dispersion at high data rates. In the 1.31-μm optical communications window used by most of the installed terrestrial base, amplifiers using praseodymium (Pr) ions in fluoride glass hosts (because Pr in silica does not emit at this wavelength) and Raman-shifted silica fiber amplifiers have been recently demonstrated in the research lab.

Fiber lasers using rare-earth ion dopants and fiber Bragg gratings as cavity mirrors were demonstrated in the early 1990s in both silica and fluoride fiber hosts. A cascaded Raman-shifted laser was demonstrated in 1990 in standard silica-germania fiber. Raman-shifted lasers eliminate the need for specialty fiber doped with rare-earths. Figure 1.8.1 shows a block diagram of a Raman-shifted laser operating at 1.3 μm. The laser consists of five stages of amplification of successively Raman-shifted light. Each stage is a cavity defined by its own tuned set of fiber Bragg grating mirrors. The Raman shift of the light in each stage is the inelastic scattering of the high-intensity light in the laser cavity from the phonons of the silica core. Higher conversion efficiencies and output power were recently obtained by sending the pump light into an optimally designed cladding that couples efficiently to the single mode waveguide core.

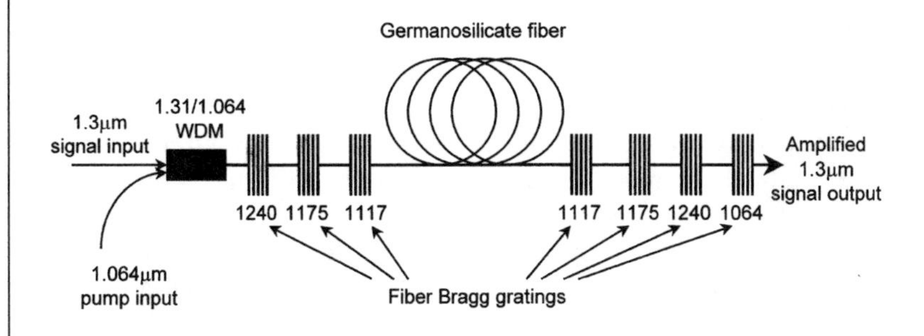

FIGURE 1.8.1 Block diagram of a Raman amplifier at 1.3 μm. (Courtesy of Bell Laboratories, Lucent Technologies.)

Future demand for broadband communications is expected to drive deployment of fiber-to-the-home communications access systems that will bring GHz bandwidths from the central office to each home. Over the next 20 years the predicted expenditure on deployment of wideband access in the United States is $150 billion. Predicted expenditures on access infrastructure globally is a factor of 3 to 5 more. The limiter for both the technology and deployment of access is low-cost, reliable components, which poses major challenges for materials and device physics. For example, lasers that can operate at elevated temperatures without active cooling are necessary, as are passive optical distribution systems and low-cost upstream communication devices such as optical modulators.

Multiple wavelength optical transmission systems for long-distance networks (DWDM, or dense wavelength division multiplexing), which use sophisticated integrated optoelectronic devices and waveguide circuits such as multiwavelength lasers, optical routers, and all optical cross-connects will be rapidly deployed. These technologies are currently in their infancy compared to their silicon microelectronics counterparts.

Optical Data Storage

The data storage market, fueled by the insatiable appetite of consumers for generating, collecting, and storing information, is growing exponentially. The total number of stored bits is expected to exceed 10^{20} by the year 2000 with the total data storage market about $100 billion. Of that, the market for optical storage is projected to be about $12 billion. In addition, the enormous consumer entertainment electronics market of digital audio and video storage will add another $20 billion in compact disk (CD) and digital video disk (DVD) technology sales and $40 billion in media sales.

The essential idea of optical storage is to store bits by forming pits or spots on a reflective surface. A spot is read by shining a focused laser beam onto the disk and measuring the intensity of the reflected light. Optical disk technologies can be classified by the number of times each portion of the surface can be written. Read-only disks, such as the ubiquitous audio CDs and computer compact-disk read-only memories (CD-ROMs) introduced in the early 1980s, are manufactured with data already recorded as pits stamped from a master onto the polycarbonate disk surface, which is subsequently aluminized and covered with a protective transparent layer. Write-once disks (also called WORM, for write once, read many) were introduced in the late 1980s. They are manufactured in an erased condition, and each sector can be written once by burning pits into the surface, such as with a laser in the disk drive. Read-write disks, also called rewritable disks, were introduced in the mid-1990s. They record data via reversible effects such as light absorption inducing a phase change between amorphous and polycrystalline states (phase-change, PC) or by flipping the direction of a macroscopic magnetic polarization in an optical medium in conjunction with an applied magnetic field (magnetooptic, MO). In PC drives data are read by monitoring the reflectivity change between an amorphous and crystalline spot, while in MO drives data are read by detecting the polarization rotation of a laser beam as it traverses the thickness of the medium.

A key enabler for today's optical recording technology was the invention of the compound semiconductor laser diode in the mid-1960s. Also necessary was the subsequent development of these tiny diode lasers into low-cost, robust, reliable, relatively high-power devices. Today's diode lasers are about the size of a grain of salt, compared to table-top-size gas-ion lasers that require expensive parts, water cooling, and high-voltage, high-current operation. Today's optical drives use near-infrared GaAlAs diodes that operate at 780 nm and 830 nm. Next-generation DVD technology will migrate to AlInGaP lasers operating near 530 nm (yellow) around 1999. Other enablers were PC media based on binary and ternary chalcogenides and MO media based on thin films of magnetic FeTb and CoPt alloys. These alloys, invented by materials physicists in the 1960s and 1970s, are still undergoing intense materials development.

Currently a typical optical drive has a slower positioning time, by a factor of

3, than a magnetic disk drive, a data transfer rate 4 to 10 times slower, and a cost per gigabyte about 5 times higher. However, the platter of an optical drive is typically removable and costs about a factor of 10 less than a typical magnetic disk of equivalent capacity. Optical drives occupy niche markets of data back-up and distribution rather than the far larger market of main drives in computers. This latter market is dominated by magnetic drives (discussed below). Future improvements in optical storage technology could potentially allow higher density optical storage with access speeds similar to those of magnetic drives, and thus become a displacement technology for magnetic storage. Development is under way in four areas for discontinuous improvements in the speed or data density of optical drives.

One development to improve storage density involves focusing a laser beam into the desired storage layer of a three-dimensional sandwich produced by stacking semitransparent layers. Data storage density is thereby increased by as much as a factor of 10.

Also under development are shorter wavelength lasers, permitting smaller spot sizes. Extensive materials research is required to make low-cost high-reliability blue-green lasers commercially available (see Box 1.9).

The third emerging development is near-field optical recording. Flying a special design optical head very close to the storage medium allows write and read spots smaller than the wavelength of laser light. This direct outgrowth of condensed-matter and materials physics and optical physics was initiated in 1928 when British scientist E.H. Synge proposed the physics of near-field optics for microwaves. The first near-field visible optical microscope was not constructed until the late 1980s, after the invention of the scanning-tunneling microscope (STM) in 1981. The STM created the technology required for scanning a small tip in a controlled manner a specified distance away from a surface with atomic-scale resolution. Near-field technology has the potential for data storage density one to two orders of magnitude higher than conventional optical and magnetic storage projections to the year 2000. Another potential advantage of this technology is the ability to use very low mass optical heads mounted directly onto sliders that have been developed for magnetic storage to reduce seek times.

The fourth development on the horizon is holographic data storage. In holographic storage a page of binary data is stored as pixels of a monochrome image. It is possible to record thousands of holograms in a spot of a storage medium with resolution on the order of the wavelength of light. Because of the three-dimensional capability, holographic storage promises a projected density two to three orders of magnitude larger than conventional optical storage. In addition, it has extremely high data rates because an entire image is transferred simultaneously. Development of low-cost, reliable, blue-green lasers and solid-state spatial light modulators, along with low-cost, robust, and reliable storage media for three-dimensional holograms, are needed to enable commercialization of the technology.

BOX 1.9 GaN Lasers

A major breakthrough in condensed-matter and materials physics in the last decade was the development of blue-green semiconducting laser diodes in the direct wide band gap GaN/AlN/InN quantum well materials system. Serious materials problems hampered research efforts concerned with the growth of quantum wells of these materials in the 1960s and 1970s, however: there was no bulk substrate crystal for lattice matching; high impurity concentrations during growth permitted growth of only very high concentration n-type material; and the relatively deep levels of p-type donors made it difficult to achieve p-type material required for current injection devices. By the mid-1970s most major U.S. industrial research efforts in this area were terminated because of the apparent intractability of these problems. Meanwhile, research in the much less robust II-VI (Zn,Cd,Mn/S,Se) materials revealed major difficulties caused by the migration of dislocation-induced dark-defect regions into the active laser area causing catastrophic damage.

Two groups in Japan, lead by Akasaki and Nakamura, continued to try to improve materials properties in the far more robust GaN system. A major breakthrough was reported in 1989 by the Akasaki group, who found p-type material after exposing as-grown GaN to an electron beam in an electron microscope. Soon thereafter Nakamura discovered that GaN grown by standard metallo-organic chemical vapor deposition was passivated by a high density of H impurities. With these breakthroughs in understanding, Nakamura was able to produce n-type material. By 1994 Nichia Chemical announced blue, and soon thereafter green, light-emitting diodes (LEDs) with extended color range. These early diodes had external quantum efficiencies higher than 5 percent, five times higher than the competing yellow-green lasers in the AlInGaP materials system. The high efficiencies of commercial LEDs in the AlGaAs and InGaN materials have opened up new markets in vehicle and brake running lights, and in highway status and traffic control signs. This market is now in excess of $1 billion per year. The addition of blue and green LEDs rounds out the visible spectrum and opens up new markets in long-life traffic lights and high efficiency, high brightness, white lighting systems. In 1996 these LEDs were already incorporated into commercial full-color displays. Figure 1.9.1 shows an outdoor full-color display incorporating blue and green InGaN LEDs and red GaAlAs LEDs by Arami Electric Co., Ltd. Several groups have subsequently demonstrated a blue-green laser in the GaN/AlInGaAs system, and a worldwide race is now under way to achieve a reliable continuous wave laser at room temperature.

Display, Printing, and Copying Technologies

Exciting recent developments in synthesis of semiconducting organic materials have enabled researchers to demonstrate a variety of optoelectronic devices based on electronically active organics. These include light-emitting diodes, thin-film transistors, photovoltaics, and nonlinear optical elements. The great potential for these devices resides primarily in the ability to process organics using cost-effective methods such as spin casting and screen printing, not in performance considerations. The potential to produce large-area devices and

FIGURE 1.9.1 A black-and-white depiction of a full-color outdoor display incorporating blue and green InGaN LEDs and red GaAlAs LEDs. [*MRS Bulletin* **22**, 30 (February 1997).]

patterns easily compatible with plastic substrates is the vision driving scientific investigation of semiconducting organics.

Application of organic materials in electrophotographic photoreceptors is already commercially successful. In that case, large-area molecularly doped polymer films with high-charge photogeneration yield have demonstrated extremely high contrast between photoactivated and dark conductivity. This property, combined with the ability to rigorously exclude deep-charge transport traps, has made organic materials superior for photocopying applications.

The new classes of semiconducting organic materials fall into two categories: discrete evaporable molecular systems and conjugated polymers. The former tend to be compositionally pure and easily suitable for layering; the latter are more thermally stable and are usually amenable to solution processing. The most mature application of these materials is light-emitting diodes; and commercial low-resolution pixelated monochrome display products based on the evaporable organics are currently available (see Figure 1.4). It is likely that analogous products based on conjugated polymers will be available in 1999. Material stability, device efficiency, and low operation voltage remain important areas of research and engineering. Further progress in synthesis of systems resistant to oxidation and electrochemical degradation, synthesis of stable electron transport materials for the polymeric devices, preparation of materials where aggregation quenching of luminescence is negligible, and development of contact modification treatments that improve injection efficiency remains a challenge for organic chemists. Meanwhile, understanding injection and transport, which determine current voltage characteristics, is crucial. Identification and passivation of luminescence-quenching sites and study of the effects of high fields present in light-emitting diodes remain important areas for basic research. A number of

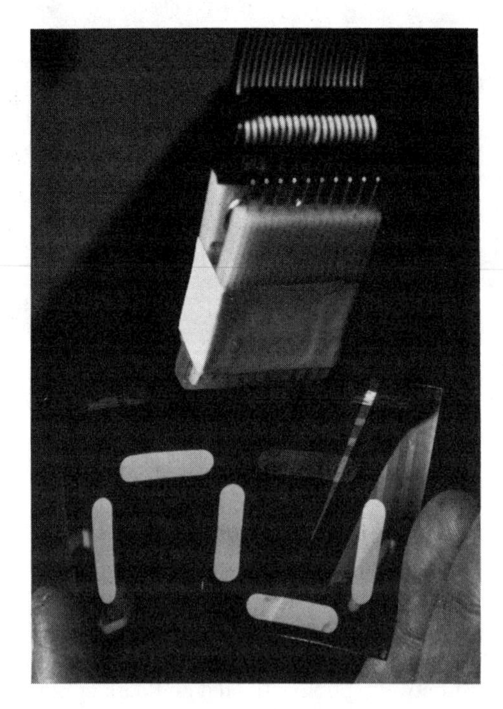

FIGURE 1.4 Flexible light-emitting diode display based on evaporated organic materials. (Courtesy of University of California at Santa Barbara.)

systems issues associated with electrical drive scheme, color fidelity, and patterning must also be resolved if organic emitters are to have wide application in display technology.

The requirements for other promising applications of light-emitting organics such as printing and lighting are substantially the same. A flurry of activity directed at making electrically pumped organic lasers, however, has raised a number of additional issues. The current injection requirements are significantly larger than what is presently achievable, and improved transport will be required. Also, design of resonator geometries suitable for retaining these large injection currents, while avoiding absorption loss associated with metallic contacts, will be required. A significant body of clever work involving distributed feedback, photonic gap, and microdroplet resonators has begun to address this problem. Hot carrier effects in the organics may also become quite important at the applied electric fields necessary to drive lasers, and "interband" emission analogous to that in inorganic semiconductors has been reported recently under conditions near dielectric breakdown. The high density of excited states required for stimulated emission gain has also spurred interesting new research on exciton-exciton interactions and cooperative emission phenomena. These rely on optical excitation of the organic films. It may turn out that the best way to make suitable solid-state organic lasers is a hybrid design using passive organic gain media that are optically pumped by electrically pumped inorganic semiconductor lasers such as those based on InGaN. For these applications, traditional laser dyes doped into inert polymers may be satisfactory.

Many of the scientific issues noted above are common to the application of organic semiconductors to transistor applications and photovoltaics. These transistor applications are likely to be further in the future because performance remains somewhat inferior to alternative technologies. Organic thin-film transistors that perform within an order of magnitude of amorphous silicon have been fabricated and may be promising as pixel-switching transistors in active-matrix liquid-crystal displays. Charge injection and transport are critical to these applications. Interface chemistry and physics at contacts remain poorly understood; control of these is essential to stable low-voltage operation. Exciting and promising results using dipole layers to reduce injection barriers have been reported. Dipole layers function much as graded gap contacts do in traditional semiconductors. Charge transport has been studied extensively, and deep traps are commonly observed that both reduce mobility for transistor applications and raise injection voltages resulting from space charge limitations on the current. Research to identify and reduce trap sites is also important to make these applications viable.

A final general class of promising applications involving active organics in electronics (see Box 1.10) relies on the conductivity of doped conjugated polymers. Doped trans-polyacetylene, although unstable in ambient conditions, has exhibited conductivities higher than metals. More stable systems, such as doped polypyrrole, that are more easily processed and are more compatible with plastic than traditional systems like indium tin oxide, show promise for application as transparent contacts for displays.

We anticipate a variety of scientific opportunities, for which the ultimate technological implications are as yet unclear, to emerge from studying electronic polymers in the next decade. Some of these opportunities may come as a result of photochemically modifying single domains or arrays of domains with a near-field optical microscope, synthesizing materials with giant optical nonlinearities, linking organic materials to inorganic quantum dot structures, examining the phase diagrams of mixed polymeric systems, and designing functional polymers that interface to biological systems. Most likely, even more exciting but unanticipated phenomena will arise from this young and vital area of research.

Although the vision for future consumer electronics as described in Box 1.10 is somewhat hyperbolic, the past decade has marked the synthesis of new active organic materials and advances in the science underlying their electronic, optical, and mechanical properties. Technologies based on these materials have the promise that comes with the processability of organic materials—namely, the potential to use inexpensive manufacturing methods like spray painting, dip coat-

BOX 1.10 A Future Vision for Organic Consumer Electronics

As you walk into the train station, you notice the large multicolor advertisement for the evening paper glowing on the electroluminescent schedule board. The large-area light-emitting diodes are made from organic conducting electrodes and luminescent polymers that have been spray painted onto the board. You want your usual sections of the local paper and then a few from different papers, so you slide your plastic profile card into the newspaper machine. The card is a few thousand transistors made from organic materials that have been printed by an ink jet printer modified to deposit organic charge transporters and electrodes. It contains your medical records, frequent flier numbers, custom newspaper preferences, and a host of other information. It also serves as a cash card. The machine asks you to put in your display. You unroll your pocket display and insert it into the machine. Several organic lasers write your customized newspaper into a thin patch of organic material about the diameter of a dime in the upper right corner of your portable display that functions as an erasable compact disk. On the train, you plug your display into the seat back. The reader is a moving organic laser whose reflection from the writing on the disk is recorded by a photovoltaic cell, which is also made from organic materials, and the display is a high-resolution luminescent color display made from organic light-emitting diodes. The active switching matrix for the display pixels is made from hybrid materials, inorganic charge transporters that have been solubilized to be processable using organic chemistry. The display was printed on a flexible polymer in a reel-to-reel manufacturing process. A tiny dot of silicon circuitry containing a microprocessor to interpret the information read from the disk and containing display drive circuitry was attached. You marvel that the display cost less than your monthly rail pass.

ing, and reel-to-reel processing. The organic materials can be designed at a molecular level to incorporate different functional entities that bring the desired optical, electrical, and mechanical properties and can then be modified to make them suitable for economical processing schemes. Active work to develop hybrid organic/inorganic structures, copolymers with regular blocks of several different types, and layered materials will likely spur development of new devices and new ways of integrating them into systems. Further research and development promises to allow us to make materials that self-organize into complex supramolecular arrays. Three-dimensional structures made in this way may be useful in fabricating photonic gap structures for telecommunications, color reflective displays, artificial photosynthetic cells, synthetic membranes, and improved electrophotographic materials.

Applying the physical science that underlyies these prospects will offer many challenges. The practical limits to charge photogeneration, injection, and transport are still poorly understood, especially at a microscopic level that would be prescriptive for synthetic chemists. The basic physics of charge and energy transfer is well established; but predictive understanding, given the complexity of morphologically disordered and compositionally impure systems such as polymers, is both an experimental and theoretical challenge. This is especially true at interfaces between materials, which are becoming increasingly important as device dimensions shrink. Overcoming challenges associated with making efficient electrical contact to organic materials is central to developing efficient light-emitting diodes, transistors, and electrically pumped lasers. Similarly, the basic science of material stability is critical in making relatively delicate organic materials commercially viable. An enormous amount of empirical progress has been made toward this end, and we expect commercial products based on organic electroluminescence to be widely available in 2000. It is likely that a deeper understanding of the chemistry of degradation will lead to even more robust and widely applicable materials.

The Physics of Optical Nanostructures and Artificially Structured Materials

As electronic devices are made smaller and faster, it will become increasingly difficult to transmit electrical signals over wire interconnects at low power consumption for chip-to-chip or, at very high speeds, on-chip communications. One possible development will be the use of miniature optical interconnects to solve the timing, power, and switching-speed limitations of electrical interconnects. This will require the development of low-power optical nanodevices that are compatible with silicon technology. Other possible applications for arrays of tiny light-emitting diodes are displays and optical correlators for producing parallel computation of images. In the last decade major advances in optical microcavity lasers, light-emitting diodes, and detectors resulted from increased understanding of the physics of quantum microcavities and advances in III-V and II-VI compound semiconductor growth and processing techniques. Simultaneously, these new materials fabrication and processing techniques have led to beautiful insights into the science of quantum optical structures, which will in turn enable more advances in technology.

By changing the dimensionality of a material by growth or processing, one can greatly alter the density of electronic states, as shown in Figure 1.5. A bulk three-dimensional distribution of electrons in a metal or semiconductor has a density of available electronic states that rises as the square root of the energy. The nature and the energies of these electronic states are greatly affected by "quantum confinement." Enclosing a thin layer (with thickness comparable to the electron's de Broglie wavelength, about 20 nm in GaAs) of lower band-gap material between two slabs of higher band-gap material, yields a two-dimensional quantum well with sharp steps in the electronic density of states (Figure 1.6a). Confining the electrons into a one-dimensional quantum wire produces a series of sharp peaks at the onset of each new quantum mode in the wire (Figure 1.6b). Confining the electrons into a zero-dimensional "quantum dot," a box comparable in size to the wavelength of the electron in all directions, produces a series of even sharper spikes that correspond to a series of confined quantum levels for the electrons in the box (Figure 1.6c). Therefore lowering the dimensionality of a structure on the atomic scale of a few nanometers will cause very large changes in the physics of electronic transport. The advances in simulations of electronic and optical processes in semiconductors, along with programmable molecular-beam epitaxy (MBE) techniques for fabrication of multiple quantum wells with atomic-scale accuracy—("quantum engineering" of structures), have led to the invention of the quantum cascade laser, which operates between confined levels of electrons in a series of tailored quantum wells (see Box 1.11), giant "pseudomolecules" with large optical nonlinearities, and resonant tunneling devices.

The optical properties of a semiconductor are greatly affected by reducing the dimensionality of its structure. For example, an exciton, an optical excitation of the system near the band edge, comprises a bound state between a conduction band electron and a valence band hole with wave functions similar to those of Rydberg atoms. Either an electron or a hole, or both, can be bound in a lower dimensional material by being confined in one, two, or three dimensions. This can also greatly affect the excitonic levels and thus the optical properties near the band gap. The dimensionality of the system is reduced if the size of a dimension (for example, the thickness of a slab) is comparable to the diameter of the exciton, which in GaAs is about 5 nm. In materials with strong coupling to light, the optical and electronic plasma modes of the material interact to form coupled exciton-polariton modes that are also greatly affected by the dimensionality of the system. Theoretically, manipulating the density of states of the optical excitations of a system can produce lasers with zero current threshold, in contrast to

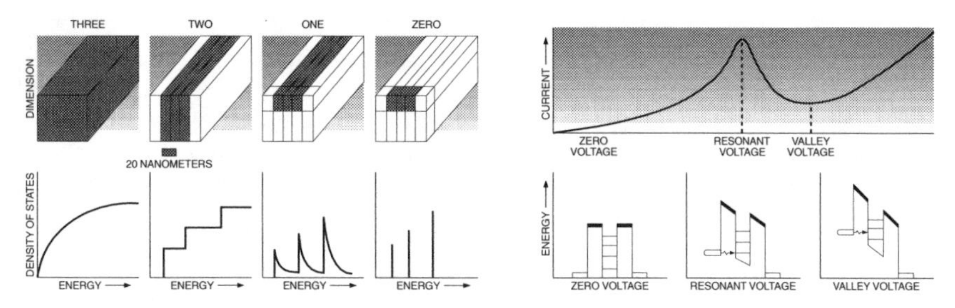

FIGURE 1.5 Illustration of the effect of quantum confinement on the density of electronic states. [Reprinted from *Scientific American* **8**(1), 26 (1997) with permission from John Deecken (artist).]

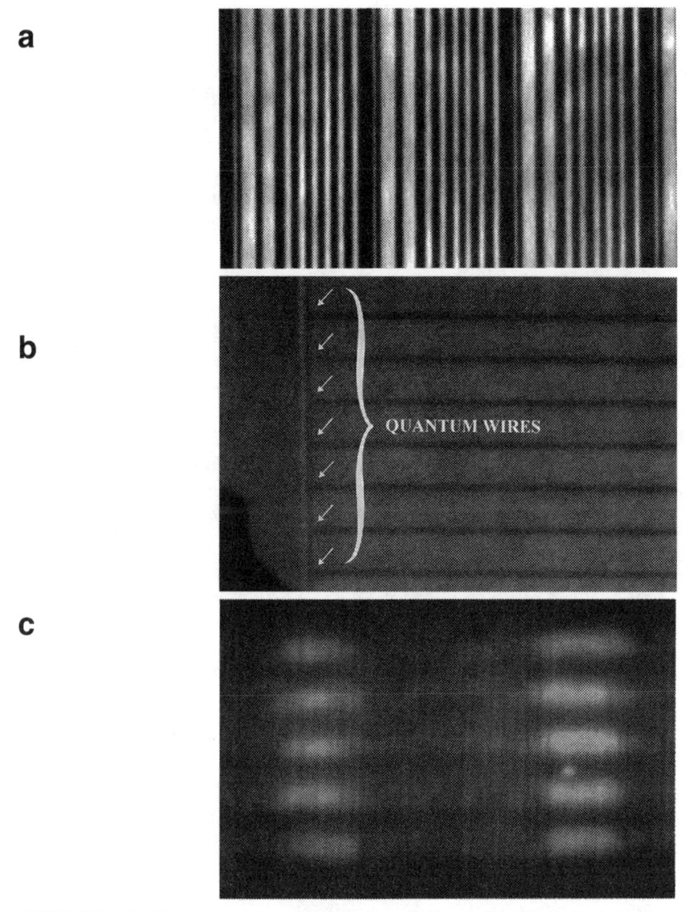

FIGURE 1.6 Two-, one-, and zero-dimensional small optical devices. [Reprinted from *Scientific American* **8**(1), 27-29 (1997) with permission from (a) S.N.G. Chu, Lucent Technologies; (b) Lucent Technologies; (c) James S. Harris, Stanford University.]

BOX 1.11 Quantum Cascade Lasers

In conventional semiconductor lasers used for applications such as fiber-optic communications and compact disk players, light is generated by the annihilation of a negative and a positive charge (an electron in the semiconductor's conduction band and a hole in the valence band) in the active region. As such, the laser wavelength is determined by the energy difference between the conduction and valence bands (the so-called energy band-gap) and its differential efficiency, in the limit of zero optical losses, cannot exceed unity, corresponding to one laser photon created per electron-hole pair annihilated. These constraints are overcome in a radically new semiconductor laser, called a quantum cascade (QC) laser, invented at Bell Laboratories in 1994. In a QC laser, light is not generated by electrons and holes recombining across the band-gap but by electrons alone as they make a transition between two excited states of ultrathin quantum wells. The QC laser is the first semiconductor unipolar laser. These energy levels arise from size quantization when their thickness becomes comparable to or less than the electron de Broglie wavelength. The wavelength is therefore not fixed by a material property (the band-gap), but can be tailored over a very wide range by changing the thickness of the layers using the same combination of materials. QC lasers are an excellent example of materials by design. The energy levels and wave functions, the optical matrix elements and the electron-phonon scattering times are designed to achieve population inversion and the desired wavelength along with other laser characteristics. The active regions of a QC laser are alternated with electron injectors from which electrons tunnel into the upper excited state of the laser transition (Figure 1.11.1 and see Figure 1.6). Electrons tumble down an energy staircase, emitting a photon in each active region. Thus in an N-stage device (where N is typically 25) N photons are created by a single electron traversing the structure. This leads to much higher optical powers than in a conventional diode laser operating at the same wavelength. The Bell Laboratories group has been able to demonstrate QC lasers based on quantum wells made of AlInAs/GaInAs material grown by molecular beam epitaxy with wavelengths spanning a large portion (3 to 13 μm) of the mid-infrared spectrum. This spectral range includes the two atmospheric windows (3 to 5 μm and 8 to 13 μm). In operating temperature and optical power these devices outperform all other semiconductor lasers emitting at these wavelengths and are the first to operate at room temperature and with powers of several hundred milliwatts in pulsed and continuous wave operation. Single-mode operation with wide wavelength tuning has also been demonstrated. QC lasers are important in the detection by absorption spectroscopy of trace gases for pollution monitoring applications. Other potential commercial uses include industrial process control, combustion diagnostics, and medical applications such as breath analyzers for the early detection of ulcers, diabetes, and various forms of cancers. Military applications include countermeasures and battlefield detection of toxic gases and biological toxins via point sensors and lidar techniques.

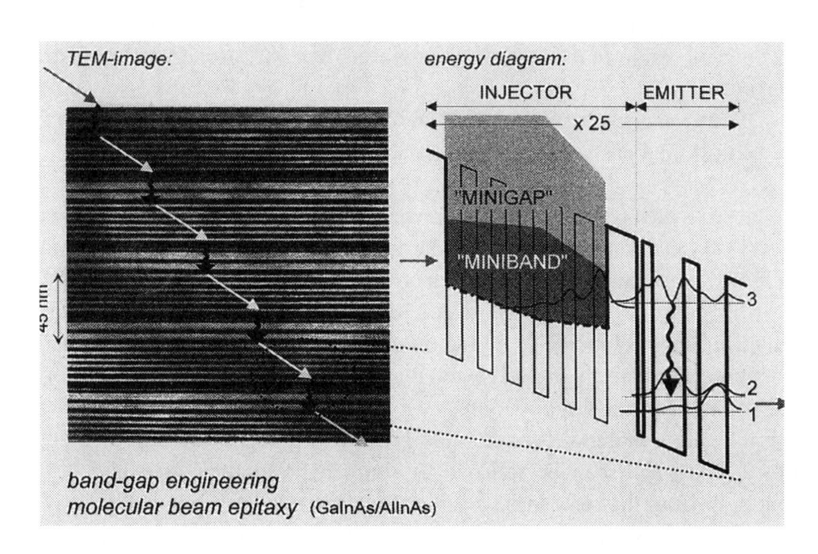

FIGURE 1.11.1 The active regions of a QC laser are alternated with electron injectors from which electrons tunnel into the upper excited state of the laser transition. (Courtesy of Bell Laboratories, Lucent Technologies.)

conventional lasers in which gain is only achieved after sufficient excitation such that the excitons have overlapped in space and have ionized into an electron-hole plasma. Because excitons are bound states of two fermions, they should therefore obey Bose statistics and can themselves exhibit Bose condensation into a macroscopic quantum ground state. The phase diagram of the excitonic matter in semiconductors and its interaction with photons, the observation of lasing between sharp excitonic levels or lasing from Bose-condensed excitons is still controversial, however, and is a field of much current interest.

Over the last decade four approaches for forming small optical devices have been used:

1. MBE has been used for growing programmed series of two-dimensional quantum wells (Figure 1.6a).

2. Cleaved-edge overgrowth with MBE has also been used to make one-dimensional quantum wires (Figure 1.6b). Quantum wells are grown by MBE, the sample is cleaved in vacuum perpendicular to the original growth direction, and new quantum wells are grown to make T-intersections of two quantum wells, forming wires of lower binding energy. Alternatively, growth in the intersection of wells forms one-dimensional wires.

3. Two successive cleaved-edge overgrowths were used to make a series of

double-T zero-dimensional quantum dots (Figure 1.6c); or, another approach, the clever use of strain in pseudomorphic overgrowth was used to produce strained dot arrays (Figure 1.3.1).

4. Solution chemistry, described in Chapter 5, was used to produce a monodisperse colloidal suspension of semiconductor quantum dots.

Selective area overgrowth has also been successfully used to tailor the quantum well thickness and composition laterally to make integrated semiconductor optical devices; examples include electro-optical modulators that consist of a ridge waveguide semiconductor laser and an optical modulator adjacent to each other on a single chip. The fabrication of novel optical nanodevices is in its infancy; many advances in design and manufacturing (such as self-assembly) are required to allow mass applications.

Enclosing a material between highly reflecting mirrors produces an optical microcavity, which can be tailored to control the angular distribution of the output light from the structure as well as the spectrum and spontaneous emission from any emitters inside. This is the principle behind vertical-cavity surface-emitting lasers, or VCSELS, which are made by enclosing an active gain medium between two highly reflective dielectric stacks of mirrors grown vertically on a substrate. Microcavities with very sharp resonances (very high Q) have been achieved by making whispering gallery mode resonators out of semiconductors or droplets of dye solution or polymer in which the index change between the material and air produces a high-Q waveguide around the outside diameter of the structure (Figure 1.7). In fact, optical nanocavities have been produced that have Q as high as several thousand. Because of the enormous field enhancement in the cavity with such Q, nonlinear effects should be observable with only a few incident photons. The intense emission of light observed, and yet to be conclusively understood, from porous silicon is to some extent both a confinement and microcavity effect. Another exciting new field involves tailoring optical materials to achieve periodic opposite polarities of the ferroelectric polarization on a length scale tailored to maximize the intensity of optical nonlinearities by efficient phase matching of coherent four-wave mixing.

Extending the concept of optical microcavities into three dimensions leads to the prediction of photonic band-gap materials, structures with periodic variations of dielectric constant on a length scale comparable to the wavelength of light. The idea is to design materials such that they can affect the properties of photons in a manner similar to the way semiconducting crystals affect electrons. In a semiconductor, the atomic lattice presents a periodic potential to an electron propagating through the electronic crystal. The geometry of the lattice and the strength of the potential are such that, owing to Bragg-like diffraction from the atoms, a gap in energy for which an electron is forbidden to propagate in any direction appears. In a photonic crystal, the periodic potential is caused by a lattice of periodic dielectric media instead of atoms. If the dielectric contrast is

FIGURE 1.7 Very high-Q microcavity. (Courtesy of Bell Laboratories, Lucent Technologies.)

sufficient, Bragg scattering of light off the lattice can also produce a forbidden band that extends over a certain energy range in which light cannot propagate in any direction. However, a defect in the periodicity will introduce localized states in the photonic band-gap in much the same way that localized states exist for electrons within the semiconducting gap. The nature and shape of the localized states will depend on the dimensionality of the defect: a two-dimensional slab or a one-dimensional line will define mirrors and waveguides in the dielectric array, and a zero-dimensional defect will define a microcavity. The design and manipulation of these defects in the photonic band-gap material promises far more control of photons.

As the technology for fabricating photonic lattices in the near infrared (IR) and visible spectral regions advances, they will offer a radically different means for controlling light. For example, Figure 1.8 shows a theoretical model of how light propagates in a periodic square lattice of dielectric rods with a waveguide produced by the intersection of two missing rows of rods. Remarkably, propagation is predicted to occur with no losses even though the bend in the waveguide is on a length scale comparable to the wavelength of light! In ordinary dielectric waveguides today, the bending losses caused by leakage from evanescent fields requires very smooth bends with bending radii of 10 cm—thus the waveguides are large, making manufacture and packaging of integrated optical structures

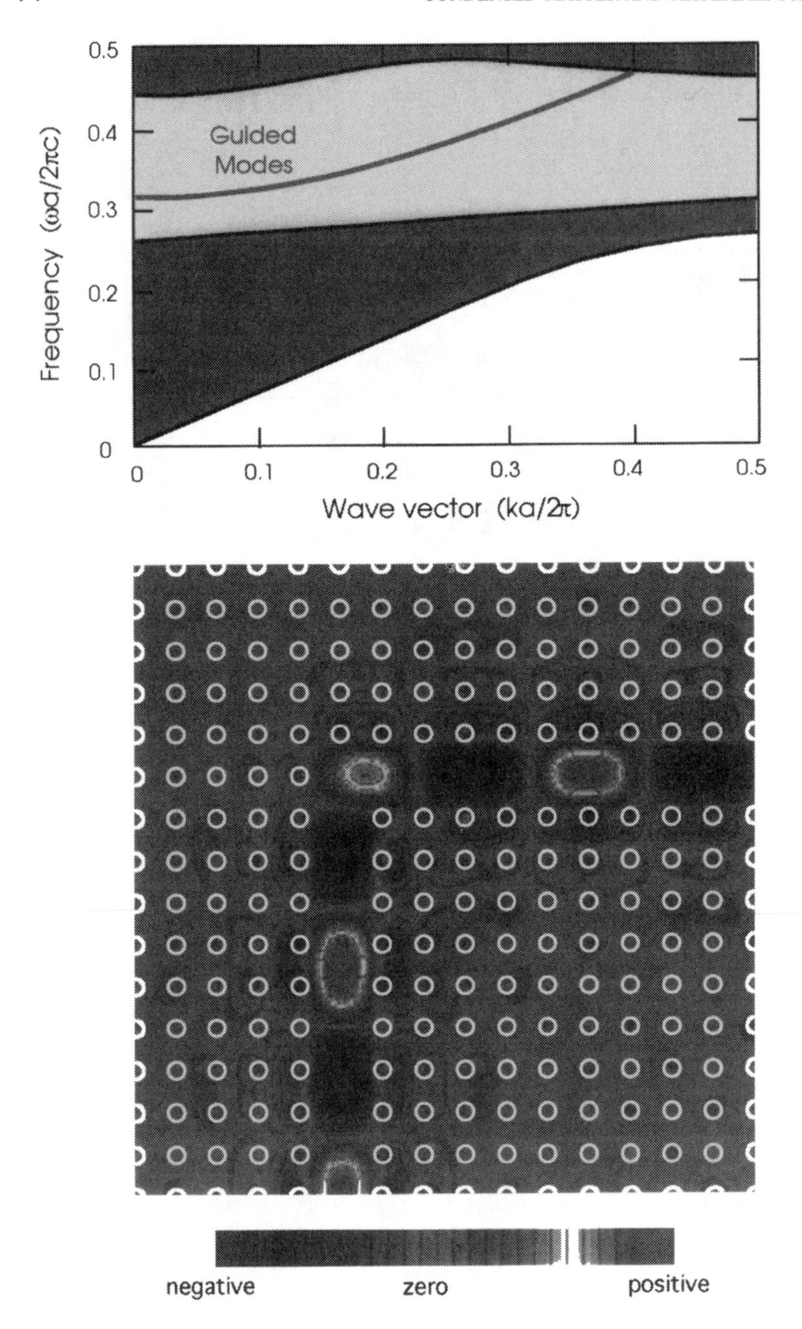

FIGURE 1.8 How light propagates in a periodic square lattice. (Courtesy of Massachusetts Institute of Technology.)

difficult. Low-cost fabrication of photonic band-gap waveguides for communications wavelengths of 1.3 and 1.5 μm, or wavelengths in the near IR and visible regions important for display, copying, and data-storage technologies, would revolutionize the field of integrated optical devices in much the same way as the integrated circuit has revolutionized electronics.

Challenges, Priorities, and Frontiers of Optical Materials and Phenomena

The frontiers in the field of optical materials and phenomena are novel manufactured materials with tailored optical properties made by self-assembly, poling, lithography, MBE and other techniques not yet invented, organics and other complex materials, nanostructures and micromachined components, biology, and ultrafast phenomena.

Some of the outstanding scientific questions about optical materials involve understanding the physics of lasing and the coupling of exciton-polariton-phonons in nanostructures and quantum cavities, fundamental understanding of the physics of electro-optical processes in organics, and understanding the physics of ultrafast nonequilibrium processes in semiconductors, metals, and biological molecules and tissues.

Some outstanding technology needs are low-cost all-optical communications network and consumer access components such as new fiber materials and devices; all-optical buffer memory, add-drop filters, amplifiers, semiconductor blue lasers, fast light switches, fast spatial light modulators; materials with tailorable optical properties such as better nonlinear optical materials, resists, and photonic-band gap materials in the near IR communications wavelength region; and low-cost assembly and manufacturing techniques for optical components such as self-assembly, stamping, and printing.

SCIENCE AND TECHNOLOGY OF MAGNETISM

Beginning with the Ancient Mariner's compass and continuing with such applications as automobile starter motors, refrigerator magnets, and computer hard disk drives, the importance of magnetic materials (Figure 1.9) in a wide range of technological uses steadily grows. Such materials display a host of fascinating properties of scientific interest. Many of these properties have proved to be useful in technological applications. The interplay of science and applications has made magnetism an extremely exciting segment of condensed-matter and materials physics. The last decade has seen an acceleration in the advances of both the technology and science of magnetism—advances that have set the stage for even more profound discovery and technological developments in the near future.

The field of magnetism enjoys an unusually strong technology pull, second in condensed-matter and materials physics only to that of semiconductors, particularly silicon. Various forms of magnetic storage technology dominate this pull. Bulk materials have long had applications in technologies such as motors, generators, and transformers. In addition, the emerging field of magnetoelectronics, with its potential impact on microelectronics technology, is stimulating a fresh wave of enthusiasm along with a broadened research agenda.

This section begins with a review of a few exciting recent developments and future directions of selected important technologies based on magnetic materials and phenomena. Against this backdrop, selected scientific accomplishments of the past decade and opportunities for the future will be presented. The section concludes by summarizing some of the major outstanding scientific questions and suggesting priority directions for future research.

Technology Pull

The industry of magnetic recording, in all its forms, constitutes an enterprise

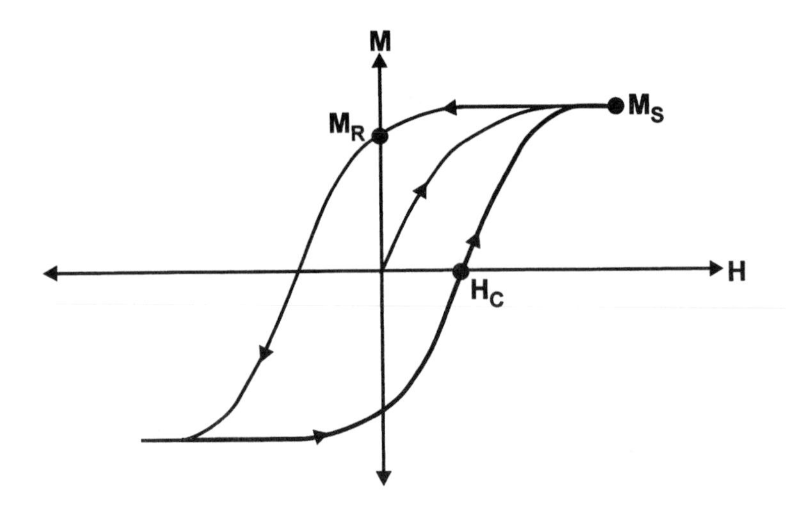

FIGURE 1.9 Magnetic response of a ferromagnetic particle. The initially unmagnetized magnetic particle exhibits a hysteresis loop in its magnetization (M) as an external magnetic field (H) is increased from zero to a maximum value, reversed through zero to a minimum value, and subsequently returned to zero. Characteristic features of hysteresis curves include the saturation magnetization, M_s (the maximum magnetization of the sample); the remnant magnetization, M_r (the magnetization that persists in the sample when the external field is zero); and the coercive field, H_c (the external field necessary to return the magnetization to zero). "Soft" magnetic materials have low H_c, high permeability (~dM/dH), and a small area enclosed by the hysteresis loop. "Hard" magnetic materials have high H_c, high M_r, and a large area enclosed by the hysteresis loop.

with annual revenues in excess of $100 billion. This consists primarily of magnetic disk, magnetic tapes, and optical disks for digital data storage together with various forms of magnetic recording for audio and video. This industry is experiencing an overall compound annual growth rate in revenue of about 10 percent per year. This growth rate is expected to continue for at least another decade, with magnetic storage playing the dominant role. The United States "owns" about 40 percent of the magnetic storage business, the largest single component of which is hard disk drives (see Box 1.12). This component alone is a $30 billion a year business.

The accelerating interplay of the science and applications of magnetism is well illustrated by the phenomenon of magnetoresistance. Lord Kelvin first observed this effect in 1856. Beginning in the early 1980s, a decade of research and development (at IBM) with this basic laboratory phenomenon perfected a product of major commercial importance. The first magnetoresistive sensor used in the recording head of a hard disk drive has an intricate structure in which data is sensed by a 20 nm thick permalloy (a NiFe alloy) layer. The useful change in resistance of this film as it passes in close proximity to a small magnetized region of a magnetic disk is about 2 percent. The time from discovery of the phenomena to a high-volume product was 135 years.

Recent research activities led to the discovery of a superior form of magnetoresistance, called giant magnetoresistance (GMR). GMR requires the interaction between at least two very thin ferromagnetic films and can register a resistance change at room temperature of about 10 percent in the same magnetic field range as permalloy. Moreover, as an interfacial phenomenon, its performance in so-called spin-valve recording sensors improves with decreasing film thickness, which also increases the storage density. In contrast to the case of magnetoresistance, only 10 years have passed since the original discovery of GMR before an initial product was produced. In the exponentially growing global hard-drive industry, GMR sensors will be needed to sustain this rate of improvement into the third millennium.

A range of magnetic tape storage systems with applications from audio and video to data storage constitute another third of the magnetic storage business. Storage densities in this arena are experiencing a single-digit compound annual growth rate with continuing cost reduction. Here, too, magnetoresistive heads play an important role in the continued scaling to higher densities. Magnetic particle tapes are still the industry standard, but thin film tapes have been introduced and will undoubtedly dominate some time in the future.

In a completely different arena, bulk magnetic materials constitute a $4 billion global market, with the United States holding approximately 20 percent of the market share. This market is projected to grow to more than $6 billion by the year 2000. "Hard" bulk magnetic materials are essential constituents in a wide variety of electric motor and generator technologies. For such applications, the strength of the permanent magnetism, or so-called "maximum BH-product," is

BOX 1.12 Magnetic Storage in Hard Disks

The basic configuration of a hard-disk drive is illustrated in Figure 1.12.1. Key aspects of this technology include the read/write head, media, head-disk interface, tracking of the head with respect to the disk, and signal processing to distinguish the ones and zeros as the disk spins beneath the head.

The evolution of the areal storage density in leading-edge products was illustrated in the introduction. This exponential growth in storage density has led to an associated growth in the number of hard-disk drive bytes shipped per year, as shown in Figure 1.12.2.

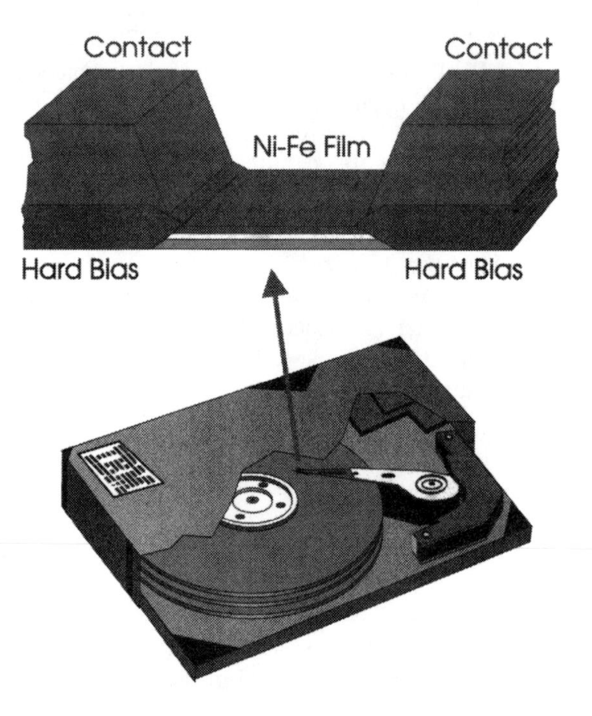

FIGURE 1.12.1 Magnetic recording disk drive assembly with a magnetoresistive head. Three disks and one of the movable arms supporting a read/write head are visible. The arrow zooms in on the magnetic sensor of a recording head. The NiFe film measures 20 nm in thickness by 1 or 2 μm in width. The permanent magnetism of the hard bias induces a reference state of the magnetization inside the NiFe film. The magnetic field emanating from small magnetized regions, representing the stored data, in the rotating disk changes the resistance in the NiFe films thus reading out the data as voltage changes across the contacts, which also supply a constant current through the NiFe film. (Courtesy of IBM Research.)

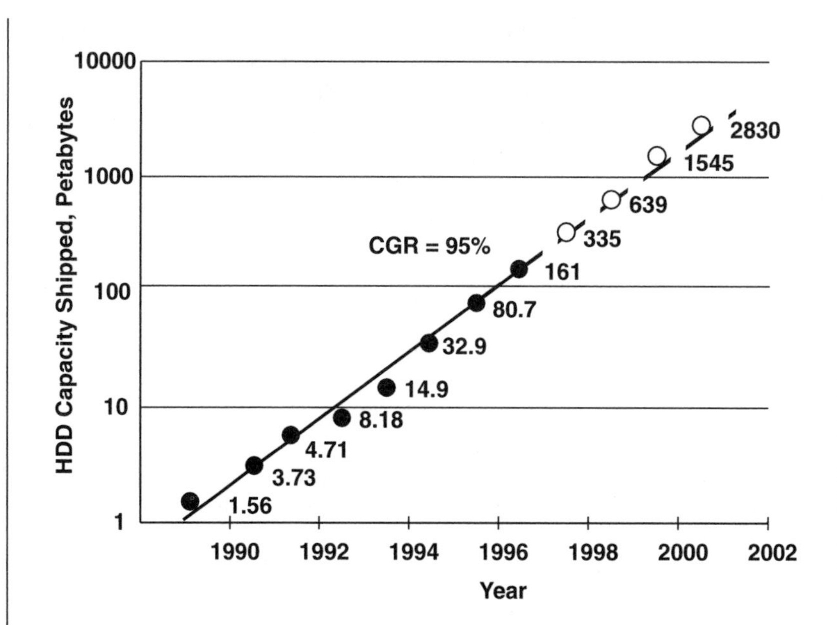

FIGURE 1.12.2 The number of hard disk drive bytes shipped in products per year. The compound annual growth rate of 95 percent is projected to continue for the foreseeable future. (Courtesy of IBM Research.)

Exponential improvement in storage density has been achieved through continued advances and scaling of all aspects of the technology. As an example, consider advances in magnetic head technology. Through the first half of the decade thin film inductive heads dominated the industry and became ever more difficult to scale. The first magnetoresistive heads were introduced in 1992 and played a pivotal role in increasing the compound annual growth rate of the areal storage density from 30 percent to 60 percent (with a 40 percent annual reduction in cost per bit!). Giant magnetoresistive heads introduced at the end of 1997 will help ensure the continuation of this trend.

the defining materials characteristic. Research and development leading to increased BH-products (Figure 1.10) has steadily decreased the costs, sizes, and weights of motors in diverse devices like auto starters, cordless shavers, hand-held drills, vacuum cleaners, washing machines, dryers, machine tools, motorized toys, and disk drives in laptop computers. Motor vehicles alone account for 70 percent of permanent-magnet usage in starters, electric windows, speakers, and cassette and CD players. A luxury car may contain 80 motors.

The most recent permanent-magnet development spurt occurred in the mid-

1980s and dramatically illustrates the unpredictable effects of the interplay between politics, economics, research, and development. Samarium-cobalt was the leading hard magnetic material in the late 1970s in spite of the high cost of samarium. The cost of samarium-cobalt magnets ballooned fivefold when the world's principal source of cobalt disappeared during the 1978 national upheaval in Zaire. Intense focused research established that samarium and cobalt could be replaced by neodymium and iron, respectively. Not known in advance was that the attainment of permanent magnetization in the new compositions hinges on complicated processing sequences including, in one case, creation of amorphous material by melt-spinning followed by crystallization through severe mechanical treatment and subsequent annealing. Equally unexpected was that the product would be stronger, both magnetically and mechanically, and less expensive. The rapidly growing NdFeB segment of the bulk magnetic materials market is projected to reach $4 billion by the year 2005.

"Soft" bulk materials play key technological roles in radio frequency (rf) and power distribution applications. At frequencies higher than 100 kHz, ferrites remain the materials of choice. They are widely used in all manner of rf and microwave elements such as antennas, filters, circulators, and insulators. Historically, advances in ferrite performance (higher permeability, higher frequency

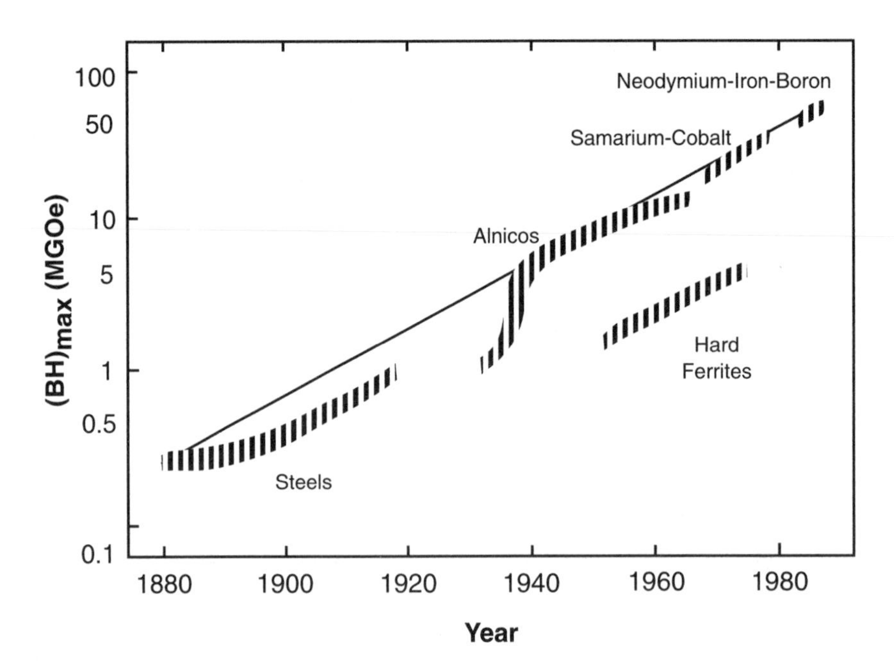

FIGURE 1.10 Chronological trend of $(BH)_{max}$ where the data represent initial demonstration in the laboratory. [*Reviews of Modern Physics* **63**, 819 (1991).]

response, and lower losses) have been driven by niche military applications, with commercialization following rapidly. At lower frequencies, soft magnetic materials are used extensively in transformers for the power-distribution industry. Here, incremental improvements can have an enormous technological and economic impact. Particularly important is magnetic metglass, an amorphous material in ribbon form prepared by ultra-fast quenching. The absence of magnetocrystalline anisotropy has the consequence that the magnetization vector can be easily rotated. Therefore, a metglass has high magnetic permeability and low losses. When used in the core of a power transformer, metglass is more expensive than crystalline materials and increases capital costs; however, this increased capital cost is often rapidly amortized by continuing savings from decreased transformer loss in electric power transmission. Worldwide savings of several billion dollars have been realized by the introduction of magnetic metglass.

Magnetostriction is another area of magnetism of considerable, and growing, interest for both military and commercial applications. Most of the materials development work in this area has been focused on rare-earth transition metal alloys. Strains of up to 1 percent, along with considerable force actuation, have been demonstrated in practical applied fields. Applications range from sonar pulse generation to high-reliability replacement of hydraulic systems in aircraft and even tanks.

A final area is magnetoelectronics (exclusive of magnetic storage), which includes a variety of devices and associated assemblies. The largest component of this industry today is sensors used for commercial, scientific, and military applications. Such sensors range from Hall effect sensors, to superconducting quantum interference devices (SQUIDs), flux gate magnetometers, search-coil magnetometers, magnetoresistance (MR) and GMR sensors, to magnetic force microscopes. Magnetoelectronics is characterized by a number of small but stable niche markets. The worldwide market for SQUID instrumentation, for example, is about $20 million per year; the market for magnetic force microscopes is similar.

The potential sleeping giant in the field of magnetoelectronics is a growing collection of novel devices and circuits that possibly can be integrated onto a high-performance chip to perform some complex function. More realistically, key elements may be integrated with high-performance semiconductor technology to produce new generations of microchips with function, density, and/or performance beyond that achievable with semiconductor technology alone. Nondestructive read out memory chips in which bits are stored in small electrically addressable magnets have been demonstrated at capacities of up to 256 kb (Figure 1.11). The nonvolatile radiation-hard nature of this memory together with potential for scaling to much higher density and performance, especially as new magnetic elements are developed, show considerable promise.

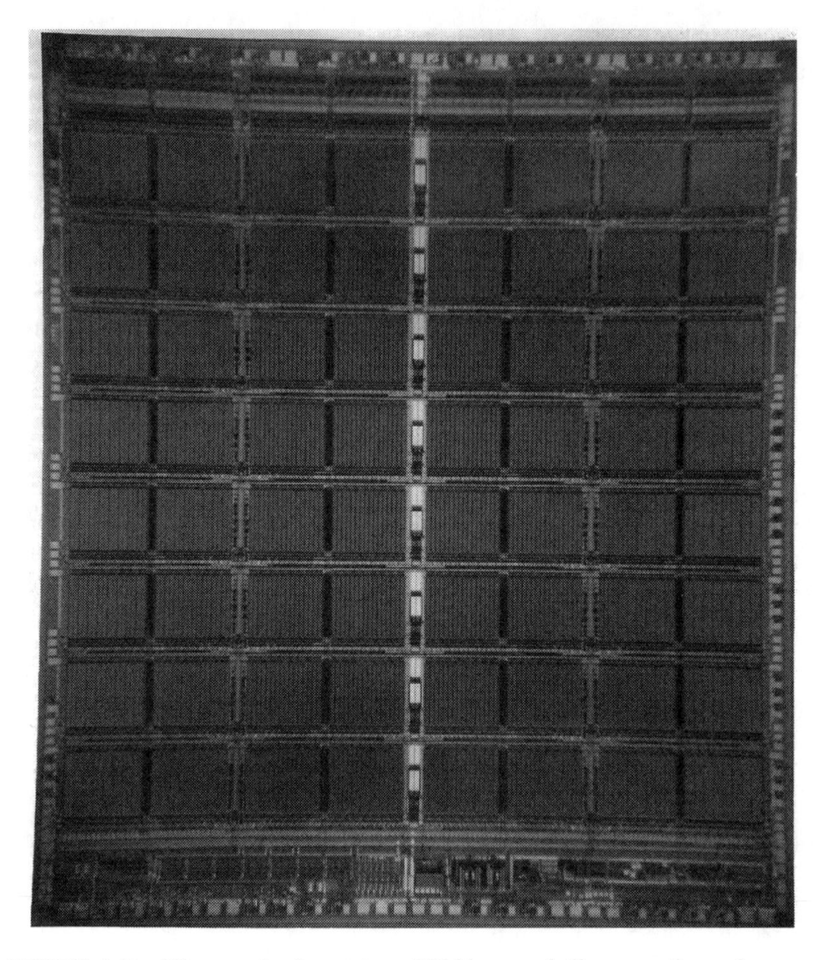

FIGURE 1.11 Micrograph of prototype 256-kb nonvolatile magnetic random access memory chip. (Courtesy of Honeywell.)

The Physics of Magnetism

It is evident from the above discussion that there is a strong technology pull for research in magnetism. The last 10 years have seen numerous exciting discoveries, a number of which have already had a direct impact on technology. Others are providing us with a better understanding of the world around us and/or are helping lay the groundwork for future technology. It is amazing that, although much is known about magnetism and while effects drive a $100 billion per year industry, our basic understanding of magnetism, even in a material such as iron, is incomplete. As an example, Figure 1.12 shows the results of a simulation of a two-dimensional array of magnetic moments approximating a 10 nm × 500 nm × 1000 nm sheet of permalloy. One might anticipate that for a system of this size, the moments would all work in unison, switching together under the influence of an applied field; however, as shown in Figure 1.12, such is not the case. The behavior of this rather elementary system exhibits a great deal of structure and complexity. There is neither experimental nor theoretical consensus on the detailed behavior of such a system. Nor is it understood how small such a system must be to ensure true, single-domain behavior in which all of the moments always remain parallel to each other.

The fundamental limit on stability of magnetic domains is an important area of basic investigation in magnetism. The advancing march of magnetic technology makes investigation of these limits inevitable, but probing these limits raises some of the most challenging questions for condensed-matter physics and materials science, such as, What is the smallest size magnetic element stable against external perturbations such as temperature fluctuations? and, Given that quantum mechanics sets bounds on the lifetime of any magnetic state, how do such bounds ultimately establish limits on the size of the smallest possible magnetic entities useful for technological applications?

Molecular magnets such as the ferric wheel shown in Box 1.13 are the subject of a wide variety of physical analyses aimed at shedding light on these and other challenging questions about magnetism in small structures. Traditional

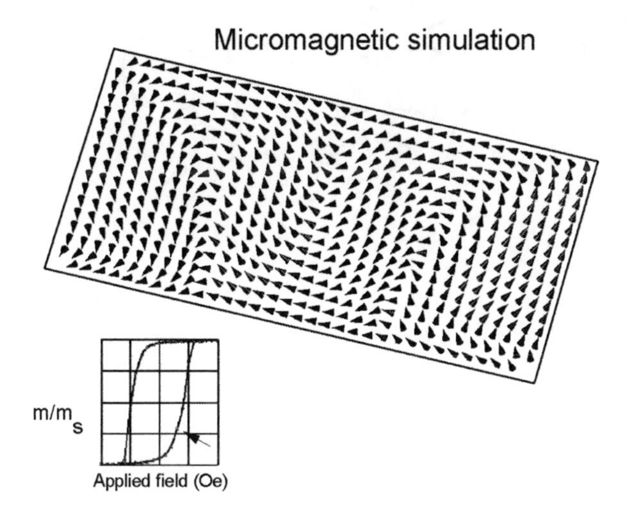

FIGURE 1.12 Micromagnetic simulation of the switching behavior of a 10 nm × 500 nm × 1000 nm permalloy dot. The arrows indicate the direction of the magnetization (m) at the point in the hysteresis cycle indicated on the curve in the lower left. (Courtesy of IBM Research.)

BOX 1.13 Nanomagnets

Fundamental questions about nanomagnets could not be addressed were it not for the remarkable advances made by materials scientists and chemists in the synthesis of exquisitely controlled nanometer-scale magnetic structures. Many "traditional" synthesis techniques have been applied successfully in this area: sub-100 nm scale permalloy particles can be made with lithographic techniques, and growth assisted by scanning-tunneling microscopes has proven that it is possible to fabricate pure iron particles with a variety of sizes and aspect ratios. The most exciting new techniques involve various "self-assembly" strategies that are emerging as extensions of chemical synthesis techniques. The relatively low-tech techniques of colloid growth have been adapted to make cobalt particles 8 nm in diameter with close to atomic control. Actual atom-by-atom control in the magnetic domain is now achieved by metallo-organic molecular synthesis; magnetic molecules containing exactly 10 iron atoms (the "ferric wheel," see Figure 1.13.1), or exactly 12 manganese ions, for example, are now routinely available.

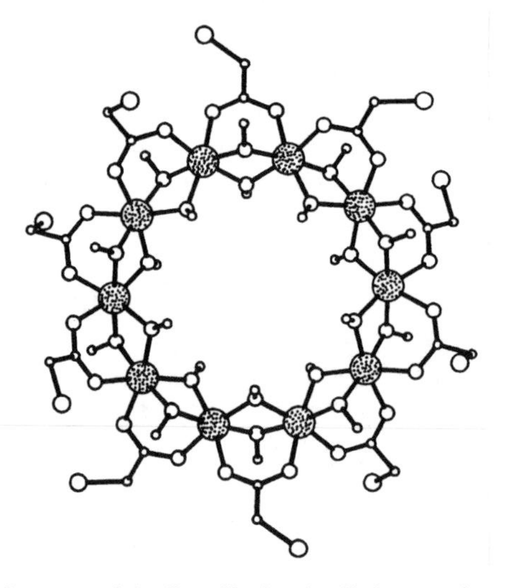

FIGURE 1.13.1 Structure of the Fe_{10} "ferric wheel" cluster, where the large solid circles represent the iron atoms and the empty circles are, in order of decreasing size, chlorine, oxygen, and carbon. The 10 Fe^{3+} ions, each with a magnetic moment corresponding to the same angular momentum or spin, are bound together into a perfectly regular ring. High magnetic field experiments have shown that the Fe^{3+} ions exhibit antiferromagnetic behavior; neighboring spins prefer to be antiparallel. The spin structure of the molecule passes through a rich sequence of phase transitions resembling those in bulk layered antiferromagnets. These experiments open the prospect of precisely controlling the structure, interactions, and dynamics of nanomagnets. [Reprinted with permission from D. Gatteschi, A. Caneschi, L. Pardi, and R. Sessoli, "Large clusters of metal ions: The transition from molecular to bulk magnets," *Science* **265**, 1056 (1994). Copyright © 1994 American Association for the Advancement of Science.]

characterization gives the basic magnetic parameters of the particle: ionic moment (i.e., the local state of spin), exchange interaction (the coupling strength between the spins), and anisotropy (the height of the energy barrier in the double-well potential separating the "up" state from the "down" state). Such characterizations demonstrate that the ferric wheel behaves very much as an atomic-scale analog of a layered antiferromagnet. Less traditional characterizations are required to understand the ultimate stability of such molecular magnets, which is determined by quantum tunneling. One particularly interesting example is the observation of what might be called "quantized hysteresis" in the Mn(12) molecule: at low temperatures, this structure shows a propensity to switch from up to down at a sequence of regularly spaced magnetic fields. Some evidence suggests that these magnetic fields coincide with resonances between quantum levels of the up and down wells, resulting in enhanced tunneling. Although some details of the tunneling mechanism remain to be understood, this is a particularly simple example of the stability (up versus down) of the moment of a molecular magnet being ultimately limited by a purely quantum effect.

Of more profound significance than the observation of quantum tunneling would be the observation of "quantum coherence" in these nanomagnets. The phenomenon is closely analogous to the microscopic quantum coherence (MQC) effect sought in small SQUIDs for years—the creation of a quantum state in a controlled superposition of up and down states. The regular advance of the phase of this superposition would result in the sinusoidal oscillation of the magnetic domain between the up and down states. Observation of such coherence oscillations would foreshadow a significant change in the role that quantum mechanics might play in the dynamics of magnetic domains. Although we might view magnetic tunneling as a nuisance, destroying the stability of a bit stored in the magnetic domain, coherence, if controllable, could be the resource needed to realize the basic element of storage and processing in quantum computing. Signs of magnetic quantum coherence have in fact been observed in a naturally occurring magnetic nanoparticle.

In the past 10 years rapid progress has been made in the characterization and understanding of magnetic multilayers, exchange coupling, and spin-dependent transport through magnetic materials and interfaces. Results from an experiment representative of this exciting, ground-breaking work is shown in Figure 1.13. This experiment measured the oscillatory exchange coupling between iron layers separated by a chromium spacer of varying thickness. The chromium wedge was grown epitaxially on the nearly perfect surface of an iron whisker crystal whose magnetization is split into two opposite domains along the [001] direction. A thin iron film was deposited on top of the chromium, and its magnetization was measured using scanning electron microscopy with polarization analysis (SEMPA). The SEMPA image, drawn on the wedge schematic, clearly shows that the exchange coupling reverses direction with almost every single monolayer change in chromium thickness. The oscillatory coupling period, which arises from nesting features in the Fermi surface of chromium, is actually slightly incommensurate with the chromium lattice, producing the phase slips observed at chromium layers 24 and 44.

The applications focus provided by GMR has helped to stimulate and invigorate the search for new magnetic heterostructures and nanostructures and new magnetoresistive materials. GMR materials consist mainly of nanometer thicknesses of interleaved metallic layers that are alternately ferromagnetic and nonferromagnetic. (Analogous nano-dispersed two-phase composites can also exhibit GMR properties.) Key to strong in-plane GMR are spin-dependent scattering at layer interfaces and an electron mean free path (approximately 10 nm) greater than sublayer thickness. The relative resistance change can be greater for current flow perpendicular to the layer planes. Resistance changes as large as 100 percent relative to the low-resistance state have recently been reported.

Work on magnetic multilayers is also stimulating new thinking concerning novel devices that can be made by integrating magnetic materials with standard semiconductor technology. An example of this is shown in Figure 1.14 where a GMR (Co/Cu) multilayer serves as the base of an n-silicon metal base transistor. Biased in the common base configuration, this device exhibited a 215 percent change in collector current in a magnetic field of 500 G at 77 K with typical GMR characteristics. The in-plane GMR of the multilayer was only 3 percent. Although by no means a practical transistor, this structure allows the study of spin-dependent scattering of hot electrons in magnetic multilayers. More practical spin transistors may be forthcoming, particularly if ways to achieve 100 percent spin-polarized injection can be devised.

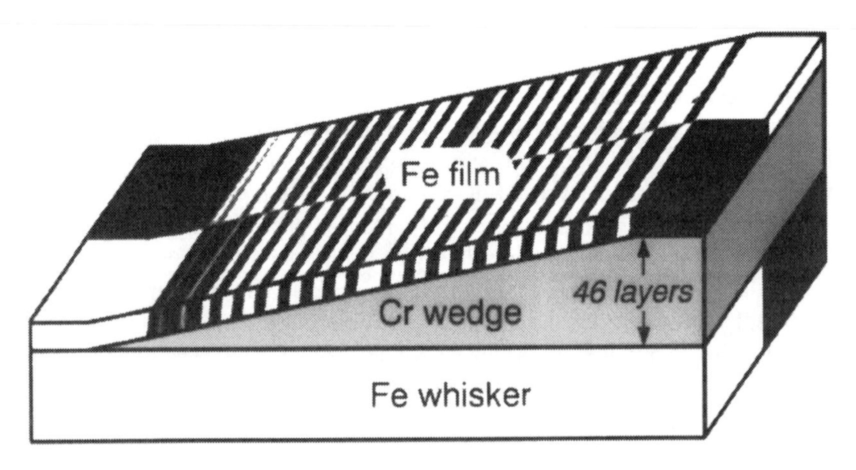

FIGURE 1.13 Oscillatory exchange coupling in Fe/Cr/Fe. [*Physical Review Letters* **67**, 140 (1991).]

Spin-polarized tunneling experiments in magnetic thin-film planar junctions are helping to elucidate novel magnetic properties as well as demonstrate features of considerable device potential. The first successes in spin-polarized tunneling between two ferromagnets through an insulating tunnel barrier occurred only very recently, even though spin-polarized tunneling was predicted more than 20 years ago. A simplified structure of this type of junction is shown in Figure 1.15. The magnetization of the ferromagnetic base electrode is pinned in the direction indicated. The magnetization of the ferromagnetic counter electrode is shown as aligned with that of the base electrode but can be reversed by application of a modest magnetic field. A proper magnetic design yields a hysteretic response curve symmetric about $H = 0$. Resistance changes of greater than 30 percent have been demonstrated. Such junctions are of considerable interest as potential storage elements in one approach to magnetic RAMs. Slightly different configurations give devices with nonhysteretic characteristics but with similar magnetoresistance. These devices are attractive candidates for sensor applications and may provide a follow-on to GMR sensors for hard-disk drives.

The enormous surge in the synthesis and study of high-T_c perovskite materials spawned a concerted effort to explore the magnetic properties of similar materials. Some of these materials exhibit what has been termed colossal magnetoresistance (CMR). The magnetoresistance of doped manganite structures such as $La_{1-x}Sr_xMnO_3$ changes by a factor of 2 or 3, although not at temperatures and magnetic fields suitable for practical device applications. These systems share much in common with high-T_c cuprate superconductors, from which dozens of new crystal structures have emerged. Replacing copper with manganese, for example, could generate a platform of new crystal chemical systems, some of which will undoubtedly exhibit promising CMR properties.

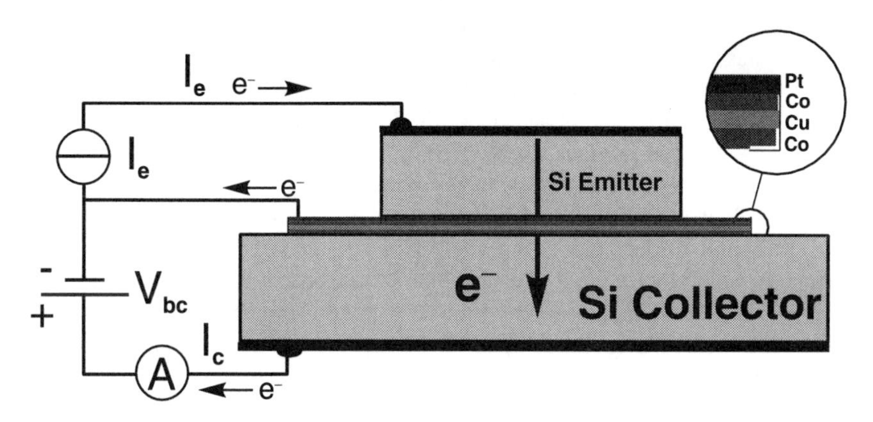

FIGURE 1.14 Schematic cross section of a prototype spin-valve transistor. [*Physical Review Letters* **74**, 5260 (1995).]

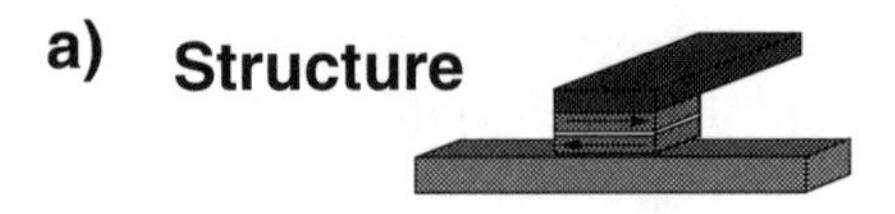

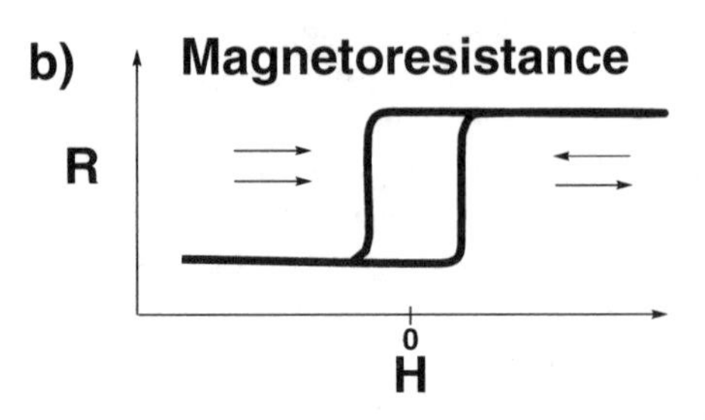

FIGURE 1.15 Magnetic tunnel junction structure. (Courtesy of IBM Research.)

Steady advances in improving the characteristics for technological applications has been realized in the more traditional bulk materials. For example, in the case of high-permeability soft materials, we have learned how to compensate for the deleterious magnetic anisotropy from one magnetic element by introducing a second element having an anisotropy of the opposite sign. Another way to cancel anisotropic effects, and thereby increase permeability, is to rapidly quench a magnetic ribbon, as with magnetic metglass, thus making its structure amorphous. The preferred axes of the atomic moments are now random and therefore the atomic contributions to anisotropy energy tend to cancel, making it easy to remagnetize. In the case of permanent magnets, the magnetic anisotropy, and therefore the coercivity of a ferromagnet, decrease steeply as its temperature approaches the Curie critical point, where these parameters necessarily vanish. We have learned how to increase the Curie temperature, and thereby the coercivity, of permanent magnet materials of the Re_xFe_y type by introducing interstitial N or C or, to lesser degrees, Ti, V, W, Mo, or Si. The complexity as well as the importance of processing details in the synthesis of high-performance magnetic materials is well demonstrated in the case of NdFeB discussed previously.

Measurement techniques are vital in the research of magnetic materials and phenomena. Experimental advances that have contributed to breakthroughs in the last decade include scanning-tunneling microscopy (STM), magnetic force microscopy (MFM), magneto-optic Kerr imaging, and scanning electron microscopy with polarization analysis (SEMPA). STM has been critical to understanding how subtle differences in physical structure can make profound differences in magnetic structure or properties. Characterization techniques based at major facilities are equally important. For CMR, as for high-T_c research, neutron diffraction is needed to obtain the position of the oxygen atoms within the unit cell as a function of temperature, field, pressure, and doping. Electron microscopy is needed to understand growth inclusions that form two-dimensional stacking faults. Synchrotron sources enable advanced spectroscopies to identify the +3 and +4 valence states of Mn and their ratio. Diffuse x-ray scattering and quasi-elastic neutron scattering are used to investigate the presence and dynamics of polaronic distortions. A number of other techniques based on such effects as spin-polarized photoemission, magnetic circular dichroism, and second harmonic generation are becoming increasingly prevalent, while many others are in the initial stages of demonstration.

Major Outstanding Materials and Physics Questions and Issues in Magnetism

Many outstanding scientific questions remain in the field of magnetism. Answers to a number of these questions will have an important technological impact and are necessary to continue the momentum and growth of the magnetics industry.

With a few notable exceptions, we lack detailed understanding of the magnetic properties of nanostructured magnetic elements and arrays of such elements. Examples include the following:

- The nature of domain structure and its influence on switching behavior;
- The dynamics and switching times in such elements or systems;
- The influence of temperature, both in the context of stability against thermally induced switching and in the context of structural change at elevated temperature;
- The nature of the interaction of spin-polarized currents with such elements, both reversible and irreversible; and
- The role and the impact of quantum coherence and macroscopic quantum tunneling in the smallest of such structures consisting of a cluster of atomic spins.

We need to understand the impact of the issues above and related issues regarding technologies such as magnetic recording and the synthesis of new materials with improved properties such as higher BH-products. Because of the resurgence of the science and applications of magnetism, it is important that we reestablish the teaching of magnetism as a priority in our universities as a whole rather than at only a few institutions that presently teach it.

Much remains to be learned concerning the nature of spin-polarized transport. Questions need to be answered about the role of structure and the relationship of surface and interface structure to magneto-transport, the scattering mechanisms at interfaces in GMR, and the physics of the temporal and spatial decay of nonequilibrium magnetism. Also required is a detailed understanding of the mechanism of spin injection, either directly or through tunneling barriers from magnetic metals into metals and semiconductors. Answers to these types of questions are needed to engineer better GMR, CMR, and magnetic tunnel junction (MTJ) materials and devices. In a particularly useful spin transistor with true on and off states separated by many orders of magnitude in conductance will require very nearly 100 percent spin-polarized current injection from one region of the device into another.

Advanced synthesis and processing techniques need to be developed to produce novel material with a high potential for scientific and technological impact. Layered structures, nanostructured three-, two-, and one-dimensional materials, and materials with higher BH-products are all attractive areas for further exploration. As in other areas of condensed-matter and materials physics, a systematic approach—technically and organizationally—is needed to explore the vast phase space of magnetic materials. As applications develop, methods to bridge the gap between fabrication techniques that serve to produce initial demonstrations and more controlled and reliable techniques that can be migrated to development and manufacturing will be needed.

Several difficult challenges remain in the measurement area. We need to understand how to magnetically probe individual electron and nuclear spins directly. Questions to be addressed include, What is the ultimate spatial resolution of magnetic measurement techniques? And, Can we fabricate a spin-polarized STM? Advances in measurement technology related to such questions will have a profound impact on our ability to understand the nature of magnetism in nanostructures.

Finally, in the technology arena, continued focus on scaling the density for magnetic storage is needed. This will involve a strong, systematic, ongoing program in both media and detectors that will draw heavily on several of the condensed-matter and materials physics areas mentioned above. In addition, there is enormous opportunity in the arena of magnetoelectronics for new magnetic effects and devices that may set the stage for magnetism to play a key role in future microelectronic chip technology. A key element will be integration of complex magnetic materials with mainstream semiconductor technology.

FUTURE DIRECTIONS AND RESEARCH PRIORITIES

Numerous outstanding scientific and technological research needs have been identified in electronic, photonic, and magnetic materials and phenomena. If those needs are met, it is anticipated that these technology areas will continue to follow their historical exponential growth in capability per unit cost for the next few years. Silicon integrated circuits are expected to continue to follow Moore's Law at least until the limits of optical lithography are reached; transmission bandwidth of optical fibers is expected to grow exponentially with advances in optical technology and the development of soliton propagation; and storage density in magnetic media is expected to continue to grow exponentially with the maturation of GMR and development of CMR and MTJs in the not too distant future. Although these changes will have major impact on computing and communications over the next few years, it is clear that extensive research will be required to produce new concepts, as will new approaches to reduce research concepts to practice, if these industries are to maintain their historical growth rate over the long term.

Continued research is needed to advance the fundamental understanding of materials and phenomena in all areas. For example, despite the extensive technological application and impact of magnetic materials and, despite more than a century of research in magnetic materials and phenomena, we lack a first-principles understanding of magnetism. By comparison, the technology underlying optical communication is very young. The past few years has seen enormous scientific and technological advances in optical structures, devices, and systems. New concepts such as photonic lattices, which are expected to have significant technological impact, are emerging. We have every reason to believe that this field will continue to advance rapidly with commensurate impact on communications and computing.

As device and feature sizes continue to shrink in integrated circuits, scaling will encounter fundamental physical limits. The feature sizes at which these limits will be encountered and their implications are not understood. Extensive research is needed to develop interconnect technologies that go beyond normal metal and dielectrics in the relatively near term. Longer term, technologies are needed to replace today's silicon field-effect transistors. One approach that bears investigation is quantum state switching and logic as devices and structures move further into the quantum mechanical regime.

A major future direction is nanostructures and artificially structured materials, which was a general theme in all three areas. In all cases, artificially structured materials with properties not available in nature revealed unexpected new scientific phenomena and led to important technological applications. As sizes continue to decrease, new synthesis and processing technologies will be required. A particularly promising area is that of self-assembled materials. We need to expand the research into self-assembled materials to address such questions as how to control self-assembled materials to create the desired one-, two-, and three-dimensional structures.

As our scientific understanding increases and synthesis and processing technologies of organic materials systems mature, these materials are expected to increase in importance for optoelectronic and, perhaps electronic, applications. Many of the recent technological advances are the result of strong interdisciplinary efforts as research results from complementary fields are harvested at the interface between the fields. This is expected to be the case for organic materials; increased interdisciplinary efforts—for example, between condensed-matter and materials physics, chemistry, and biology—offer the promise of equally impressive advances in biotechnology.

In conclusion, the committee identifies a few major outstanding scientific and technological questions and research and development priorities.

Major Outstanding Scientific and Technological Questions

- What technology will replace normal metals and dielectrics for interconnect as speed continues to increase?
- What is beyond today's FET-based silicon technology?
- Can we create an all-optical communications/computing network?
- Can we understand magnetism on the meso/nano scales needed to continue to advance technology?
- Can we fabricate devices with 100 percent spin-polarized current injection?

Priorities

- Develop advanced synthesis and processing techniques, including those for nanostructures and self-assembled one-, two-, and three-dimensional structures.
- Pursue quantum state logic.
- Exploit physics and materials science for low-cost manufacturing.
- Pursue the physics and chemistry of organic and other complex materials for optical, electrical, and magnetic applications.
- Develop techniques to magnetically detect individual electron and nuclear spins with atomic-scale resolution.
- Increase partnerships and cross-education/communications between industry, university, and government laboratories.

2

New Materials and Structures

Our ability to make new materials and structures—both in bulk and in reduced dimensions or length scales—is inextricably linked to the advancement of our understanding of fundamental phenomena in condensed-matter and materials physics. This chapter describes some of the past decade's advances in inorganic materials and structures. Some of the advances and promising new areas in organic materials are discussed in Chapter 5. As described in Box 2.1, an astonishing array of new materials with unexpected properties has come over the horizon. Improvements in synthesis and processing have led to dramatic improvements in the properties of established materials and our ability to exploit these properties. As a result, we can now fabricate new combinations of materials, features of reduced dimensions, and other characteristics that differ in significant ways from previous possibilities. Some of these developments have provided fertile ground for condensed-matter and materials physicists to explore novel fundamental phenomena; others show promise for finding applications quickly; some have the potential to change our lives.

New materials underlie the science and technology described throughout this report. Beyond condensed-matter and materials physics, they enable both science and future technologies. In some cases, entirely new and unexpected phenomena appear in a class of new materials. Layered cuprate high-temperature superconductors are a new class of materials that has kept experimentalists and theorists alike searching to understand the physical basis of high-temperature superconductivity. New materials sometimes allow entirely new device concepts to be realized or lead to a dramatic change in their scale, such as single-molecule wires made of carbon nanotubes; and new forms of already known materials can possess different properties. Semiconductor nanoclusters, which emit light whose wavelength depends on cluster size, offer the possibility of tailoring material properties to suit a particular need. Even mature techniques, such as those for bulk crystal growth, demand continuous improvements in process control to produce the size or quality of material required for either technological applications or fundamental studies.

Better understanding of the mechanisms at play in materials that have been known for decades can lead to new approaches that alleviate detrimental properties. An excellent example is the introduction of metallic oxide electrodes in ferroelectric devices, which reduces aging effects dramatically. Better understanding of the details of materials preparation can give rise to improvements in processing. Improved insight into the kinetics of epitaxial growth can dramati-

BOX 2.1 Additions to the Zoo: New Materials and Structures of the Past Fifteen Years

There have been far too many new developments in the past 15 years or so to document them all in detail, but all these developments have been made possible by advances in two intertwined areas: complexity and processing. Many of the new materials and structures are dramatically more complex, compositionally or structurally, than have been studied previously. In general, this trend has required advances in processing to allow control of the increased complexity. In other cases, the final product may not be much more complex than other well-known materials or structures, but the processing itself may need to be altered to achieve more control over the growth process in order to obtain the new material.

Advances giving rise to new materials and structures fall into three categories. Some involve the synthesis of an entirely new compound or material. The advance may have been revolutionary, meaning that the properties of the new material (or in some cases its existence) could not have been predicted. In other cases, advances in processing have allowed fabrication of new or modified materials or structures whose properties were suspected before the material was actually made. This may allow a well-known compound to be remade in a new form with different properties. Third, well-known materials are sometimes found to exhibit new (in some cases unexpected) properties that appear when the ability to process them is improved. The new property may be found in a known material simply by looking at it in a new light, which shines on it as a result of insight gained from another materials system.

The materials advances listed in Table 2.1.1 were driven by different motivations. Many addressed a technological need, such as the need to transfer or store information. Others were driven by scientific curiosity. Although the driver can be clearly identified in each case, the two sets are not mutually exclusive. Many discoveries that result from pure scientific curiosity ultimately find their way into products. For example, low-temperature superconductors are now used in magnets for magnetic resonance imaging. Other discoveries, though originally motivated by a technological need, give rise to very beautiful and fundamental insights.

For example, the fractional quantum Hall effect was first observed in high-mobility semiconductor structures now used in high-frequency applications.

TABLE 2.1.1 Some New Inorganic Materials of the Past Fifteen Years

Advance	Driver	Nature of Advance
New compounds/materials		
High-temperature superconductors	Science	Revolutionary
Organic superconductors	Science	Revolutionary
Rare-earth optical amplifier	Technology	Evolutionary
Intermetallic materials	Technology	Evolutionary
High-field magnets	Technology	Evolutionary
Organic electronic materials	Technology	Evolutionary
Magnetooptical recording materials	Technology	Evolutionary
Bulk amorphous metals	Technology	Evolutionary
New structures of known materials		
Quasicrystals	Science	Revolutionary
Buckyballs and related structures	Science	Revolutionary
Nanoclusters	Science	Evolutionary
Metallic hydrogen	Science	Evolutionary
Bose-Einstein condensates	Science	Evolutionary
Giant magnetoresistance materials	Technology	Revolutionary
Porous silicon	Technology	Evolutionary
Diamond films	Technology	Evolutionary
Quantum dots	Technology	Evolutionary
Foams/gels	Technology	Evolutionary
New properties of known materials		
Gallium nitride	Technology	Revolutionary
Silicon-germanium	Technology	Evolutionary

cally lower the growth temperature in semiconductor processing. Understanding and exploiting fundamental growth mechanisms can lead to previously unattainable structures as in the use of strain to induce the self-assembly of quantum dots.

Many advances in condensed-matter and materials physics are the direct result of the availability of materials and structures of a quality not previously attainable. These materials and structures in turn exist because of improvements in the technology used to make, study, measure, and see them. The impetus for improvements in the materials is often a technological need, not the search for new knowledge. The new knowledge generated, however, in some cases itself becomes the enabler of revolutionary technology. This interplay is explored in Box 2.2.

BOX 2.2 The Science—Technology Circle

Tremendous advances in compound semiconductor devices were enabled by dramatic improvements in the growth of thin films that began in about 1970 with the invention of molecular beam epitaxy (MBE, see Figure 2.2.1). The direct antecedents of MBE were developments in vacuum technology beginning in the 1960s and continuing into the 1970s, driven by accelerator development and space physics. As the attainable vacuum improved, it became possible to keep a surface atomically clean for long enough to study it. Surface probes such as Auger spectroscopy and electron diffraction techniques allowed the clean surfaces to be studied.

MBE enabled the controlled, layer-by-layer growth of compound semiconductors. The composition of the film could be changed abruptly. Extremely high mobility was achieved in GaAs-GaAlAs heterostructures through "modulation doping." Research into these structures was pursued because of their utility in high electron-mobility transistors (HEMTs, see Figure 2.2.2) which are used today in high-speed electronics.

Study of these layers at low temperatures in extremely high magnetic fields led to the discovery of the quantum Hall effect, which takes place in a two-dimensional electron "gas" produced in a transistor-like device. Under these conditions, electron correlations dominate, leading to precise quantization of the Hall conductance. As the quality of the layers was improved further, the mobility also improved, and the fractionally quantized Hall effect (FQHE) was discovered, in which the quantum number describing the system is a fraction rather than an integer (see Figure 2.2.3). The FQHE has subsequently been used for unprecedentedly accurate measurements of the fundamental quantity h/e^2 (Planck's constant divided by the square of the charge of the electron).

FIGURE 2.2.1 Molecular beam epitaxy was invented at Bell Laboratories in about 1970 as an outgrowth of advances in vacuum technology and surface science techniques. (Courtesy of Bell Laboratories, Lucent Technologies.)

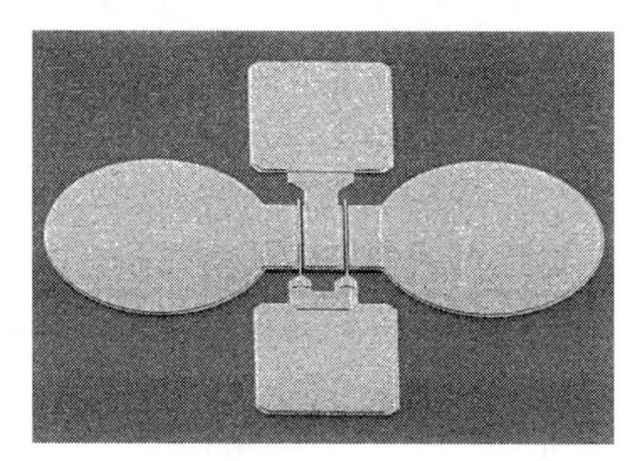

FIGURE 2.2.2 A high electron-mobility transistor (HEMT) such as those used in cellular telephones. The round bonding pads are 100 μm in diameter, roughly the size of a human hair. The gate of the transistor, just 0.05 μm across, appears as the two narrow lines in the center of the scanning electron micrograph. (Courtesy of Sandia National Laboratories.)

FIGURE 2.2.3 A pictorial representation of the many-particle state that underlies the fractional quantum Hall effect. The height of the landscape represents the amplitude of the quantum wave of one electron as it travels among its companions (shown as balls). The arrows indicate the vortices induced by the magnetic field, which attach themselves to the electrons to form composite particles. (Courtesy of Bell Laboratories, Lucent Technologies.)

The remainder of this chapter examines a few of the past decade's most impressive advances in materials and structures. The selections emphasize a number of themes that have emerged in materials research. Some of the discoveries have been completely unexpected. Others were predicted, although the experimentalists did not always know of these predictions when they did their work. Our thinking about new materials has changed fundamentally; we now consider dramatically more complex possibilities in our search for new materials than we did a decade ago. In some classes of materials that have been studied for many decades, we have achieved a much deeper understanding of physical and chemical mechanisms that govern their properties. This understanding in turn has led to improvements in the properties of the materials, either through elimination of problems inherent in existing materials by improved processing or by the introduction of new materials. Even in a material as thoroughly studied as carbon, a myriad of new forms has been discovered, exhibiting a wide range of properties. Shrinking the dimensions of well-known materials such as semiconductors has led to properties dramatically different from those of the bulk. New concepts in thin-film growth have led to improved film properties by changing the growth and processing windows. Finally, there has been a change in the attitude toward strain in heteroepitaxial systems that allows strain to be used to tailor the morphology as well as the electrical properties of the layers. The culmination of this effort is in the use of strain to induce self-assembly of quantum dots.

COMPLEX OXIDES

Surely one of the most surprising developments since the publication of the Brinkman report[1] has been the discovery of high-temperature superconductivity in complex oxide materials, beginning in 1986 with the observation by Bednorz and Müller of superconductivity near 30 K in $La_{2-x}Ba_xCuO_4$. This discovery was rewarded with the 1987 Nobel Prize in Physics (see Table O.1). The field exploded with the discovery of superconductivity at temperatures in excess of the boiling point of liquid nitrogen (77 K). The family of known high-temperature superconducting materials now numbers near 100, with the highest superconducting transition temperature (T_c) above 130 K. High-temperature superconductivity has significantly altered the direction of condensed-matter and materials physics in several ways. The excitement generated by this totally unexpected discovery attracted researchers from throughout the field of condensed-matter and materials physics and beyond to the study of these fascinating materials. More recently, the principles that have been successful in the study of these materials have proven valuable in the study of other areas of condensed-matter and materials physics, most notably other sorts of oxides.

---

[1]National Research Council [W.F. Brinkman, study chair], *Physics Through the 1990s,* National Academy Press, Washington, D.C. (1986).

High-temperature superconductors are much more complex than many of the materials that have occupied the attention of condensed-matter and materials physicists for many decades (see Figure 2.1). This complexity, however, is a two-edged sword, giving rise to a richness in the possible structures and properties of the materials, but also making the materials extremely challenging to produce, control, and understand.

The crystal structures of these materials are dramatically more complicated and have lower symmetry than those of low-temperature superconductors or semiconductors. The physical properties are similarly anisotropic. This makes the control of crystallographic orientation extremely critical. The unit cell is large. A large unit cell and low symmetry offer many opportunities for the formation of defects during materials preparation, either in individual atomic sites or in long-range crystallographic perfection.

The superconducting coherence length is of the order of the interatomic spacing in some crystallographic directions. This makes the materials exquisitely sensitive to defects—from atomic-scale defects, such as vacancies, interstitials, and substitutional atoms, to grain boundaries and other larger-scale imperfections. Separating the intrinsic properties of such materials from artifacts caused by defects is critical to gaining full understanding of the high-T_c phenomenon. It places extreme emphasis on materials preparation and serves as an example of the true collaboration that must exist between those who seek to understand and control the growth of the materials and those who probe their underlying physics, as illustrated in Box 2.3. Conversely, the carefully controlled introduction of

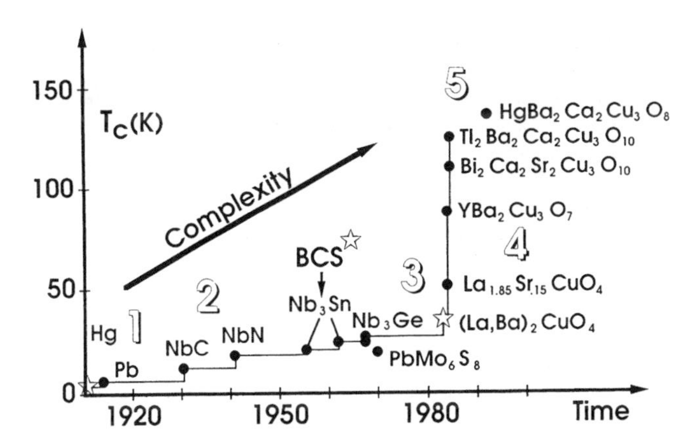

FIGURE 2.1 Historical development of inorganic superconductors, with Nobel prizes indicated by stars. Increasing superconducting transition temperature correlates with increased chemical complexity and more constituent elements, as shown by the numbers 1 to 5. (BCS stands for the Bardeen-Cooper-Schrieffer theory of classical superconductors.) [*MRS Bulletin* **19**, 26 (1994).]

BOX 2.3 Vortex Matter: A Novel Window on Materials Physics

In the past 5 years, a new area of condensed-matter investigation has emerged, based on the remarkable behavior of vortices in superconductors. For decades, type-II superconductors in a magnetic field have been understood as arrays of quantized tubes of magnetic flux, each surrounded by a circulating "vortex" of supercurrent that defines its interaction with its neighbors and the outside world. In traditional superconductors, thermal energy is limited to about 20 K by the superconducting transition temperature, and the vortex tubes form an elastic solid. High-temperature superconductors offer a new possibility: thermal energies up to about 100 K may melt the vortex solid, creating a novel liquid state with dramatically different properties that arise from the relative motion of vortices. As early as 1988, motion of vortices below T_c was found to create undesired dissipation in high-T_c cuprates. The thermodynamic nature of the melting phase transition and of the resulting vortex liquid has been vigorously debated. Theorists soon realized that vortex phases and phase transitions embody many fundamental features of condensed-matter physics, including reduced dimensionality, entanglement of flexible line objects, and the role of disorder on elastic media. Studies of vortex matter provide new insight into these basic materials physics issues in other condensed-matter environments. The diversity of equilibrium vortex phases is illustrated in the phase diagram of Figure 2.3.1.

Although theoretical analysis of vortex liquids and solids abounded, experimentalists were frustrated by the quality of the available high-temperature superconducting materials. Real materials contain defects like impurity clusters, dislocations, twin boundaries, and rough surfaces. In superconductors, these defects generate pinning sites that immobilize vortices and remove them from participation in equilibrium behavior. The experimental observation of vortex phase transitions had to await more perfect materials with dramatically reduced defects and vortex pinning.

In 1992, the first indications of vortex lattice melting were observed in electrical transport experiments. These measurements accurately located the melting line in the H-T plane and gave tantalizing but indirect evidence that the transition was first-order in clean materials. Further transport experiments suggested that first-order melting was destroyed by controlled pinning disorder and suggested the existence of a critical point in the melting line. These and other experimental observations created new interest and activity in the field.

Experimentalists then sought the next level of fundamental information—thermodynamic characterization of the order and entropy changes on vortex melting—with magnetization and specific heat experiments. Such experiments require an even higher level of sample perfection to ensure thermodynamic equilibrium in the solid phase, where pinning effectiveness is significantly enhanced by shear elasticity. Sample size was a second serious problem: the most perfect crystals are also the smallest, making it extremely difficult to resolve the tiny magnetic and thermal signatures of melting from the much larger background. Nevertheless, sample preparation techniques continued to improve with better understanding of the roles of composition, growth rates, and annealing procedures. Improved materials enabled several landmark thermodynamic experiments, which have now settled the question of the thermodynamic order of the transition and raised new questions about critical points, vortex entanglement, and the dimensionality of the liquid.

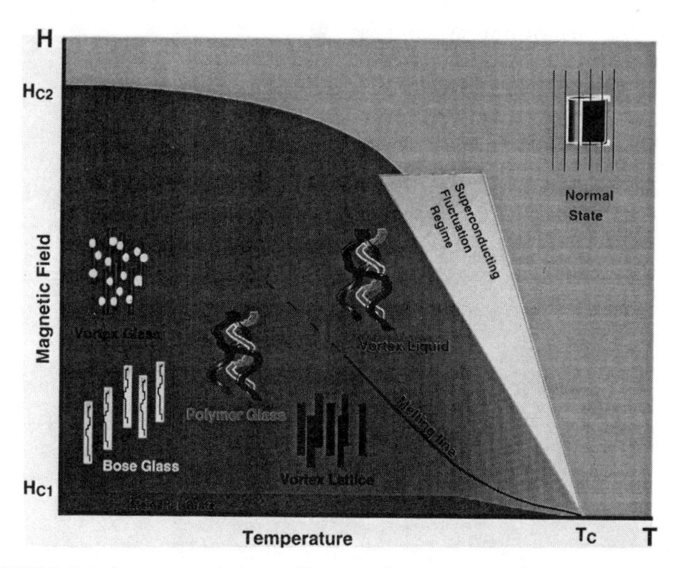

FIGURE 2.3.1 A suggested phase diagram of vortex matter in the magnetic field-temperature plane. Several vortex liquid and solid phases are illustrated, including a liquid of entangled vortex lines, a perfect hexagonal lattice, a polymer glass of entangled lines, and solid phases disordered by point pinning defects (vortex glass) or by line pinning defects (Bose glass). The melting transition is first-order from a lattice and proposed to be second-order or continuous from a glass. A critical point may occur on the melting line, where the first-order character disappears. The normal and vortex liquid states are separated by a fluctuation dominated by crossover rather than by a true phase transition. (Courtesy of Argonne National Laboratory.)

Vortex matter has emerged as a vital field, with its own developing issues and international community of researchers. It extends traditional studies of atomic matter in several ways. For example, vortex density is linear in the applied magnetic field, so it can easily be changed by an order of magnitude with the twist of a dial. Experimental access to such a large density range is unheard of in atomic matter. The interactions among vortices are well-known Lorentz forces, which can be treated analytically or in simulation with no uncontrolled approximations. Advanced materials development has produced clean crystals with few pinning defects, revealing intrinsic thermodynamic behavior and its evolution under controlled disorder induced by electron or heavy ion irradiation. Finally, vortices can be set in motion by the Lorentz force from an externally applied transport current, enabling studies of driven phases, steady-state motion, and the new area of dynamic phase transitions. This remarkably rich microcosm of condensed-matter physics owes its existence to two materials developments: the landmark discovery of high-temperature superconductors, which introduced large thermal energies into the vortex phase diagram, and dramatic improvements in materials perfection, which enabled experimental studies of the delicate thermodynamics of collective vortex behavior.

defects of a particular type into the material by, for example, ion irradiation or judicious atomic substitution allows the properties to be adjusted.

The superconducting oxides with T_c above 77 K all contain at least four elements, two of which are copper and oxygen. Oxygen moves readily in these materials, during both sample preparation and subsequent processing. Changing the oxygen content by just a few percent can determine whether a material is a superconductor or an insulator. It can also govern the symmetry and crystal structure of the material, resulting in phase transformations during specimen preparation that, to date, have been unavoidable. Precise control of the stoichiometry of the metal constituents is also required to optimize the superconducting properties, although the consequences of deviations from ideal stoichiometry are not nearly as critical for the metals as for oxygen.

Current interest in the high-temperature superconducting materials centers around two general areas: superconducting electronics and the carrying of large currents. The electronics applications can be further subdivided into logic and high-frequency applications. Electronics applications require thin films, generally in combination with films of other materials. The fabrication of reproducible tunnel junctions with useful properties for logic applications has been very challenging because of the incompatibility of high-temperature superconducting materials with most nonoxide barrier materials and the extremely short coherence length of the superconductor. Quite a few metallic oxides with compatible crystal structures have been identified and studied as a result of considerable research into suitable barrier materials. A promising area of application is in components for communications, particularly in the gigahertz frequency domain. The major issues are the surface resistance of the material and electrical nonlinearities at high frequencies. Though there has been considerable progress in improving surface resistance in the past few years, detailed understanding of the relationships between this and other relevant properties and the structure of the materials is still emerging.

Technological applications demand large-area films that can be deposited fast enough to be economically viable. There has been dramatic progress, with high-quality films of $YBa_2Cu_3O_{7-x}$ (see Figure 2.2) now available on substrates several hundred square centimeters in area.

Current-carrying applications require bulk material or thick films. Grain boundaries, especially those with significant misorientation between grains, are extremely detrimental to high critical currents because of both the extreme anisotropy of the materials properties and the properties of the grain boundaries themselves. The most successful approach for bulk materials with properties of potential technological interest has been the use of drawn, multifilament wires, especially in the bismuth system. The drawing induces alignment of the grains in the filaments and increases the critical-current density. More recently, biaxial orientation has been achieved in thick $YBa_2Cu_3O_{7-x}$ films deposited on metal substrates, either coated with an aligned buffer layer fabricated by ion beam-assisted deposition or with strong crystallographic alignment induced in the substrate by rolling.

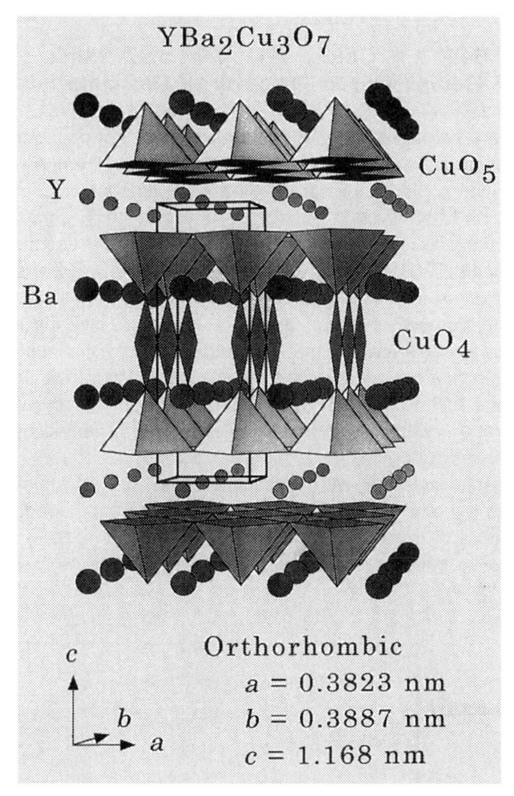

FIGURE 2.2 The orthorhombic crystal structure of superconducting $YBa_2Cu_3O_{7-x}$. The superconducting transition temperature of this material is above 90 K. Note the presence of four elements in the compound and the low symmetry of the structure. These characteristics make materials synthesis challenging and give rise to dramatic anisotropy in the physical properties of the material. (Courtesy of Princeton University.)

It has proven very fruitful to apply the principles discovered and techniques developed for high-temperature superconductivity to other classes of complex oxides. In some cases, this research has been driven by the need for materials with specific electronic or magnetic properties that are chemically and structurally compatible with high-temperature superconductors. These materials are typically needed as buffer or barrier layers. Compatible materials with other properties could be needed in the future if high-temperature superconducting devices are to be successfully integrated with devices having other functionality, such as memory and optical devices. Perhaps the most impressive demonstration of the application of lessons from high-temperature superconductivity has been the recent interest in colossal magnetoresistance in $LaMnO_3$-derived materials (see Box 2.4).

BOX 2.4 Colossal Magnetoresistance: A Rediscovered Property of Old Materials

The discovery of high-temperature superconductivity changed our thinking about complexity in materials composition and structure. It emphasized that truly different properties could be obtained when new dimensions of complexity are considered. A particularly good example of the change in paradigm that high-temperature superconductivity has caused is the observation and study of colossal magnetoresistance (CMR). CMR was first observed in a perovskite material, lanthanum manganate ($LaMnO_3$) in which some of the lanthanum is substituted by an alkaline-earth element: calcium, strontium, or barium (see Figure 2.4.1). The manganates have been known for many decades and had previously been studied for their promising catalytic properties. Presumably, more recent advances in magnetic storage technology sensitized researchers to rediscover the CMR effect and to pursue it with the vigor and determination sparked by the potential applications.

As innumerable materials with perovskite-based crystal structures received new attention in the aftermath of the high-temperature superconductivity discovery, the alkaline-earth-substituted manganates were found to have magnetoresistance effects up to three orders of magnitude larger than the previously known

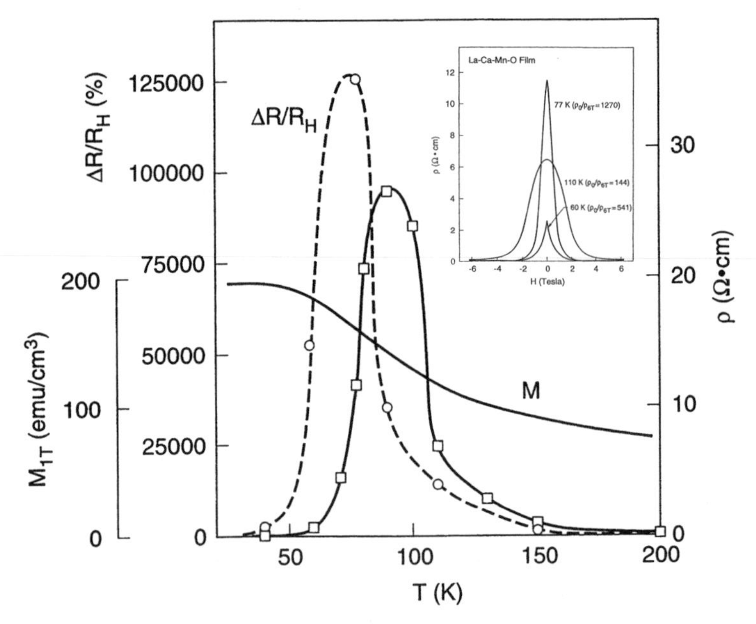

FIGURE 2.4.1 Resistivity (ρ), magnetoresistance ratio ($\Delta R/R_H$), and magnetization (M) as a function of temperature curves for an epitaxial La-Ca-Mn-O film. [Reprinted with permission from S. Jin, M. McCormack, T.H. Tiefel, R.M. Fleming, J.M. Phillips, and R. Ramish, *Applied Physics Letters* **64**, 3045 (1994). Copyright © 1994 American Institute of Physics.]

giant magnetoresistance metal multilayers. This observation was not predicted, and an understanding of the phenomenon is even now just being developed. The current consensus is that the relevant magnetic behavior of these compounds results from the motion of electrons between adjacent manganese ions via the intervening oxygen ion. This "double-exchange" interaction favors parallel (ferromagnetic) spins of neighboring Mn^{3+} and Mn^{4+} ions produced by the alkaline-earth substitution. This interaction competes with the antiferromagnetic coupling of manganese ions in the absence of the mobile electrons.

Ultimately, the symmetry of the spin ordering on neighboring manganese sites determines the electrical resistivity: parallel spins lead to low resistivity while antiparallel spins give high resistivity. Thus the variation of resistivity with temperature can be understood in terms of the transition between a semiconducting paramagnetic state above the Curie temperature, T_c, and a metallic ferromagnetic state below T_c. Above T_c, a magnetic field enhances spin alignment and reduces resistance, but it has little effect below T_c. Therefore the largest bulk magnetoresistance effect is observed in the temperature region near T_c.

Magnetoresistive effects, T_c, and the mobility of electrons between manganese ions are thus closely connected. This is one of the key ingredients of manganite physics, and it has been tested recently in experiments in which the manganese-manganese overlap has been systematically varied through controlled bond-angle variations induced by substitutions of ions with various sizes. Because the richness of physical phenomena derives from the interplay among these electronic properties, local lattice strains, and other parameters of comparable magnitude, control and modification of materials properties are of paramount importance to exploration and exploitation of the manganites.

Although the bulk properties of manganites are of great interest in their own right, the recent discovery of spin-dependent electron transport across grain boundaries holds further promises. These phenomena occur in fields low enough for possible sensor applications in magnetic storage. Active research has now started on the development of small-area thin-film devices that rely on the spin-dependent resistance across interfaces, and this work will benefit from the many advances in materials engineering developed in contact with high-temperature superconductors.

Progress in these materials demonstrates once again the complex interplay among the detailed chemistry, materials structure, and electronic and magnetic properties, which provides a fruitful research field for fundamental studies and hope for future technology.

Although ferroelectric materials have been a topic of considerable research for a long time, developments in high-temperature superconductivity within the past decade have aroused new interest and insight, leading to improved electrode materials and better control over the structure and properties of the ferroelectrics themselves. An outgrowth is the current interest in high dielectric constant materials, generally complex oxides, for use in high-density semiconductor memories.

Research on complex oxides in general and high-temperature superconductors in particular has spawned new growth and processing techniques as a result of the unusual properties of these materials. This research has given rise to materials of ever-improving quality, which allow physicists to probe the fundamental mechanisms at work and technologists to explore more fully their promise for applications. The area is vital and will be a source of exciting physics and materials research for the foreseeable future.

ELECTROCERAMICS

Electroceramic materials have been studied and used for many decades because of their interesting and in some cases novel properties, such as ferroelectricity, piezoelectricity, pyroelectricity, and electro-optic activity. Current interest in the ferroelectrics centers on their potential in nonvolatile memories and high dielectric constant capacitors. Micromachines, such as accelerometers, displacement transducers, actuators, and so on, require piezoelectric materials. Room-temperature infrared detectors make use of pyroelectric properties. Electro-optic properties enable color filter devices, displays, image storage systems, and the optical switches required in integrated optical systems.

Electroceramics can serve as "smart" materials, functioning as both sensors and actuators (see Box 2.5). All smart materials have at least two phase transitions (e.g., crystallographic and electronic), and their synthesis and processing must be carefully controlled to regulate their excursions through phase space. The complexity of these phenomena and the materials that display them has made this an exciting area. There has been dramatic progress in the control of electroceramic materials properties, in understanding the relationships between properties of interest and the underlying microstructural mechanisms that control them, and in integrating various materials to give improved properties or even new behavior.

Progress has been especially impressive in the ferroelectric materials. Extensive research has focused on understanding the mechanisms responsible for the degradation of ferroelectric and high-permittivity perovskite thin films with time, temperature, and external field stress. The three most important degradation phenomena are ferroelectric fatigue, ferroelectric aging, and resistance degradation.

Ferroelectric fatigue, the loss of switchable polarization with repeated polarization reversals, is caused by pinning of domain walls, which inhibits switching of the domains. Elimination of fatigue is critical for nonvolatile memory applications. Recent results have shown that charge trapping at internal domain boundaries is the primary fatigue mechanism. Fatigue also induces changes in the oxidation states of isolated impurity point defects, which are much more stable than optically generated ones in unfatigued samples.

Fatigue can be largely eliminated in some ferroelectric systems [e.g., lead

BOX 2.5 Electroceramics in Composite Smart Materials and Biomimetic Systems

As the sensitivity and selectivity of a sensor increases, its structure generally becomes more complex. This often involves moving from single-phase materials to composites in which the connectivity, symmetry, or scale of the composite is designed to give a desirable field concentration or composite symmetry. Each constituent material has an associated phase transition. For example, polymeric materials with phase transitions in which the elastic properties change dramatically may be combined with ferroelectric materials in which the dielectric properties have an associated instability. The different types of instability allow fabrication of structures especially good for sensing and actuating. Different connectivity patterns optimize the tensor coefficients that contribute to the figure of merit for the application. Figure 2.5.1 shows the figure of merit for composite piezoelectric materials with different connectivities. The largest figure of merit by far is for the "moonie" structure, which consists of either a piezoelectric ceramic disk or a multilayer stack, sandwiched between two specially designed metal end caps. This design provides a sizable displacement, as well as a large generative force.

The evolution of these sensors and actuators moves the composites progressively closer to the configurations adopted by biological systems that perform the same function. The 1-3 composite hydrophones, for example, mimic the geometry and sensing function of the lateral line of the North Atlantic cod, a series of fibrous sensors spaced along the length of the fish. Similarly, the air space in the moonie transducer is a resonant cavity that corresponds to the fish swim bladder. By vibrating the bladder wall, a fish emits a low-frequency grunt that propagates well through the ocean. The same cavity also makes the moonie and the fish more sensitive. As more complex sensing and actuating functions are designed, inspiration can be gained by studying the analogous biological "composite" systems.

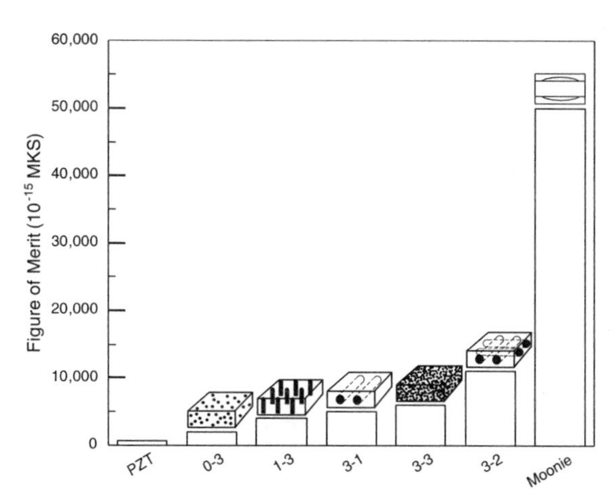

FIGURE 2.5.1 Hydrophone figure of merit for composite piezoelectric materials. The connectivity is indicated schematically above each bar. (Courtesy of Pennsylvania State University.)

zirconate titanate (PZT)] by using electrodes of metallic oxides, such as (La,Sr)CoO_3 (LSCO) and (RuO_2). The interaction between the ferroelectric material and its electrode plays a critical role in determining fatigue performance, perhaps because of the accumulation of oxygen vacancies near the electrodes during cycling. This suggests that electrodes with a large tolerance for oxygen deficiency, such as some of the metallic oxides, should offer better fatigue characteristics than those that serve as ineffective sinks for oxygen vacancies. The improvement in fatigue characteristics offered by this approach is shown in Figure 2.3.

Layered perovskite materials such as $SrBi_2Ta_2O_9$ have recently received attention because of their lack of polarization fatigue even with simple metal electrodes. The emerging picture attributes this to less oxygen vacancy accumulation at the electrodes caused by either a smaller vacancy population or reduced

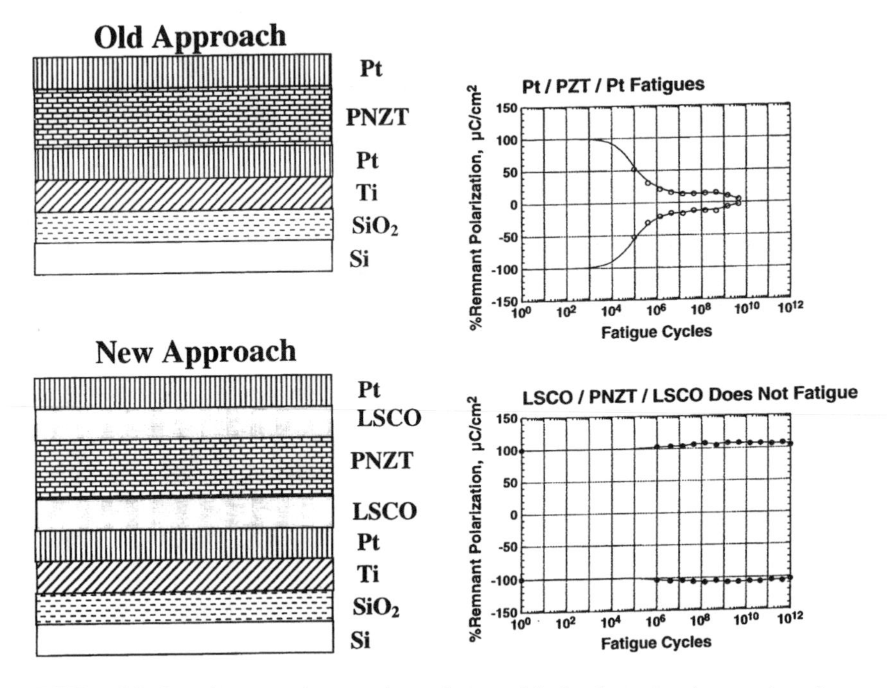

FIGURE 2.3 Developments in new electrode materials for ferroelectric capacitors have reduced the fatigue in these devices by more than 6 orders of magnitude. The upper image shows a "conventional" capacitor structure of the ferroelectric material Pb((Nb,Zr)Ti)O_3 (PNZT) with unbuffered platinum electrodes, along with the fatigue of the remanent polarization. The polarization decays to half its initial value after 10^5 cycles. The lower image shows a capacitor structure with the platinum electrodes buffered by $La_{0.5}Sr_{0.5}CoO_3$. This capacitor shows no fatigue even after 10^{12} cycles. (Courtesy of the University of Maryland.)

mobility. Smaller polarization in these materials should also result in weaker pinning and thus higher unpinning rates.

In addition to their ferroelectric and dielectric properties, many of the same materials also exhibit piezoelectric properties. PZT is the most widely used. On cooling from high temperature, the crystal structure of PZT undergoes a displacive phase transformation, and the point symmetry changes from cubic to tetragonal. To make use of piezoelectric ceramics, compositions near a second phase transition are chosen. At the Curie point, PZT converts from a paraelectric state with the ideal cubic perovskite structure to a ferroelectric phase located near a morphotropic phase boundary between the tetragonal and rhombohedral states. Very large piezoelectric coupling between electric and mechanical variables is obtained near this phase boundary. Much of the current research in this field involves looking for other morphotropic phase boundaries to further enhance the electromechanical-coupling factors. Ferroelectric thin films have been successfully used in a variety of microelectromechanical systems applications, including accelerometers, microvalves, pressure sensors, and infrared detectors. Microactuators and microsensors are designed to make use of the strong piezoelectric response of ferroelectrics such as PZT and to ease the fabrication and incorporation of on-chip electronics. The development of fabrication methods such as surface micromachining, low-stress silicon nitride deposition, and solution deposition of ferroelectric thin films has been essential.

Control of the growth of ferroelectric films has been a prerequisite for the progress that has been made. Film crystallinity has improved dramatically, and techniques such as ion bombardment have allowed the growth temperature to be lowered, improved the selection of the desired perovskite phase over the pyrochlore phase, improved the degree of preferred alignment in the films, and resulted in denser, smoother films. The use of oxide electrodes and templates for growth has helped eliminate unwanted orientations and dramatically improved the electrical properties. Control of the film microstructure has led to improved leakage current.

Nanoscale force microscopy has begun to allow examination of the switching of individual grains in a polycrystalline matrix with resolution of about 10 nm, as shown in Figure 2.4. The ability to follow localized processes will be important in unraveling the physical phenomena that govern these complex materials.

NEW FORMS OF CARBON

The first documented conjecture of the existence of hollow-cage molecules of carbon appeared in 1966. Four years later, the existence of C_{60} (buckminsterfullerene) was predicted theoretically, with more rigorous treatments appearing in subsequent years. Nevertheless, the molecule itself was not observed until 1985, when it appeared serendipitously during a series of graphite laser vaporiza-

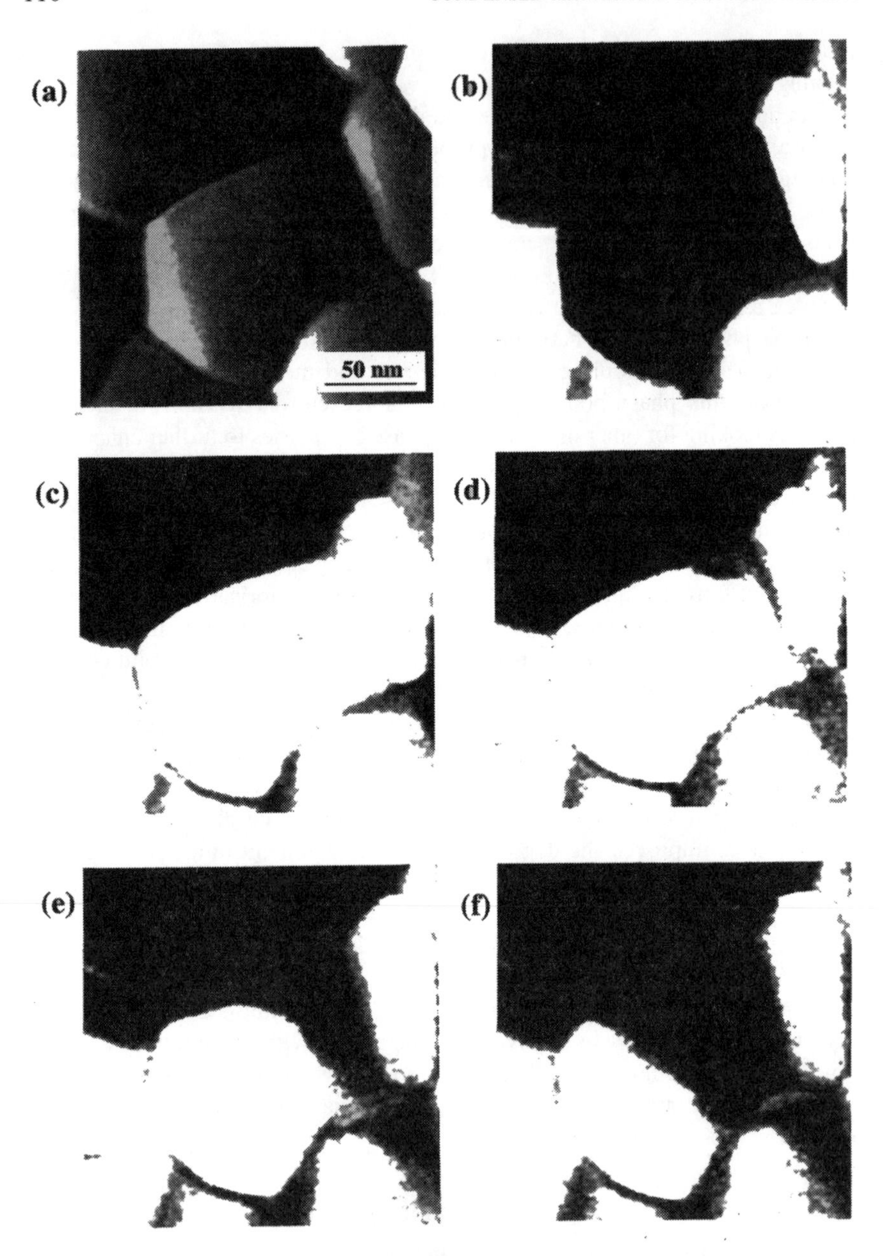

FIGURE 2.4 Scanning-force microscopy of topographic (a) and piezoresponse (b-f) images of a PZT film grown on a LSCO/TiN/Si substrate. The central grain was switched completely from the polarization direction down (dark) to up (white). The switch back of the central grain into the polarization down direction starts mainly at the grain boundaries with the surrounding grains. [*MRS Bulletin* **23**, 39 (1998).]

tion experiments designed to simulate the chemistry in a red giant carbon star. Finally, in 1990, a method was developed to produce macroscopic quantities to allow the intense investigation of this and related compounds that have been a focus of research in the present decade. As suggested in Figure 2.5, C_{60} turns out to be just one of a veritable menagerie of three-dimensional closed carbon molecules: spheres, tubes, particles, and combinations thereof, with one or multiple layers. The discovery of fullerenes by Curl, Kroto, and Smalley was recognized with the 1996 Nobel Prize in chemistry (see Table O.1).

The remarkable geometry of these molecules is enabled by slight deviations from the hexagonal bonding configuration found in graphite resulting from the desire to eliminate energetic dangling bonds at the edges of graphite sheets. The addition of twelve pentagons to the hexagonal array transforms the open graphite structure into any of the observed closed molecules that have only positive curvature. Heptagonal rings give rise to a saddle-shaped surface when buried among hexagons.

Carbon nanotubes, which were originally grown as a by-product in the fullerene-generating chamber, are quasi-one-dimensional structures with a simple and well-understood atomic structure. A chemist might think of a carbon nanotube as a monoelemental polymer. The nanotube is an ideal model for quasi-one-

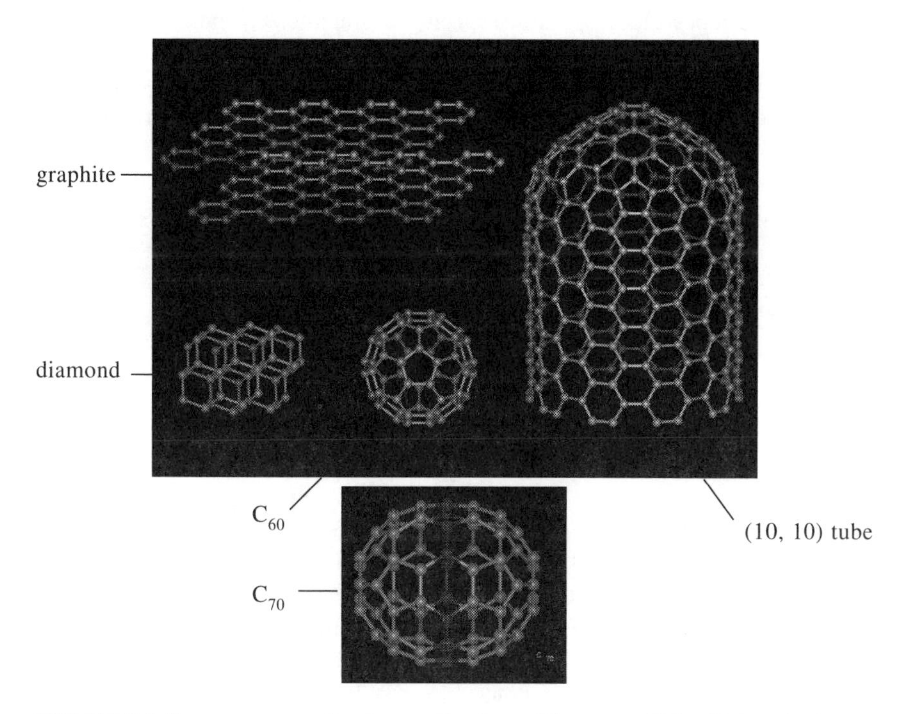

FIGURE 2.5 Some representative forms of carbon. (Courtesy of Rice University.)

dimensional structures because its known atomic structure makes computer simulations more reliable. Nanotubes can be as much as several microns long, and tube diameters range from one to a few tens of nanometers. A metal serves as a catalyst for nanotube formation, preventing the growing tubular structure from wrapping around and closing into a smaller fullerene cage. Nanotube growth is believed to take place at the open ends of the tubes. During growth, the open tubule end required to fabricate long, single-wall nanotubes can be maintained by a high electric field, by the entropy opposing orderly cap termination, or by the presence of a metal catalyst. The tube ends tend to close quickly when the growth conditions become inappropriate, for example, when the temperature drops or when the carbon atom flux is too low. As long as the tube end is open, carbon atoms can be deposited on the tube-end peripheries, and they can grow. When pentagons are formed for some reason, the tubes will be capped. If the axial growth rate is dominant over the radial one, the tubule will become a single-shell tube. A comparable growth rate in both the axial and radial directions will form spheroidal particles.

Characterization of carbon nanotubes has been slow compared to fullerene research activity, partly because of the inability to synthesize macroscopic quantities of the tubules and to refine them. Many nanotubes are in the form of a multiple-shell structure of nested cylindrical tubes separated by about 0.34 nm, which is the same as the d_{0002} lattice spacing of graphite. Cylindrical crystals are often seen in biological protein crystals but rarely in inorganic materials. Recent measurements on single-wall carbon nanotubes have shown that they do indeed act as genuine quantum wires, confirming theoretical predictions, as shown in Figure 2.6.

Electronic and mechanical properties of nanotubes deviate from those of a bulk graphite crystal. Depending on tubule diameter and helicity, both of which affect the band gap, the behavior can range from metallic to semiconducting. Because it makes for a more symmetrical structure, less helicity leads to better conductivity. This leads to true molecules that are also true metals, something chemistry has never had before. Because of the quasi-one-dimensionality of these nanotubes, conduction is quantized.

One unexpected phenomenon in nanotubes is the ability to fill them with a material. Nonhexagonal carbon rings in the hexagonal network are responsible for tubule morphologies and presumably local strain. After deposition of a small amount of lead on tubule surfaces and heating, some of the metal clusters move to heptagon sites. Nanotubes can be opened by mild oxidation at the reactive site at the closed end. On heating, some of the lead is transported into the central hollow in the tubule. The intercalated material is crystalline and not pure lead but lead carbonate or oxide. This finding suggests that the tubule tips react selectively in air at elevated temperature, but the rest of the tubules do not react. Strain induced by including pentagons in the tubule tips may be responsible for the selective reaction. Carbon onions have also been stuffed with metals and metal carbides.

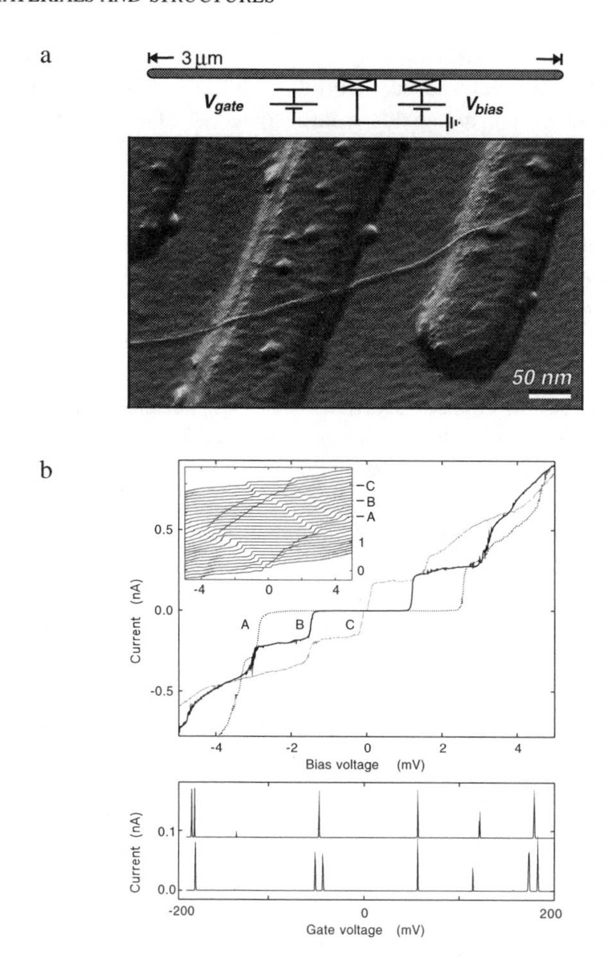

FIGURE 2.6 (a) Atomic-force microscope tapping-mode image of a carbon nanotube on top of a Si/SiO_2 substrate with two 15-nm-thick platinum electrodes and a corresponding circuit diagram. This single-wall nanotube has a diameter of ~1 nm. Its total length is 3 mm, with a section of 140 nm between the contacts to which a bias voltage (V_{bias}) is applied. A gate voltage (V_{gate}) applied to the third electrode in the upper left corner of the image is used to vary the electrostatic potential of the tube. (b, top) Current-voltage curves of the nanotube at a gate voltage of 88.2 mV (trace A), 104.1 mV (trace B), and 120.0 mV (trace C). The inset shows more I-V_{bias} curves with V_{gate} ranging from 50 mV (bottom curve) to 136 mV (top curve), with vertical offsets for clarity. The variation with V_{gate} of the gap around V_{bias} = 0 implies Coulomb charging of the tube. The stepwise increase of the current at higher voltages may result from an increasing number of excited states entering in the bias window. (b, bottom) Current versus gate voltage at V_{bias} = 30 mV. The two traces shown were performed under the same conditions. [Reprinted with permission from S.J. Tans, A.R.M. Verschueren, and Cees Dekker, "Individual single-wall carbon nanotubes as quantum wires," *Nature* **386**, 474 (1997). Copyright ©1997 *Nature*.]

A large family of structures can be generated in the high-temperature regimes of the arc experiments used to produce fullerenes and nanotubes. In addition to nanotubes, the arc yields a large quantity of polyhedral graphitic particles with well-defined faceting and a wide-ranging size distribution (8 to 60 nm). If electrodes incorporating a metallic salt or oxide are used, a small percentage of encapsulated metallic particles within closed graphitic shells is produced. The quasi-spherical onionlike structure (buckyonion) is the exclusive result of the irradiation process. The same effect may be obtained from diverse starting carbon materials (polyhedral graphitic particles, buckytubes, and even disordered forms of carbon). The onions grow by a kind of internal epitaxy as the layers reorganize, progressing from the surface to the central shell. The model structure for the final product is the concentric arrangement of spherical fullerenes, formed by $60n^2$ carbon shells. Small buckyonions (2 to 4 shells) are very stable under intense electron bombardment, suggesting that they may be the most stable forms of carbon cluster.

By virtue of their unique structures, fullerenes exhibit novel chemical transformations. These molecules are spherical or nearly so. Molecules with high point-group symmetry, which are not bound strongly in the solid state, tend to crystallize into structures with long-range periodicity of the molecular centers of mass, but the molecular orientations are random or even dynamically disordered. Because C_{60} has high electron affinity, it forms anion salts with alkali and alkaline-earth metals as well as with strong organic molecules. K_3C_{60} and other alkali fulleride salts exhibit superconductivity with T_c above 30 K, as discussed in Box 2.6.

The organic molecule tetrakis dimethylaminoethylene (TDAE) is known to be an effective electron donor. A C_{60}-TDAE salt has been formed that exhibits a ferromagnetic state below 16 K. This material, which has a low-symmetry monoclinic structure because of the highly nonspherical nature of TDAE, holds the record for the highest Curie temperature of any purely organic molecular solid.

NANOCLUSTERS

Ever since the early 1980s, when scientists began discovering the various potentially advantageous properties of ultrasmall grains of material—nanoclusters—there has been tremendous activity as researchers strive to create and control new types of particles. Much of the recent research has been directed at finding ways to make small clusters of uniform size with common optical, electrical, and mechanical properties. These efforts have already begun to have commercial payoffs, as in the case of ceramics and chemical catalysts that have increased efficiency because of their high surface-to-volume ratio.

In any material, substantial variation of fundamental electrical and optical properties with decreasing size will occur when the electronic energy-level spacing exceeds the temperature. The variation is especially pronounced for semi-

BOX 2.6 Superconductivity in Alkali and Alkaline-Earth-Doped C_{60}

Because fullerenes act as electron acceptors, they can form different types of salt. Those of the alkali and alkaline-earths have particularly interesting electronic properties. Photoemission studies probing the occupied electronic states, as well as inverse photoemission measurements probing the unoccupied electronic states, have allowed direct monitoring of the nature of electron doping into the C_{60} levels. The valence band of solid C_{60} is derived from a fivefold degenerate h_u orbital. The stability of this orbital makes it difficult to remove electrons from C_{60}. On the other hand, C_{60} is a good acceptor because of the threefold degenerate t_{1u} and t_{1g} levels. Exposing C_{60} to alkali vapor results in electron filling of the t_{1u} level. Because the Fermi energy is pinned to the top of the filled level, with increased filling, the spectral manifold is shifted to lower energy. The threefold degeneracy of the level means that half-filling corresponds to three electrons. K_3C_{60} is a metal that becomes superconducting at temperatures lower than 19 K. Further filling leads to the compound A_6C_{60} (where A is an alkali element). Because this material has a fully filled t_{1u} lowest unoccupied molecular orbital, it is insulating. The structures of C_{60}, K_3C_{60}, and Cs_6C_{60} are shown on the right in Figure 2.6.1. The structures are cubic but are represented in tetragonal form. The alkali metal atoms sit in the tetragonal and octahedral voids of the fcc-C_{60} structure. With alkaline-earth metals, the t_{1g} orbital derived band is also partially filled. Thus Ca_5C_{60}, Sr_6C_{60}, and Ba_6C_{60} are also superconducting.

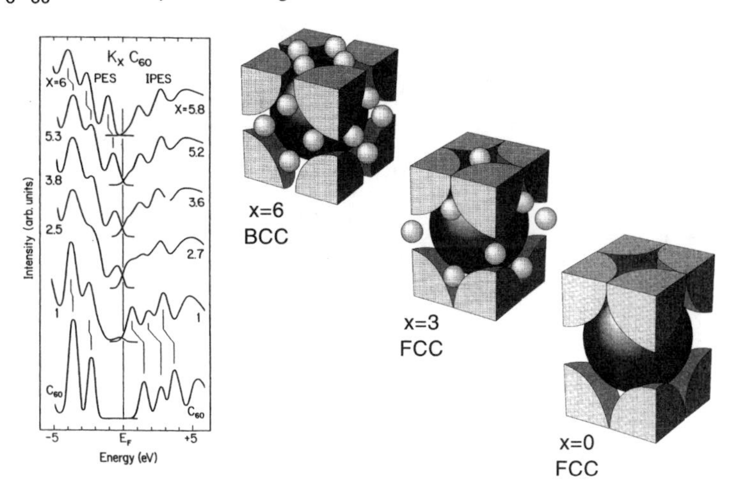

FIGURE 2.6.1 Normal (PES) and inverse (IPES) photoemission density of states of C_{60} as a function of exposure to potassium vapor. The gradual filling of the C_{60} t_{1u} lowest unoccupied molecular orbital is clearly seen. The spectral manifold shifts to lower energy with increasing exposure because of Fermi level pinning. On the right, body-centered tetragonal representations show the structures of Cs_6C_{60} ($x = 6$), K_3C_{60} ($x = 3$), and C_{60} ($x = 0$). [Left: *Journal of the Physics and Chemistry of Solids* **53**, 1433 (1992); Right: *MRS Bulletin* **19**, 28 (November 1994).]

conductors, in which size dependence emerges at relatively large size compared with metals, insulators, or molecular crystals. This arises because the bands of a solid are centered about atomic energy levels, with the width of the band related to the strength of the nearest-neighbor interactions. In the case of van der Waals or molecular crystals, which have weak nearest-neighbor interactions, the bands in the solid are very narrow, giving little size dependence of optical or electrical properties in the nanocrystal regime. As the cluster grows, the center of a band develops first and the edges last. Thus, in metals, for which the Fermi level lies in the center of a band, the relevant energy-level spacing is small, and at temperatures above a few Kelvin, even small clusters (10 to 100 atoms) have electrical and optical properties that resemble those of a continuum. Because the Fermi level lies between two bands in semiconductors, the band edges dominate the low-energy optical and electrical behavior. Optical excitations across the gap thus depend strongly on size even for clusters as large as 10,000 atoms (see Figure 2.7). Electrical transport also depends heavily on size, mainly because of the large variation in energy required to add or remove charges on a nanocrystal. As a consequence, many useful size-dependent phenomena are observed in clusters characterized by an interior that is structurally identical to the corresponding bulk solid and a surface layer that contains a substantial fraction of the total number of atoms in the cluster.

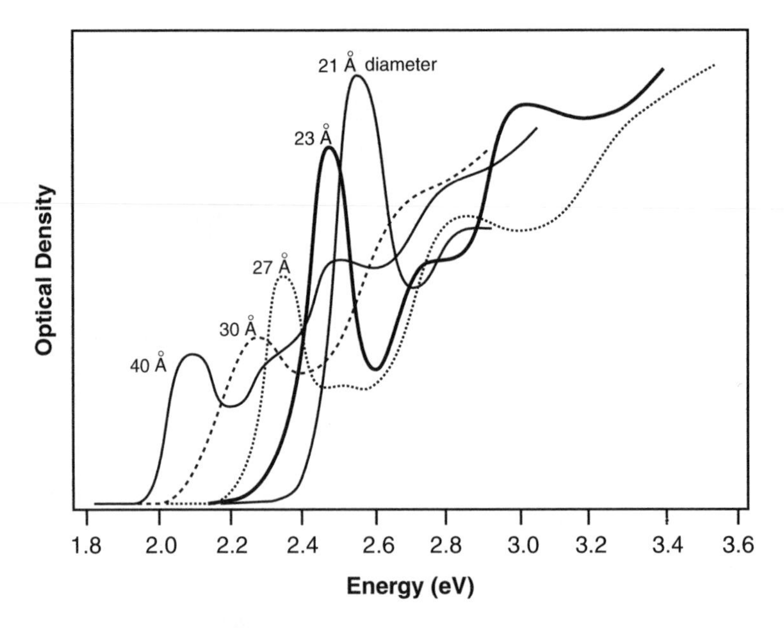

FIGURE 2.7 Quantum confinement causes the optical spectra of CdSe nanocrystals to sharpen and move to higher energy as the size of the particle shrinks. [*MRS Bulletin* **20**, 23 (1995).]

The surface of a semiconductor exerts a critical influence on optical and electrical properties. Passivation is critical to the control of these properties, and clusters are no exception. As in the bulk, the ideal termination removes the surface reconstruction, leaves no strain, and simply produces an atomically abrupt jump in the chemical potential for electrons or holes at the interface. A great deal of current research into semiconductor clusters is focused on the properties of quantum dots with the bulk bonding geometry and with surface states eliminated by immersion in a material of larger gap.

There have been two approaches to the fabrication and investigation of quantum dots. The top-down approach involves the gradual reduction of the extent and dimensionality of solid matter. The quantum dots thus produced are between 1 μm and 10 nm in size. They are well passivated and immobilized on a substrate where they may be investigated optically and electrically. The bottom-up approach views quantum dots as extremely large molecules or colloids. Nanocrystals vary in diameter from 1 nm to about 20 nm. Their surfaces are derivatized with organic molecules, which prevent them from aggregating and render them soluble. Whether these organic molecules provide electronic passivation as well is an open question. These samples may be manipulated chemically in a wide variety of ways, yielding entirely new sample configurations (see Figure 2.8).

Much of the work on semiconductor nanocrystals started with the realization that it is possible to precipitate a semiconductor out of an organic liquid. A set of precursors are injected into a very hot liquid. Upon injection the temperature immediately rises above the nucleation limit so that nucleation occurs, and then the temperature quickly drops. The concentration drops quickly because dilution occurs, resulting in crystallites in a fluid. The crystallites become encapsulated in a layer of organic material so that they do not collide and fuse. The crystallites are each single-crystalline, and they have a preponderance of low-energy, low-index facets with very few high-index surfaces. One issue that quickly emerges in the study of nanocrystals is that approximately half the atoms or more are on the surface of the crystal, making control of the surface even more important than in bulk materials or conventional films.

Although the high-pressure behavior of semiconductor nanocrystals is ultimately the same as that of the bulk, the details differ rather dramatically. For example, CdSe nanocrystals 4.2 nm in diameter require almost three times greater pressure than does bulk material to transform from the wurtzite to the rock salt crystal structure. When the nanocrystal does transform, it remains a single crystal, indicating that only a single nucleation event has occurred per crystallite. The smaller the crystallite, the higher the transformation pressure. This difference has been explained by noting that the transformation involving the least atomic motion requires transforming from a wurtzite crystallite with only low-index surfaces to a rock salt structure with numerous high-energy faces. This consideration increases in importance with decreasing crystallite size, as more and more atoms in the crystallite occupy its surface.

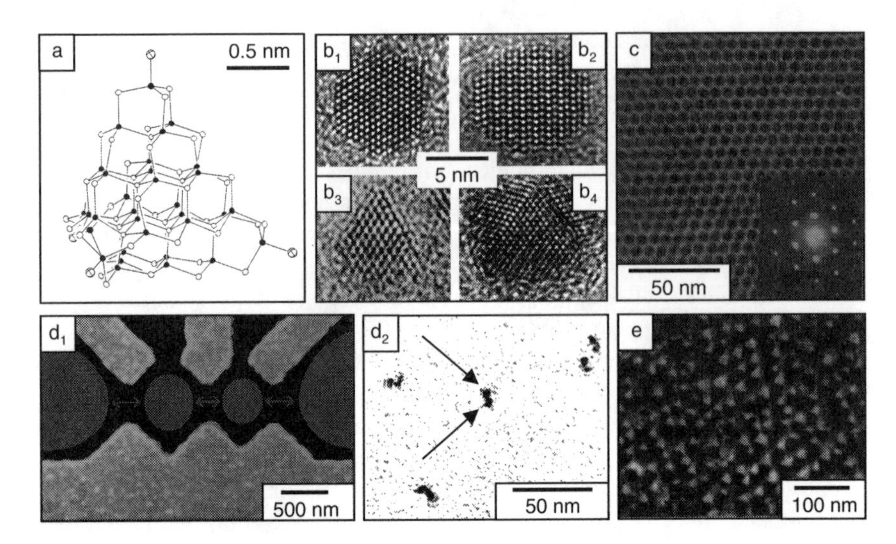

FIGURE 2.8 Gallery of quantum dot structures: (a) Positions of cadmium and sulfur atoms in the molecular cluster $Cd_{32}S_{55}$, as determined by single-crystal x-ray diffraction. This cluster is a small fragment of the bulk CdS zincblende lattice. The organic ligands on the surface are omitted for clarity. (b_1) and (b_2) Transmission electron micrographs of CdSe nanocrystals with hexagonal structure, as viewed down different crystallographic axes. These nanocrystals were prepared colloidally and exhibit well-defined facets. The surfaces are passivated with organic surfactants. (b_3) and (b_4) Transmission electron micrographs of CdS/HgS/CdS quantum dot quantum wells. The faceted shapes show that epitaxial growth for passivation is possible in colloidally grown nanocrystals. (c) Transmission electron micrograph of a CdSe quantum dot superlattice. (d_1) Scanning electron micrograph of two coupled GaAs quantum dots about 500 nm in diameter. The strength of the coupling can be adjusted by adjusting the gate voltage. (d_2) Transmission electron micrograph of coupled CdSe nanocrystal quantum dots 4 nm in diameter. These crystallites are joined by an organic molecule. The coupling can be tuned by changing the linker length. (e) Transmission electron micrograph of InAs quantum dots in a GaAs matrix, prepared by molecular beam epitaxy. [Reprinted with permission from A.P. Alivisatos, "Semiconductor clusters, nanocrystals, and quantum dots," *Science* **271**, 934 (1996). Copyright © 1996 American Association for the Advancement of Science.]

The main reason for the high level of interest in semiconductors of reduced dimensionality results from their large quantum-size effects. The band gap in cadmium selenide can be tuned between 4.5 and 2.5 electron volts (eV) as the size is varied from the molecular regime to the macroscopic crystal, and the radiative lifetime for the lowest allowed optical excitation ranges from tens of picoseconds to several nanoseconds. The energy above the band gap required to add an excess charge decreases by 0.5 eV. The melting temperature increases from 400 to 1600°C, and the pressure required to induce transformation from a four- to a six-coordinate phase decreases from 9 to 2 GPa. There are many questions in the literature about what would happen to an indirect-gap material, such as silicon, as the nanocrystallite size decreases. As silicon crystallites become smaller and smaller, they become more emissive because of the quantum-size effect. However, the fundamental matrix element does not change; it is only a density-of-states effect. Silicon will never become a direct-gap material in the relevant size range. On the other hand, a direct-gap semiconductor like cadmium selenide already has an allowed electronic transition, and as it shrinks, it is more allowed. Direct-gap semiconductors will retain their advantages over indirect-gap materials proportionally. As they are made smaller and smaller, they will continue to radiate more efficiently than indirect-gap semiconductors. One recent development with potentially far-reaching impact is the development of ultra-small, highly efficient semiconductor lasers, known as quantum dot lasers. Because of quantum confinement, controlling the size of the nanocluster leads to control of the color of light emitted (see Figure 2.9). Quantum dots are so small that they tightly confine normally mobile electrons, so the charges spend less energy on their wanderings. Thus more energy is released when the electron and

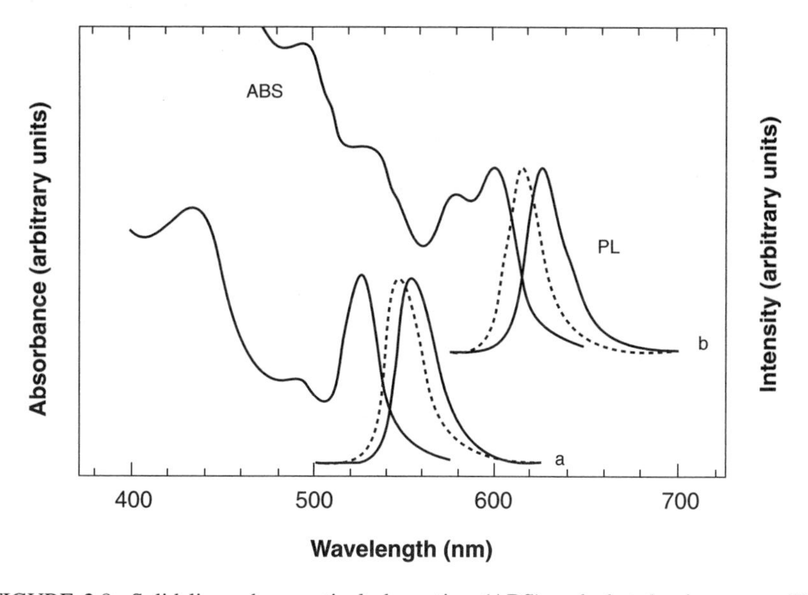

FIGURE 2.9 Solid lines show optical absorption (ABS) and photoluminescence (PL) spectra at 10 K for close-packed solids of CdSe quantum dots 3.85 nm (curve a) and 6.2 nm (curve b) in diameter. Dotted lines are photoluminescence of the same dots but in dilute form dispersed in a frozen solution. [Reprinted with permission from C.B. Murray, C.R. Kagan, and M.G. Bawendi, "Self-organization of CdSe nanocrystallites into three-dimensional quantum-dot superlattices," *Science* **270**, 1336 (1995). Copyright © 1995 American Association for the Advancement of Science.]

hole combine, resulting in a shorter wavelength. The smaller the dot, the greater the frequency shift. Making a true quantum dot laser has proven difficult. It is not straightforward to make the dots the same size, and the result has been that the devices emit a range of light frequencies. Very recent work in this area has yielded dots of more uniform size, with characteristics more indicative of true laser activity.

THIN FILMS, SURFACES, AND INTERFACES

The revolution in the control of the growth and properties of thin films, surfaces, and interfaces traces its origin to advances in vacuum technology in the 1960s and 1970s and to the development of surface-sensitive probes such as surface spectroscopies and electron diffraction techniques. Although the field is no longer new, there have been impressive gains in the past decade, both in the enabling technologies for thin-film growth and in the insights that growth and surface studies have provided. The result has been more control over and understanding of the growth process, and hence the properties of the resulting film. Defects can be placed as desired or eliminated altogether. More complex materials can be grown with acceptable quality as a result of the increased control.

The base vacuum available in molecular beam epitaxy systems has improved by more than an order of magnitude to $\sim 10^{-12}$ torr. Such improvement allows the surface contamination prior to growth to be lowered and also results in less contamination in the growing film. Alternatively, the growth rate can be lowered while preserving low levels of contamination, improving control over the process. Various growth techniques have been developed or refined, as discussed elsewhere in this report. Of particular relevance are pulsed-laser deposition, which has proven particularly useful for the deposition of complex materials such as high-temperature superconductors; ion beam-assisted deposition, which induces crystallographic alignment in a growing film independent of the crystallography (or lack thereof) of the substrate; and various refinements of traditional techniques such as molecular beam epitaxy, chemical vapor deposition, and sputtering.

The ability to image surfaces and films in real space, as discussed in Chapter 6, has revolutionized studies of film growth. Most of the earliest studies using the new scanning probe microscopies corroborated surface structures previously arrived at through tortuous interpretation of surface spectroscopic and diffraction data. It quickly became apparent, however, that the ability to observe surfaces and films in real space with atomic resolution could enable far more understanding of surfaces than could be derived through more indirect techniques. Nucleation sites, evolution of surface morphology, etc., can now be observed directly, as discussed in Box 2.7. Another advance in the area of monitoring that has had a significant impact on thin film studies is the ability to monitor film growth in real time. Perhaps the most impressive demonstration of the power of this technique is the in situ studies of chemical vapor deposition growth performed at

BOX 2.7 Early Stages of Film Growth

The study of film growth has been increasingly characterized by the application of surface science methods to understanding growth at the atomic level. Both technology and the desire for fundamental knowledge at the atomic level are driving the search for atomic-level control of the fabrication processes for novel materials and new devices.

Growth of thin films from atoms deposited from the gas phase is intrinsically a nonequilibrium phenomenon, governed by a competition between kinetics and thermodynamics. Precise control of the growth and thus of the properties of thin films becomes possible only through an understanding of this competition. Experiment and theory have both made impressive strides in exploring the kinetic mechanisms of film growth, including adatom diffusion on terraces, along steps, and around island corners; nucleation and dynamics of the stable nucleus; atom attachment to and detachment from terraces and islands; and interlayer mass transport. The synergism between experiment and theory has tremendously improved our understanding of the kinetic aspects of growth.

The diffusion of an adatom on a flat surface or terrace is by far the most important kinetic process in film growth. Despite the vital importance of surface diffusion, accurate determination of the surface diffusion coefficient in a broad range of environments has been a major challenge. Scanning-tunneling microscopy (STM) has improved the situation considerably. STM can image a vastly broader range of surfaces than can field ion microscopy, which has traditionally been used for such studies. Atom-tracking STM has been especially valuable because it allows an atom or cluster to be followed as it migrates. Information from such experiments can then be fed into theories to provide deeper understanding of the mechanisms at play in adatom diffusion.

The availability of new probes of the initial stages of nucleation and growth has meant that even well-studied systems have continued to yield new insights. Much recent attention has focused on the possible pathway for nucleation of a silicon addimer, the stable nucleus for a wide range of growth conditions for homoepitaxy on Si(100) (see Figure 2.7.1). A silicon adatom may have multiple diffusion pathways on the surface before finding a partner, as all calculations have suggested. Recent experiments have focused on determining the relative stability of different dimer orientations and have been able to distinguish slight differences. Studies have also focused on the preferred locations of dimers, where there are still significant differences between experiment and theory. Experiments have revealed some surprisingly large anisotropies in larger islands as they grow.

As islands grow, specific island shapes or morphologies develop. One class is compact, whereas another is fractal-like, with rough island edges or highly anisotropic shapes. Recent studies of two-dimensional island formation in metal-on-metal epitaxy have identified several aspects of atom diffusion along island edges that are important in controlling the formation of fractal islands. Fractal island growth is very dependent on bonding geometry, having been reported only on face-centered cubic (111) or hexagonal close-packed (0001) substrates, both of which have approximately triangular lattice geometry. Growth on face-centered cubic (100) surfaces with square lattice geometry has so far resulted only in compact islands. This observation has required modification of the classic diffusion-limited aggregation model.

BOX 2.7 Continued

Explorations seeking improved understanding of the relative importance of the various atomistic rate processes important in the initial stages of film growth have led to discoveries of various ways of rate manipulation to improve the quality of films grown by vapor-phase epitaxy. For example, any enhancement of downward diffusion of adatoms below their landing site would improve layer-by-layer or two-dimensional growth. Increasing the ability of an atom to cross a corner of an island at which two edges meet would lead to more compact islands. Larger surface diffusion would lead to earlier achievement of step-flow, hence layer-by-layer growth. Such insights are already leading to greater control over the precise morphology of thin films to achieve desired structures. The rate of progress in this area will surely increase as our understanding continues to grow.

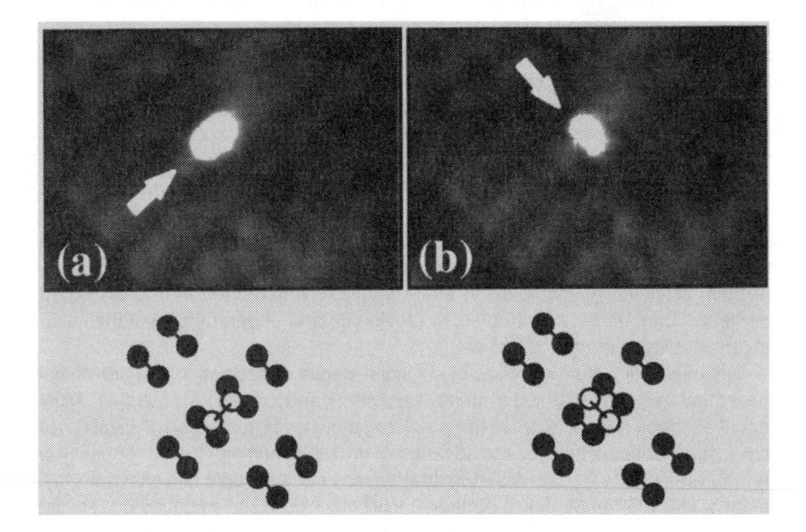

FIGURE 2.7.1 Scanning-tunneling microscope images showing the rotational dynamics of a silicon ad-dimer formed on top of a dimer row in Si(001). (a) The bond of the ad-dimer (arrow) is parallel to the substrate dimer rows in the 2×1 reconstruction, as schematically shown below the image. (b) The same ad-dimer has rotated by 90 degrees. The two images were taken 40 seconds apart at room temperature. The orientation in (a) is energetically slightly more stable. [Reprinted with permission from Z. Zhang and M.G. Lagally, "Atomistic processes in the early stages of thin-film growth," *Science* **276**, 377 (1997). Copyright © 1997 American Association for the Advancement of Science.]

synchrotron sources, as described in Chapter 6. The power of this technique has just begun to be tapped.

There have been major strides in understanding the kinetics of epitaxial growth. In silicon molecular beam epitaxy, it has been shown that the notion of a minimum temperature for epitaxial growth is incorrect, even at a fixed growth rate. Instead, for all temperatures of the regime studied (30 to 300°C), the growing epitaxial film becomes amorphous above a limiting thickness. The epitaxial thickness for the epitaxial-amorphous transition follows an Arrhenius temperature dependence, depending on the deposition rate. Other systems, including gallium arsenide, also exhibit this behavior. Such improved understanding is relevant to the problem of dopant incorporation in silicon epitaxy, in which controlled growth of a highly doped layer of arbitrary thickness can be achieved at a temperature low enough to avoid dopant segregation. This low growth temperature approach allows very sharp doping profiles.

The ability to control the structure of inorganic thin films on an unprecedented scale has recently been demonstrated using a technique called "glancing angle deposition," which maximizes atomic shadowing and minimizes adatom diffusion. By making use of extremely high adatom angles of incidence, coupled with substrate rotation (or other motion) around the substrate normal, slanted, zigzag, helical, or other morphologies have been demonstrated, as shown in Figure 2.10. The helically structured films have been shown to rotate the plane of polarization of light in a manner analogous to other chiral media; pitches between 50 and 2000 nm have been demonstrated.

Major progress has been made in substrate engineering and control, along with understanding of the role of surface properties in film growth, in the last decade. Compliant substrates allow control of the strain induced by the constraint of epitaxy so that the electronic and structural properties of the film can be tailored to fit the needs at hand. If the substrate is sufficiently thin, or if the top layer is sufficiently weakly bonded to the rest of the substrate, the substrate can readily deform elastically when a lattice mismatched material is grown on it. The extent to which this approach can be reduced to practice is as yet unclear. Epitaxy has been shown to be strongly affected by the presence of an adsorbate (surfactant), which changes the surface properties because of its bonding arrangement, usually accompanied by a change in the surface reconstruction (see Figure 2.11). The saturation of the dangling bonds yields a chemical passivation of the surface, a change in electronic structure, and a reduction of the surface free energy, which causes the strong segregation of the surfactant layer. The adlayer floats on the growing film, and only a small fraction is incorporated. These properties can have a profound effect on the growth mode of the film, converting a film that normally grows in an island fashion to layer-by-layer growth. In heteroepitaxy, interfaces prepared by growth with an adsorbate present, either hydrogen or a dopant species, remain sharper at a higher growth temperature than interfaces prepared by growth on a bare surface.

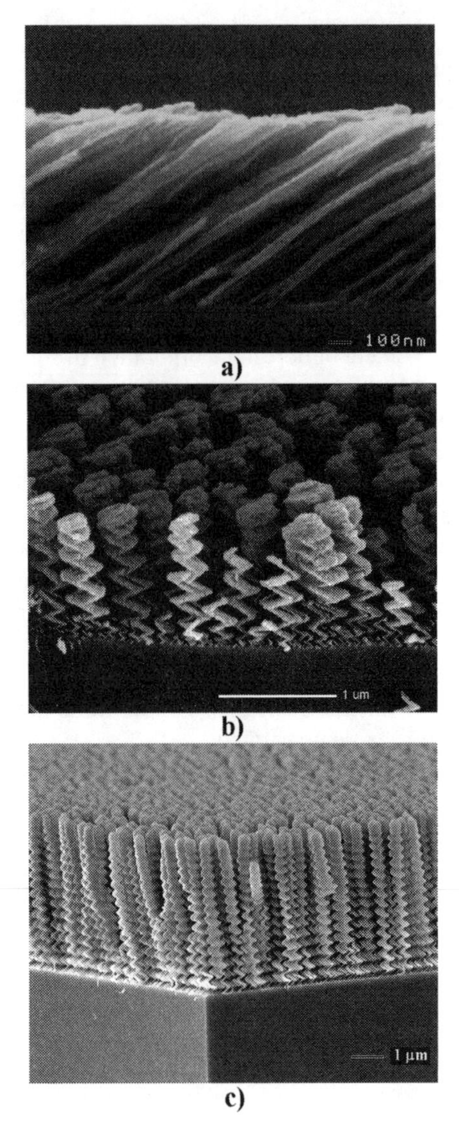

FIGURE 2.10 Films deposited by glancing angle deposition (GLAD): (a) oblique evaporated flux at 85 degrees from the substrate normal produces a slanted, porous microstructure; (b) periodically alternating the oblique flux from angles of 85 degrees to –85 degrees produces a porous film composed of isolated "zigzags"; (c) rotating the substrate about an axis normal to the wafer center while maintaining obliquely incident (85 degree) flux produces isolated helical structures on the substrate. [Reprinted with permission from K. Robbie and M.J. Brett, "Sculptured thin films and glancing angle deposition: Growth mechanics and applications," *Journal of Vacuum Science and Technology A* **15**, 1460 (1997). Copyright © 1997 American Vacuum Society.]

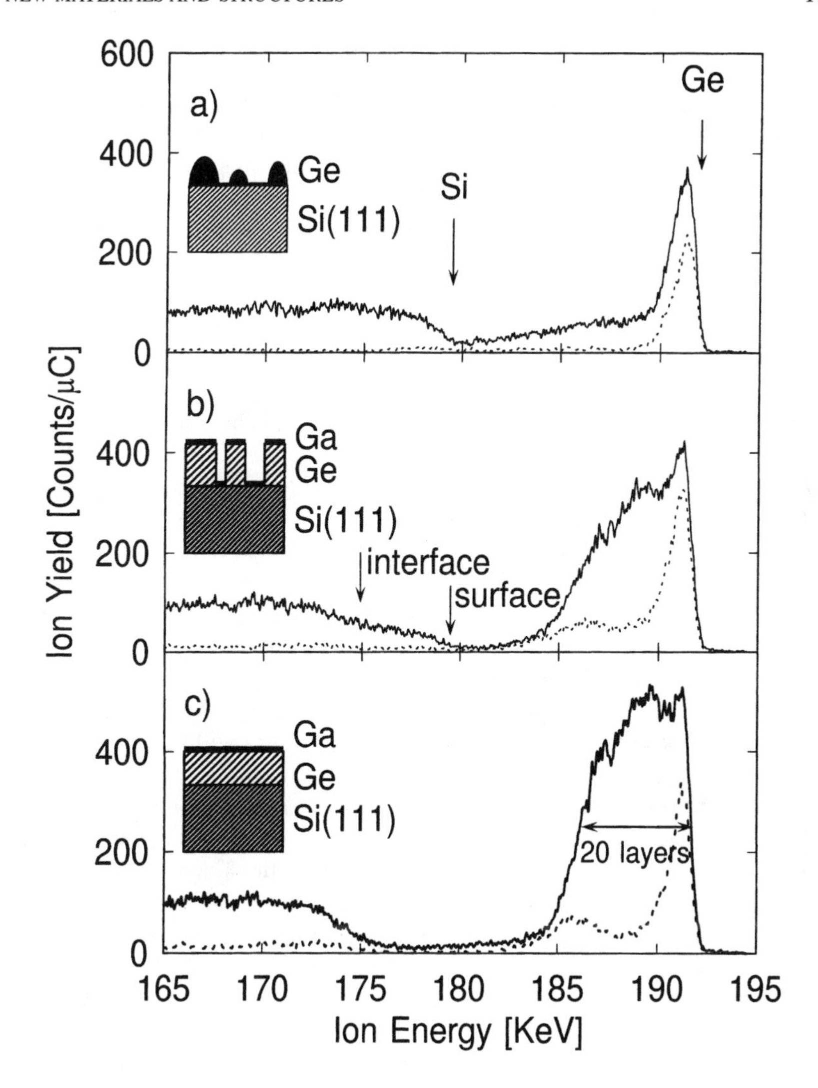

FIGURE 2.11 The use of a surfactant dramatically alters the morphology of a growing film. In this figure are medium energy ion scattering (MEIS) spectra for germanium films on Si(111) at 470 °C. Both random (solid line) and channeling (dotted line) data are shown. (a) Ten monolayers of germanium deposited with no gallium. Note the island morphology. (b) Twenty monolayers of germanium deposited with one-third of a monolayer of gallium as a surfactant. Note the columnar morphology. (c) Twenty-eight monolayers of germanium with one monolayer of gallium surfactant. Note the smooth morphology. [Reprinted with permission from J. Falta, M. Capel, F.K. Legroves, and R.M. Tromp, *Applied Physics Letters* **62**, 2962 (1993). Copyright © 1997 American Institute of Physics.]

ARTIFICIALLY STRUCTURED MATERIALS

Artificially structured materials—materials whose structure or composition differs in some intentional way from materials available in nature—have enabled many of the advances in condensed-matter and materials physics over the past decade that are described in this report. Such materials are frequently dominated by interfaces, a feature that often leads to properties very different from those of bulk materials. Because artificially structured materials are frequently prepared far from thermodynamic equilibrium, they can exhibit phases or properties that are not otherwise achievable. Their multilayer structural length may be on the order of the length-scale characteristic of nonlocal physical phenomena in solids, making such materials ripe for fundamental investigations. The physical limits to their fabrication have been pushed to the greatest extent for semiconducting materials. The coming decade will undoubtedly see the same limits pushed for other classes of materials as well: complex oxides, polymers, biological materials, and composites are a few of the most exciting.

Many if not most artificially structured materials involve heteroepitaxy, the crystallographically oriented growth of one material on a dissimilar one. In nearly all cases, heteroepitaxy involves a lattice mismatch between the different materials, which produces strain in the initial epitaxial layer. The strain is relieved as the epitaxial film thickness increases, by a roughening of the surface of the epitaxial layer or by the introduction of defects such as dislocations into the epitaxial layer or both. Controlling when and how strain relief occurs is a key issue in heteroepitaxy.

The strained films discussed here are grown near the thermodynamic limit. Sputter-deposited films, such as those discussed in Box 2.8 and in the section on magnetic multilayers in Chapter 1, are deposited in the kinetic limit, which is required for alternating layers of extremely disparate materials.

One distinct trend has been toward the use of more highly strained heteroepitaxial combinations, such as InGaAs/GaAs (see Figure 2.12) and SiGe/Si. Such systems must be approached with great care in order to achieve the optimum structural (and consequently electrical or optical) quality. Morphology-related strain relief is not a new phenomenon. Mounding in a film may partially relieve elastic stress of the epitaxial material within each mound. Even though there is additional compression of lattice planes at the grooves between the mounds, the roughened morphology is energetically favorable because the volume of material subjected to additional stress is much less than the volume experiencing partial stress relief. Another very important factor is the surface free energy. Roughening generally increases this energy, so that roughening is suppressed until the free-energy reduction in the system by stress relief is greater than the free-energy increase caused by surface area increase and step formation.

Strain-induced roughening can be problematic in the fabrication of coherently strained device structures, for which it is important to understand the early

BOX 2.8 Multilayers for X-Ray and Extreme Ultraviolet Optics

Multilayers are artificially structured materials that are periodic in one dimension in composition or both composition and structure. These layered materials are, if perfect, equivalent to single crystals in one dimension. Thus the multilayer acts as a superlattice, diffracting longer-wavelength radiation in a manner directly analogous to the diffraction of x-rays by crystals. This application of multilayer structures as dispersion elements for soft x-rays and extreme ultraviolet radiation was the impetus for the first attempts to synthesize multilayer materials. Many factors determine the character of the multilayer response to an incident spectrum. The important parameters are the substrate quality (roughness and figure), the uniformity and thicknesses of the component layers, the x-ray optical constants of the component elements, the number of layers in the structure, the interfacial width between layers (i.e., interfacial abruptness in atomic position and composition), and roughness at layer interfaces. Many of these factors depend in turn on the synthesis process and the materials. Therefore, understanding of multilayer performance depends on a knowledge of the relationships among synthesis process, resultant microstructure, and properties for these engineered microstructure materials.

The individual layers of the optics have a specific set of properties related to bulk forms of the materials. Primary issues include the compositions and structures of the layers, the x-ray optical properties of the layers, and uniformity of the areal density of atoms in the layers (see Figure 2.8.1). Specific synthesis questions relate to the film nucleation and growth behavior because deposition of material A onto a substrate or layer B may differ substantially from deposition of material B onto a substrate or layer A. Interfaces within the multilayer must also be controlled to an excruciating degree. They must be compositionally abrupt, smooth, clean, and flat.

Recent work has shown that precise control of sputtering parameters during multilayer deposition allows control of individual layer thicknesses to an accuracy of better than ~0.01 nm, which greatly enhances reflectivity for both nickel-carbon and tungsten-carbon multilayers. Sputter deposition of multilayers typically produces higher quality structures than thermal source techniques. This has been attributed to ion bombardment by the sputter plasma resulting in smoother interfaces and higher reflectivities. Results of ion beam-assisted deposition support this proposal. Thermal-evaporation-source synthesized rhodium-carbon multilayers with and without argon ion bombardment (300 eV) at an incidence angle of 10 degrees show the effect. A gain of more than a factor of two in reflectivity was found for the samples "polished" by the incident ion beam. This increased reflectivity is attributed to smoothing of the interfaces between the carbon and rhodium by a factor of 30 percent by the ion bombardment. Combining these two improvements in control are likely to facilitate fabrication of higher quality multilayer structures, particularly of smaller periods.

Multilayer structures may be optimized and engineered for specific spectral ranges by an analysis for optimum materials on the basis of their x-ray constants and an assessment of their suitability for multilayer microstructure synthesis. As an example, there are difficult spectral regions in which the lowest absorption materials useful as spacer layers are either toxic, such as beryllium, or unstable, such as lithium. Candidate materials such as magnesium are difficult to deposit as

BOX 2.8 Continued

a uniform thin film because of its low melting point and high vapor pressure. Mg_2Si has been identified as a possible new material for this application, and W/Mg_2Si multilayers have reflectivities that are the highest reported in the 800 to 1300 eV range.

Multilayer x-ray optics and instrumentation are now mature enough to be both an enabling technology and an area of scientific investigation in their own right. The promise that was held for soft x-ray and extreme ultraviolet multilayer optics is now coming to fruition, and many of the advanced optical systems envisioned in the late 1970s are becoming reality. Such x-ray optics will likely form the critical element of vacuum ultraviolet optics for the next generation of lithography in the semiconductor industry.

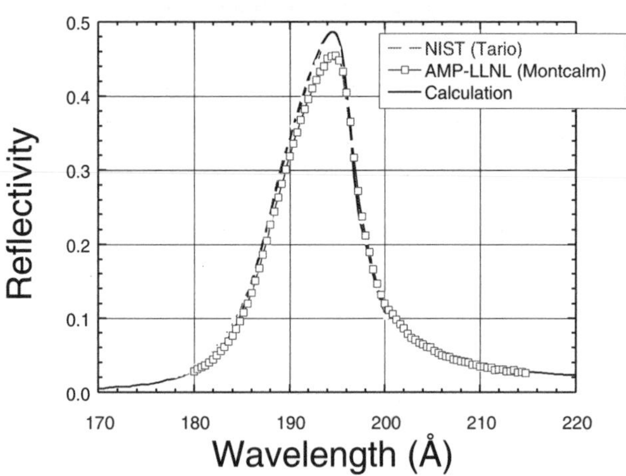

FIGURE 2.8.1 Transmission electron micrograph of a 6.9 nm period Mo_2C/Si multilayer x-ray mirror (top) and the experimental and calculated reflectivity as a function of x-ray wavelength (bottom). The experimental reflectivity is 93.5 percent of the calculated values. [Reprinted with permission from T.W. Barbee, Jr., and M.A. Wall, "Interface reaction characterization and interfacial effects in multilayers," *Proceedings of the SPIE* **3113-20**, 204 (1997). Copyright © 1997 SPIE.]

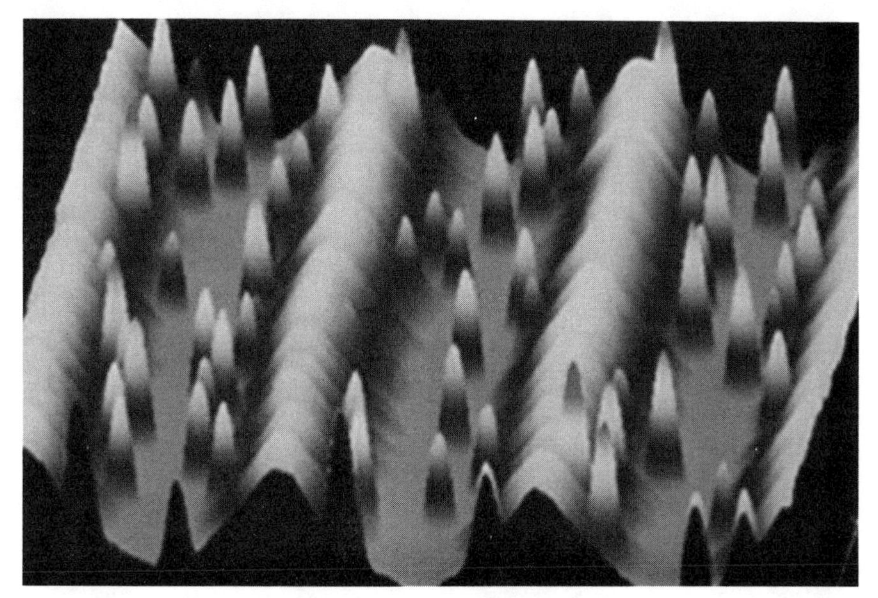

FIGURE 2.12 Atomic force micrograph of self-assembled InAs islands deposited by molecular beam epitaxy on a patterned GaAs(001) surface. The valleys and hills, which have been defined by optical lithography and etching, have a period of ~240 nm. The InAs islands, which are formed by deposition of about 1.5 monolayers of InAs on the corrugated GaAs surface, are preferentially located in the valleys of the surface. The height and diameter of the InAs islands are 10 nm and ~20 nm, respectively. These islands will be transformed into quantum dots by in situ overgrowth of a GaAs cladding layer. (Courtesy of the University of California at Santa Barbara.)

stages of the transition in order to avoid or suppress three-dimensional growth. On the other hand, the strain-driven transition is beneficial for the self-assembly of quantum dots, in which it is necessary to control the size distribution and self-organizing behavior of the islands. It is critical to understand the kinetic pathways to island formation.

Recent theoretical investigation suggests that systems with tensile stress could be more resistant to roughening than those with compressive stress. Recent work has supported this prediction in, for example, the Si-Ge system. Molecular dynamics modeling has attributed this observation to an increase in the energy of certain types of surface steps under tensile strain, which makes it energetically favorable for the surface to remain planar.

One of the major contributions of scanning-probe microscopies to heteroepitaxy has been improved understanding of morphological evolution in heteroepitaxy. A general trend in all experiments is the decreasing size of typical morphological features with increasing misfit stress. In many highly mismatched systems, surface ripples exhibit a strong tendency to facet along inclined planes,

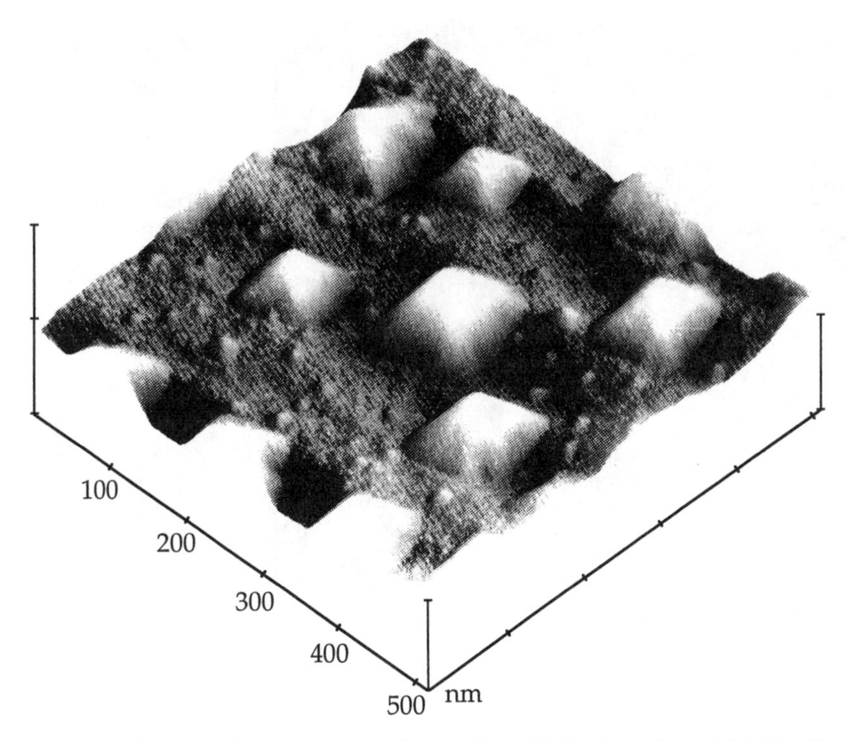

FIGURE 2.13 Atomic force microscope image of an initially planar 2-nm thick $Si_{0.5}Ge_{0.5}$ alloy layer on Si(001) after annealing to produce hut-shaped islands caused by strain in the layer. [*MRS Bulletin* **21**, 31 (1996).]

as in the epitaxial germanium "hut" clusters in Figure 2.13. The presence of {501} facets appears to be a general feature of strained-layer growth in the Si-Ge system. It is not understood, however, why such facets are stable and what role they play in the growth of coherently strained islands. The picture is emerging that {501} facets are the natural result of the desire to release as much elastic energy as possible without unduly creating energetically costly surface-step configurations.

During growth, strain relaxation in heteroepitaxial systems can lead to changes in island shapes. In semiconductor systems, coherent islands are often faceted and characterized by large aspect ratios [up to a 10:1 base-to-height ratio in the case of the germanium "huts" and up to 50:1 in silver on Si(100)]. Recent calculations have shown that strained islands are likely to undergo a shape transformation during growth. Below a critical island size, the energy balance favors compact, symmetric islands; for large islands, elongated shapes with high aspect ratios are preferred. This suggests one approach to the challenge of producing quasi-one-dimensional quantum wire structures.

Turning now to the other major strain-relief mechanism—defects such as dislocations—we enter an area of research with much longer antecedents. Traditionally, misfit dislocations have been avoided simply by keeping the film thickness at less than the "equilibrium critical thickness." More recently, other strategies have been introduced, such as reduced temperature growth or substrate patterning. When substrate patterning is used to isolate regions of the sample, relaxation is greatly retarded because most isolated regions contain no heterogeneous nucleation sites. These regions can be remarkably stable during thermal anneal, which shows that controlling nucleation may be the key to controlling, or even suppressing, relaxation.

One of the most successful methods to date of fabricating quantum dots uses self-assembly that results from growth kinetics controlled by strained-layer epitaxy. A strain-induced transition from two- to three-dimensional growth results in the formation of coherently strained islands on the surface of the semiconductor. Using these principles, islands <20 nm in diameter can be fabricated with size distributions within ± 10 percent. The resulting islands are pseudomorphically strained and dislocation free.

The random distribution of islands can be modified by appropriate control of their nucleation and growth, kinetics. It was readily apparent early on that preferential nucleation of islands takes place at surface steps. This effect could be used to order islands. Recent progress in making extremely perfect kinkless steps over micron distances on Si(111) offers hope of ultimately achieving ordering of island assemblies by this approach.

Another strategy for ordering the islands is based on the very sharp transition from two- to three-dimensional growth. A corrugated substrate with concave and convex areas will tend to flatten and minimize its surface energy during epitaxy through faster growth of the convex areas. Thus, when depositing a strained layer over such a surface, the critical thickness will be reached sooner in these areas, and quantum dots will nucleate preferentially in these regions.

Another promising approach based on strain-induced nucleation allows for regulating the size and ordering of the islands in the growth direction. If two or more layers of quantum dots are grown sufficiently close to each other (closer than 10 nm for InAs/GaAs), it is possible to obtain self-alignment of the islands in the growth direction.

FUTURE DIRECTIONS AND RESEARCH PRIORITIES

The examples of new materials and structures presented in this chapter point out the major themes in the search for new and improved materials and properties that have characterized the past decade. The themes include the discovery of new and unexpected materials with novel properties and the use of new tools to provide improved understanding and control in well-known materials. These developments foreshadow many of the advances that are likely in the coming years. It is critical to emphasize, however, that many of the most exciting developments have been complete surprises; this will almost certainly be true for the foreseeable future as well.

Materials Properties by Design: Complexity

The ability to tailor materials and structures to obtain a desired set of properties is in its infancy (see Box 2.9). Band-gap engineering, which has been achieved by judicious use of strain, alloying, and quantum-size effects, is one area that has had considerable impact. Control of the microstructure of a material, through processing and judicious choice of geometry or neighboring materials, has been used to control the physical properties of some materials. Artificially layered materials have been given unusual dielectric properties and have contributed to the search for new high-temperature superconductors.

Future progress will build on these recent accomplishments. Eventually, we can expect to be able to tailor materials at the molecular level, building up materials molecule by molecule in three dimensions. This capability will enable truly three-dimensional designs with as much or as little symmetry as needed. In the future, a design will be able to incorporate structure at multiple length-scales, enabling the optimization of multiple properties that involve phenomena operating at very different dimensions. Today, the most advanced artificially structured materials utilize individual material constituents of the same or very similar classes: III-V semiconductors, for example. We can look forward to being able to use a much more colorful palette, not limited to a single class of materials or even just to inorganic materials. Polymers, organic molecules, and even biological molecules are likely to become integral parts of increasingly complex structures as we learn more about how to manipulate molecules individually. A glimpse into the possibilities is described in Box 2.10.

As the structures that we are able to build become more complex, we will need the tools to be able to see, characterize, and manipulate them. We will need to be able to work with these structures on all of the length scales that are relevant for the properties we desire. The scanning-probe microscopies are an important step in this direction, but more, equally revolutionary advances will be required to truly take advantage of this new regime of materials design.

Finally, we will need to be able to make predictions about these new materials. Our ability to predict the existence of new materials with interesting properties is extremely limited. Witness the complete surprise presented by the discovery of high-temperature superconductors or the lack of theoretical guidance concerning other interesting avenues of inquiry in the search for other high-temperature superconducting families. Theoretical guidance regarding promising synthetic routes for fabricating new materials would be most useful to experimentalists trying to prepare them. Finally, improved communication between theorists and experimentalists might help to shorten the gap between prediction and fabrication—several decades in the case of C_{60}, for example.

BOX 2.9 Combinatorial Chemistry and the Search for New Materials

Systematically varying the composition of a multicomponent system to optimize its properties is a time-honored empirical method in materials synthesis. For example, "phase spreads" of thin films proved powerful in the study of metal-insulator transitions almost 2 decades ago. Scaling up the approach to allow the fabrication and testing of tens or hundreds of compositional variants requires the ability to prepare small volumes of precisely controlled composition under known preparation conditions as well as the capability to test these miniature samples for desired properties. Known more recently as combinatorial chemistry, this approach is used to make a large number of chemical variants in parallel, to screen them quickly and reliably for chemical activity, and to build a library of information about the resultant chemical diversity. Until recently, combinatorial chemistry has been used primarily to transform the way new drugs are discovered, but in the coming decade, it may have an impact on the search for other classes of new material as well. The most notable forays of combinatorial chemistry into non-medical arenas are in the areas of superconducting compounds and phosphors (see Figure 2.9.1). The applicability of the technique to the search for catalysts is also being investigated. All these materials share the property of being very complex, containing many elements and eluding prediction of their properties or even existence using any currently available theories or models. As the entire field of condensed-matter and materials physics moves toward increasingly complex systems, techniques such as combinatorial chemistry are likely to make a home for themselves alongside more traditional techniques such as bulk crystal growth or physical and chemical vapor deposition. For this promise to be realized, however, new tools need to be developed that can analyze and sort the large number of samples that are produced by this powerful technique.

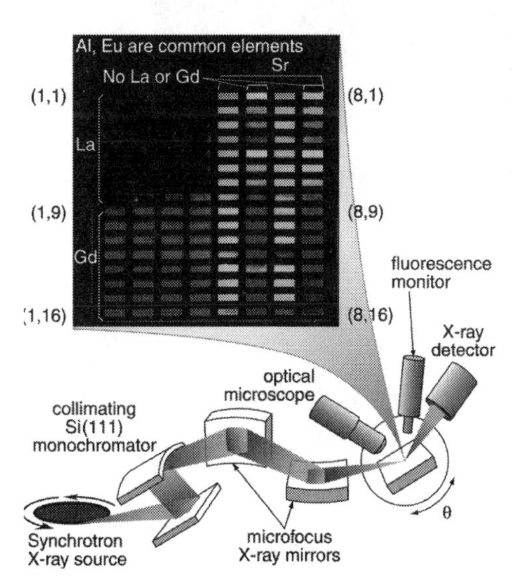

FIGURE 2.9.1 An array of different combinations of phosphors being screened for brightness in ultraviolet light. [Reprinted with permission from E.D. Isaacs, M.A. Marcus, G. Aeppli, X.-D. Xiang, X.-D. Sun, P. Schultz, H.-K. Kao, G.S. Cargill III, and R. Haushalter, "Syncrotron x-ray microbeam diagnostics of combinatorial synthesis," *Applied Physics Letters* **73**, 1820 (1998). Copyright © 1998 American Institute of Physics.]

BOX 2.10 Polymers Enable Porous Inorganic Materials Synthesis to Order

Block copolymers self-assemble into ordered nanostructures consisting of three-dimensional arrays of spheres, cylinders, lamellae, and even bicontinuous domains depending on the lengths of the blocks. Polymer physicists have achieved a remarkably detailed understanding of the interplay between chain architecture and thermodynamics that leads to these nanostructures. As useful as this insight has been for all-polymeric materials, it now promises to be equally important for the synthesis of inorganics. A long-standing need has been for porous ceramics that have well-defined but large pore sizes (>5 nm). These can now be achieved using block copolymers as a template. A copolymer with hydrophilic and hydrophobic blocks is ordered into a nanostructure in which the hydrophobic block forms aligned cylinders. A ceramic precursor is absorbed preferentially into the hydrophilic matrix surrounding the cylinders and allowed to condense into a robust inorganic oxide network. The block copolymer can be extracted, leaving a ceramic matrix surrounding ordered uniform cylinderical pores, as shown in Figure 2.10.1. Pore size can be controlled between 5 and 30 nm simply by changing the length of the copolymer, leaving the ratio of blocks the same.

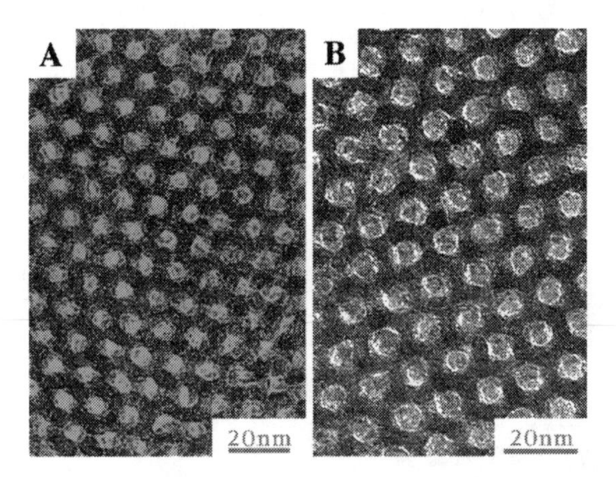

FIGURE 2.10.1 Transmission-electron microscope micrographs of mesoporous silica with pores sizes 6 and 9 nm. [Reprinted with permission from D. Zhao, D. Feng, Q. Huo, N. Melosh, G.H. Fredrickson, B.F. Chmelka, and G.D. Stucky, "Triblock copolymer syntheses of mesoporous silica with periodic 50 to 300 angstrom pores," *Science* **279**, 550 (1998). Copyright © 1998 American Association for the Advancement of Science.]

Synthesis and Processing: Control

To make the increasingly complex materials and structures of the coming decades, tremendous advances in processing will be required. One can look forward to the day when arbitrarily complex materials and material combinations can be made with the same level of control as is possible for semiconductors today. One of the first classes of material that is likely to see the benefits of improved processing is the complex oxide family, including high-temperature superconductors. The complexity of these materials, however, pales in comparison with some of the other structures involving vastly different classes of material such as biological and inorganic materials. Although demonstrations of increasingly complex structures designed on the molecular level may be made using scanning-probe techniques, fabricating structures that can be studied intensively will require faster techniques that can make multiple samples. This almost certainly calls for a dramatic increase in our understanding of and ability to use self-assembly and biomimetic techniques to produce and process materials.

In the past decade there has been impressive progress in the understanding and control of defects—what they are, where they come from, and how to eliminate them or control their placement when they serve a defined purpose—in some materials systems, especially semiconductors. For other materials to reach the same level of perfection and processing control, the same level of understanding will be required.

Nanoscale fabrication and processing, wherein molecular chemistry and condensed-matter physics merge, will be key to achieving the level of control that will be needed to realize many of the exciting possibilities posed in this report.

Physics: Understanding

The materials and structures on the horizon offer rich possibilities for condensed-matter and materials physicists. More perfect materials will enable us to move toward developing a full understanding of the relationship between the detailed structure of a material and its properties. The ability to control defects will enable them to be studied themselves—how they interact with the material they inhabit and even how judiciously assembled collections of them interact with one another and with different defects. Advances of the past decade in probing surfaces and interfaces on the atomic scale offer the possibility that a full understanding of the initial stages of growth in systems more complex than silicon may one day truly be possible. Control of the structure of materials on various length scales simultaneously offers the opportunity to look for effects that result from the interplay of structure on these different length scales.

Technology: Relevance

Advances in new materials and structures have dramatically improved our lives in the past, and there is every reason to expect that new advances will have comparably great impact in the years to come. For this to happen, sustained research will be needed over many years. This research will need to have a balance between fundamental investigations into the physical mechanisms at play and research and engineering aimed at investigating the numerous questions that must be answered before a material can enter the technological mainstream: What can the material be used for? Is there a potential market of sufficient size to pay for the needed research and development? Is the advance so revolutionary, with improvements in customer capability so great, that it can found a new industry? If the improvement is in an area already occupied by an existing technology with significant infrastructure, can the material be integrated with the existing technology? And if so, is the improvement worth the development cost?

Just as revolutionary advances in new materials and processes enabled the transistor, the optical fiber, the solid-state laser, and many other technologies that have improved our lives and strengthened the economy, new developments in materials and structures hold out the promise of revolutionary breakthroughs in the twenty-first century.

Outstanding Scientific Questions

- Can we complement empiricism with predictability in our search for new materials and structures with desired properties? Can we predict the composition and structure of a new material, its properties, and how to synthesize it?
- Can we develop a full understanding of the initial stages of growth?
- Can we develop a full understanding of the relationship between the detailed structure of a material and its properties? Can we truly control defects?

Research Priorities

- Tailor materials at the molecular level.
- Use more complex combinations of materials: polymers, organic molecules, biological molecules, etc.
- Develop new tools to synthesize, visualize, characterize, and manipulate new materials and structures.
- Make increasingly complex materials and combinations with as much control as is currently possible in the making of semiconductors.
- Increase our understanding of and the ability to use self-assembly and biomimetic techniques to produce and process materials.
- Merge molecular chemistry and condensed-matter and materials physics to understand and control fabrication and processing on multiple length-scales.
- Integrate processing of new materials and structures with existing technologies.

3

Novel Quantum Phenomena

One might imagine that because condensed-matter and materials physics deals with known objects (atoms) interacting via well-defined and well-understood forces (the Coulomb interaction among charged particles) there are no surprises and no fundamental intellectual challenges left to be discovered. Nothing could be farther from the truth. Quantum mechanics is a strange business, and the quantum mechanics of large collections of atoms and molecules can be stranger still. It inevitably happens that when assembling a collection of atoms to form a material, the whole is greater than the sum of the parts in the sense that "emergent phenomena," such as spontaneously broken symmetries and quantum or classical phase transitions, often appear in large collections of atoms. For example, a set of widely spaced copper atoms has an energy gap for charge excitations and thus is an insulator. This is because the atoms are largely independent of each other and retain the discrete spectra of isolated atoms. Even at absolute zero, compressing the atoms into solid copper causes the electrons to "melt" into a new "liquid" phase that has no excitation gap and is an excellent electrical conductor. If the same experiment is carried out with aluminum atoms, very subtle differences in the atomic properties lead not to an ordinary metal but to a superconductor. There exist vast families of complex materials, with many atoms per unit cell, whose surprising properties are still extremely difficult to predict a priori. This is an especially important challenge for theoreticians in the coming decade.

The idea of emergent phenomena teaches us that, even though we understand and can relatively easily compute the properties of individual atoms, materials constructed from large collections of atoms will routinely surprise us with completely unexpected properties. Living matter and life itself are perhaps the most spectacular examples of emergent phenomenon; no matter how much we learn about individual atoms, life cannot be understood or explained in this purely reductionist manner. One of the biggest surprises of the last decade was high-temperature superconductivity. It is hard to imagine a less likely candidate for a superconductor than an insulating ceramic compound with properties similar to those of a china coffee cup. Yet when chemically doped to introduce charge carriers, such compounds not only superconduct, they do so at record high temperatures.

The characteristic energy scale for individual atoms is 1 to 10 electron volts (eV). However, as we look on larger length scales at collections of atoms, characteristic energies become smaller and smaller, and excitations become more and more collective. At low energies, the effective elementary degrees of freedom may be collective objects very different from individual electrons and atoms, and their effective interactions may be very different from the original "bare" Coulomb interactions. These collective effects are the source of the surprises that emerge.

It is instructive to compare this situation with that in high-energy elementary particle physics. There we know the effective degrees of freedom and their interactions at low energies—it is the world of atoms around us. The intellectual challenge is to understand degrees of freedom at shorter and shorter length scales and higher and higher energy scales. This is done by constructing high-energy particle accelerators to act as microscopes with ever greater magnification, or by studying extreme conditions in astrophysical systems and the early universe. This approach is just the reverse of what is done in condensed-matter physics, where we strive to understand collective effects at longer and longer length scales. The analog of the particle accelerator is the refrigerator, which lowers thermal energy scales and increases the distance over which particles suffer inelastic collisions. The analog of an extreme astrophysical system is a sample in a dilution refrigerator. The intellectual challenge is the same in the two fields: to find correct descriptions of the physics that work over a wide range of scales.

Fifty years ago understanding a novel quantum object known as a "hole" (see Box 3.1) led to the invention of the transistor. In the past decade there has been tremendous progress in the discovery and study of a variety of novel quantum phenomena. This chapter presents brief descriptions of a few examples drawn from superfluidity, superconductivity, Bose-Einstein condensation, quantum magnetism, and the quantum Hall effect. It cannot cover many other fascinating areas of development in the last decade, including significant advances in our understanding of quantum critical phenomena, non-Fermi liquids, metal-insulator and superconductor-insulator transitions in two dimensions, quantum chaos and the role of interactions, coherence, and disorder in mesoscopic systems.

There has been particularly significant progress in this last area, both technologically and theoretically. For example, electron "wave guides" have been constructed, and the quantization of their conductance in units of e^2/h has been

BOX 3.1 Exotic Quantum Objects in Today's Technology

Fifty years ago at the time of the invention of the transistor, the hot topic in condensed-matter physics was an exotic quantum object, the "hole," whose predicted existence was one of the great early triumphs of quantum mechanics. The ability to create and manipulate these "holes" is crucial to the operation of diodes, transistors, photocells, light-emitting diodes, solid-state lasers, and computer chips.

To understand the concept of the hole, consider the fact (illustrated schematically in Figure 3.1.1) that when atoms are assembled into a solid, the discrete quantum energy levels of the individual atoms smear out into bands of quantum levels. The Pauli exclusion principle tells us that each band state can hold no more than one electron. In a semiconductor, the highest occupied band (the valence or "bonding" band) is separated by a small but crucially important energy gap from the lowest unoccupied band (the conduction or "anti-bonding" band).

The Pauli exclusion principle has the important consequence that a filled band is inert. It is impossible to excite the system by moving an electron to a new state within a band that is already entirely filled up, so the only way to achieve the lowest energy excitation that can be made in a semiconductor is to lift an electron from the valence band across the gap to the conduction band. It is easy to visualize the electron in the conduction band as a particle that can move around, carry current and so forth. Paradoxically, quantum mechanics also teaches us that the absence of an electron in the otherwise-filled valence band should be viewed as a hole that behaves like a kind of anti-particle (much like the positron, which is the anti-particle of the electron in high-energy physics). Without the hole, the valence band is inert and carries no charge or current. The electron that was removed had negative charge and carried some particular current. Hence, we must assign the hole a positive charge and the opposite current. Without quantum mechanics guaranteeing that a filled band is inert, this assignment would not be meaningful.

Introduction of chemical dopants into semiconductors can produce an excess of electrons (n-type material) or an excess of holes (p-type material). Remarkable materials physics advances in purification and doping control of silicon now allow routine inexpensive construction of the special types of junctions between p- and n-type material that play such a crucial role in today's solid-state technology. So next time you turn on your computer, remember quantum mechanics is at work!

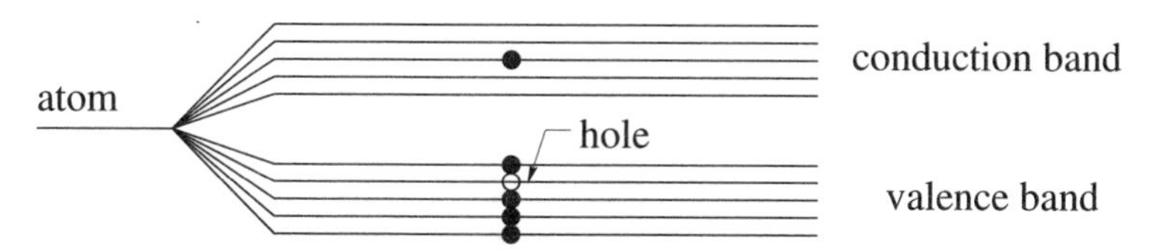

FIGURE 3.1.1 Energy bands in solids. [Reprinted with permission from S.M. Girvin, "Exotic quantum order in low-dimensional systems," *Solid State Communications* **107**, 623 (1998). Copyright © 1998 Elsevier Science.]

observed. It is also now possible, for example, to construct mesoscopic grains of metal large enough to be superconducting but small enough that significant changes in transport properties are observed depending on whether the number of electrons on the grain is even or odd. In addition, we now have observations of interference effects for electrons diffusing in a mesoscopic normal conductor bounded by segments of superconductors with variable relative phase.

One of the significant characteristics of condensed-matter and materials physics that is worth emphasizing is the close interaction between experimentalists and theorists. Unlike many other areas of physics, theorists often collaborate closely with experimental groups and they frequently jointly publish papers. The field owes its intellectual vitality to this close collaboration and to a continuous and exciting stream of quite unexpected experimental discoveries.

Newly uncovered electronic phenomena in complex and strongly interacting materials have challenged fundamental theoretical paradigms such as the Fermi liquid. In some cases, such as for the fractional quantum Hall effect, entirely new concepts (for example, fractionally charged quantum vortices) have been developed that enriched the foundations of the field and beautifully explained the new phenomenology. High-temperature superconductivity and heavy-fermion systems present a bewildering array of paradoxes that have spawned many new paradigms whose validity is still being sorted out. This challenge will require, and will lead to, fundamental experimental and theoretical advances in the next decade.

SUPERFLUIDITY AND SUPERCONDUCTIVITY

Superfluids and superconductors have the remarkable property of carrying matter or charge currents completely without friction. In helium-4, the atoms undergo Bose-Einstein condensation (see next section) and become superfluid a few degrees above absolute zero. In a superconductor, pairs of electrons join together to form an effective boson-like degree of freedom. In an ordinary low-temperature superconductor, these Cooper pairs of electrons have a diameter much larger than the spacing between the electrons. Hence, it is not usually appropriate to view the superconducting phase transition as Bose-Einstein condensation, though it is closely related.

The 1996 Nobel Prize in Physics was awarded to Lee, Osheroff, and Richardson for the discovery of superfluidity in helium-3 (see Table O.1). This isotope of helium is a fermion, and it is Cooper pairs of helium-3 atoms that just barely condense at exceedingly low temperatures. Unlike the electrons in an ordinary superconductor, which form pairs in a state of zero relative angular momentum, helium-3 atoms pair in a p-wave ($\ell = 1$) angular momentum state. This feature gives superfluid helium-3 many novel properties because the state of the system is determined not just by the complex phase of the condensate wave function, but also by the local orientation of the pair angular momentum vector.

The exotic pairing state of helium-3 is naturally connected with high-temperature superconductors for two independent reasons. First, it has recently been established via several ingenious experiments that the pairing state in high-temperature superconductors is d-wave ($\ell = 2$), rather than the usual s-wave ($\ell = 0$). Unlike the case of helium-3, however, the direction of the angular momentum is not free to change but is fixed by the underlying lattice. In fact, as illustrated in Figure 3.1, the d-wave is actually a standing wave with the angular positions of its antinodes parallel to the axes of the square copper oxide planes.

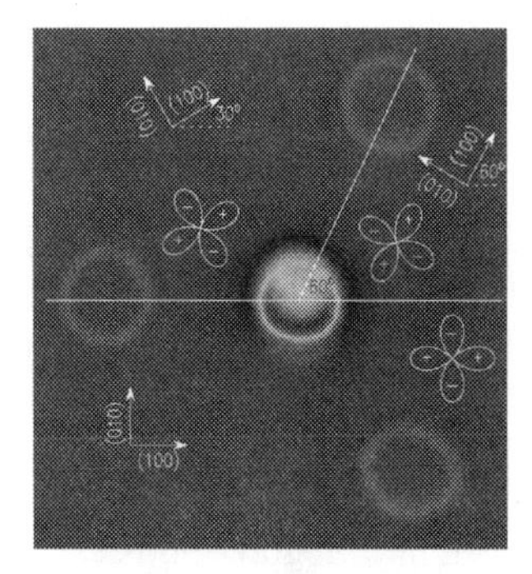

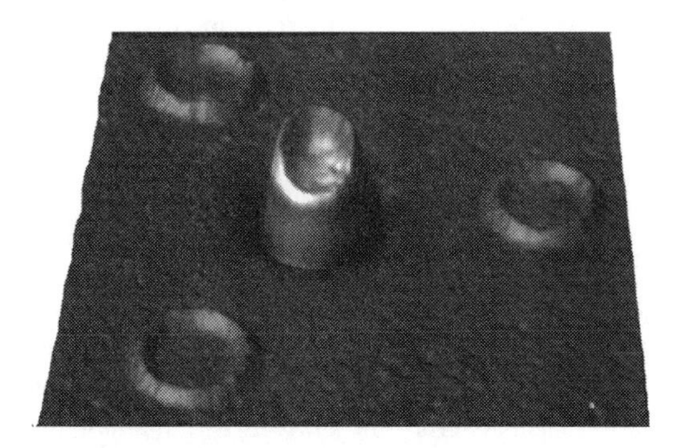

FIGURE 3.1 (Left) The long straight lines mark the boundaries of three substrate strontium titanate films with different crystallographic orientations. The four circles are rings of the high-temperature superconductor yttrium barium copper oxide, grown epitaxially on the substrate. The substrate orients the d_{x2-y2} Cooper pair wave function as indicated by the four-leaf clovers. At the point where the ring crosses from one substrate to the next it has a grain-boundary Josephson junction. Cooper pairs moving from the lower section to the upper left suffer an orientation change of less than 45 degrees and hence have a positive overlap with their new state. The same is true for tunneling from the upper left to the upper right. However, the orientation change from the upper right to the lower section exceeds 45 degrees and hence gives a negative overlap. A Cooper pair traveling around the central ring thus picks up a net minus sign (a phase shift of p), which results in destructive interference that raises the energy. The other three rings remain unfrustrated. The frustration of the central ring can be alleviated if a spontaneous current begins to circulate, which produces a half quantum of magnetic flux, because the Aharonov-Bohm effect would then introduce an additional ±p phase shift. This saves more than enough energy to pay for the cost of producing the current. Thus, if the superconductor is d-wave, the system is unstable to producing half a flux quantum trapped in the central ring. The bright spot in the central ring is an image based on a scanning-probe measurement, using superconducting quantum interference devices, of the local magnetic field. [Reprinted with permission from Barbara Levi, *Physics Today* **49**, 19-22 (1996). Copyright © 1996 American Institute of Physics.] (Right) The same data are shown but in a three-dimensional representation. The total integrated flux in the central ring is very close to half a flux quantum, providing proof that the system is frustrated and is almost certainly d-wave. [Reprinted with permission from *Nature* **373**, cover page (January 9, 1995). Copyright © 1995 Nature.]

One novel feature of d-wave superconductivity is that scattering at interfaces and nonmagnetic scattering by disorder can be pair-breaking. This effect has come to the fore recently because of evidence that has been obtained suggesting that certain crystal faces of high-T_c materials spontaneously break time-reversal symmetry by nucleating an additional pairing channel (for which surface scattering is not pair-breaking) to form a complex gap function.

The second similarity between the oxides and superfluid helium-3 is that both are strongly correlated systems. The oxides, however, are unusual in that the coherence length (size of a Cooper pair) is small, comparable to the spacing between the particles, which means that the transition bears a closer resemblance to Bose-Einstein condensation. Fluctuation effects are far more important than in an ordinary superconductor because mean field theory relies for its validity on there being a very large number of particles within the volume occupied by each Cooper pair.

Understanding strong correlation effects is an important challenge for both the superconducting state and the unusual normal state of these materials. At present there is no clear understanding of the mechanism for high-temperature superconductivity. We do not know, for example, whether the superconductivity occurs because of, or in spite of, antiferromagnetism. Indeed, it is not unreasonable to ask whether the phrase "*the* mechanism" is even meaningful in this case.

Superfluid helium-3 is a strongly correlated Fermi liquid that just barely manages to form Cooper pairs. It is not yet clear how to describe the corresponding strong correlations in high-temperature superconductors. The strange properties of the normal state—for example, the extremely linear temperature dependence of the resistivity—may mean that the standard theory of Fermi liquids cannot be used to describe them. If so, a totally new paradigm will have to be developed. Ideally, it will be possible to develop a simple picture that captures the essential physics and allows us to construct new materials with even higher critical temperatures, perhaps even higher than room temperature.

The small size of Cooper pairs in high-temperature superconductors has the benefit that, at least naively, it increases their tolerance for very strong magnetic fields. However, it may also be one of the many factors that limit the critical currents in these materials. Despite the technological problems caused by this latter effect, and despite the difficult materials problems, progress is being made toward practical applications of high-T_c materials.

Study of the short coherence length and associated strong fluctuations of the order parameter in high-T_c materials has led to some interesting and fundamental advances in statistical mechanics. The theoretical ideas that have been developed are directly relevant to technological problems presented by the strong suppression of the critical current by magnetic fields. Because of their short coherence length, high-T_c superconductors are strongly type-II. This means that an applied magnetic field penetrates the sample relatively uniformly, inducing a high density of vortex lines.

Application of a current produces a driving force that pushes the vortices sideways, leading to dissipation of energy. Naturally occurring or artificially introduced disorder produces random traps that tend to impede the motion of the vortices. One of the deep questions about this random statistical mechanical system is whether, in the limit of a weak driving force, the vortices are perfectly pinned. That is, is the linear response resistivity ever truly zero at any finite temperature? More colloquially, is a superconductor in a magnetic field really a superconductor?

For many years it was thought that the answer to this question was no. The rate of vortex "creep" was known to become extremely small at low temperatures, but it was believed to be thermally activated, giving a never-vanishing resistivity of the form $\rho \sim e^{-\varepsilon/k_BT}$. The physical picture behind this is that there is a characteristic energy barrier ε, associated with the random pinning, which is finite and can be overcome by thermal fluctuations.

In low-temperature superconductors this question, although important in principle, is nearly moot in practice because the barrier ε tends to be large relative to typical thermal energies at T_c. The pinning is thus extremely effective at all temperatures where superconductivity exists. In high-T_c materials, the pinning barrier ε is smaller (another side effect of the short coherence length) and T_c is much larger. Hence, magnetic fields induce high dissipation rates, which allows the temperature dependence to then be followed over a significant range below the zero-field T_c.

It is now understood that the correct answer to our question—Is a superconductor in a magnetic field really a superconductor?—is yes. As the temperature is lowered, the highly fluctuating "tangled spaghetti" of vortex lines begins to exhibit collective correlations over a length scale ξ that diverges at a characteristic "vortex glass" temperature T_g.

Associated with this diverging length is a divergence in the effective collective pinning barrier $\varepsilon \sim |T - T_g|^{-\theta}$. Below this temperature the barrier is infinite and cannot be overcome by (equilibrium) fluctuations. The linear-response resistivity is thus truly zero, not merely small.

The existence of a distinct vortex glass phase, as shown in Figure 3.2, makes clear the point that the upper critical field H_{c2} is a purely mean-field concept. In mean-field theory, the sample is "normal" for $H > H_{c2}$ and "superconducting" for $H < H_{c2}$. We now understand that H_{c2} is merely a crossover scale below which the order parameter rapidly becomes large. Because of the strong fluctuations in high-T_c materials, there is a "vortex liquid" regime in which the resistivity is finite even though the superconducting order parameter is large. Only when the system enters the frozen vortex glass phase does a true phase transition occur. Only when the vortices are frozen in place does the dissipation vanish.

Because of the "floppiness" of vortex lines in high-T_c materials (because of the short coherence length and the extremely weak coupling between copper oxide planes along the c axis), random point defects are not very effective at

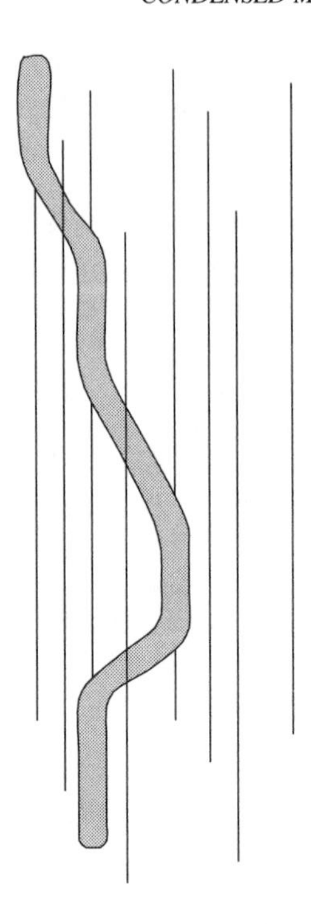

FIGURE 3.2 One can interpret classical superconductor vortex lines as world lines of quantum bosons moving in two space dimensions and one time dimension. In the Feynman-path-integral formulation of quantum statistical mechanics, the trajectories of the bosons must be followed over a time interval h/k_BT. Hence the thickness of the superconducting sample determines the inverse "temperature" of these fake quantum bosons. Notice that the columnar defects constitute a potential for the bosons that is random in space but constant in time. Thus we have the quantum mechanics of bosons in two dimensions in the presence of a static random potential. There has been a great deal of interest in the last decade in this "dirty boson" problem as a model for helium in porous media (or adsorbed on surfaces) and for modeling the superconductor-insulator transition in metallic films (viewing the Cooper pairs as bosons). There are two different phases at zero temperature: at low densities or strong disorder, the bosons are localized in an insulating "Bose glass" phase; at high densities or weaker disorder, the bosons are condensed into a superfluid state. We can now map these pictures back onto the vortex system: The "insulating Bose glass" phase is superconducting because the vortices are localized. The "superfluid" state of the bosons means that the vortices are freely moving (even though they are highly entangled). Thus the "vortex liquid" has dissipation and represents the nonsuperconducting state. This "Bose glass" generalization of the vortex glass idea can be pursued further to include analogs of the Mott-Hubbard Bose insulator, Mott variable range hopping, and boson tunneling between localized states. Furthermore, our knowledge of the dynamical structure factor for quantum bosons makes predictions for the static structure factor for the vortex fluctuations that have been confirmed by small-angle neutron scattering experiments.

pinning. The vortex glass phase does not exist in two dimensions, and it exists in three dimensions only in the absence of magnetic screening (so that collective effects are sufficiently strong).

The pinning efficiency for extended columnar defects is much better. These can be constructed using the linear damage tracks produced by fast-moving heavy ions from an accelerator. The statistical mechanics of fluctuating vortices in this situation has an elegant interpretation by means of an analogy with the quantum mechanics of a Bose liquid (see Figure 3.2).

This quantum boson analogy clearly demonstrates the existence of a phase transition in which the vortices can become localized by columnar pins leading to a state with truly zero resistivity in linear response.

The relatively small size of the Cooper pairs in high-temperature superconductors puts the superconducting transition in a new regime, closer to the Bose-Einstein condensation limit.

The most extreme regime of Bose-Einstein condensation has recently been achieved with the creation of condensates in gases of alkali metal atoms held in atom traps and cooled to nanokelvin temperatures. These are analogous to helium-4 in the sense that the particles are bosonic, but in this case the gas is dilute—the spacing between particles is much larger than the scattering length—and, hence, represents the nearly ideal case of pure Bose-Einstein condensation.

BOSE-EINSTEIN CONDENSATION IN ATOM TRAPS

Tremendous excitement has been generated by the recent success of the atomic physics community in creating Bose-Einstein condensates (BECs) in dilute gases of alkali atoms. A BEC is a coherent state of matter (see Figure 3.3) in which a finite fraction of the particles occupy the same quantum state. The indistinguishability of the particles and the fact that below the critical temperature their thermal de Broglie wavelengths exceed their spacing are central to the phenomenon. Using a two-stage process of laser cooling followed by evaporative cooling to temperatures on the order of 100 nanokelvin, it is now possible to produce condensates containing a few million atoms in an atomic trap.

BECs have been important in condensed-matter and high-energy physics for many years. BECs were first observed in superfluid helium in 1938. More recently, evidence has been obtained for condensation of excitons in semiconductors. BEC is also believed to provide the mechanism for generating the masses of the hadronic elementary particles.

What is new here is the parameter range; atomic vapors represent an unprecedented low-density regime in which the scattering length (the effective size of the particles) is much smaller than the typical particle spacing. This opens up a new range of possible experiments in which one can control the number of particles, the density, and the atomic species, thereby giving the possibility of varying, and therefore studying, effective interaction strengths in ways that have

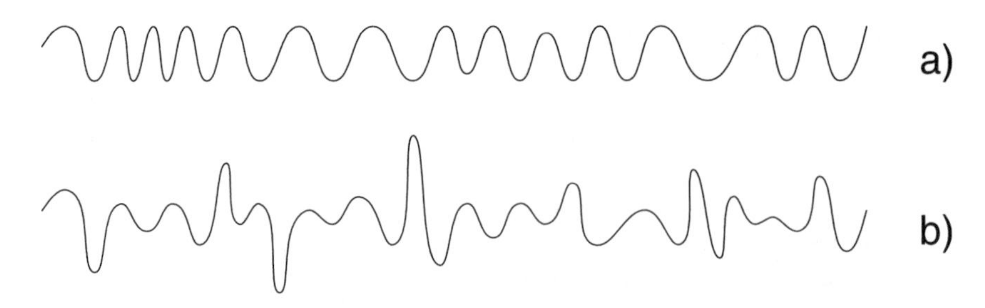

FIGURE 3.3 A laser produces coherent light, which we can think of as Bose condensate of photons. Understanding what this means can help us understand Bose-Einstein condensates (BECs) and atom lasers. Laser light wave oscillations are very similar to the waves produced by an ordinary radio transmitter, as shown in a. Because nothing fixes the phase of an oscillator, the phase undergoes a slow random walk, but the amplitude fluctuates very little. The random walk of the phase introduces a finite correlation time and hence a small but finite spectral bandwidth. A thermal source of the same bandwidth could be obtained by passing black-body radiation through a narrow filter. The resulting wave would look something like that shown in b. The large amplitude fluctuations are a result of interference among the different Fourier components, each of which has a random amplitude and phase. If the intensity and bandwidths are the same, the autocorrelation functions will be very similar. To distinguish coherent light from incoherent, we therefore have to look at fluctuations. For a perfectly coherent state there are no fluctuations: $\langle\psi^\dagger\psi^\dagger\psi\psi\rangle - \langle\psi^\dagger\psi\rangle^2 = 0$. The quantum mechanical interpretation of this is that the photons are Poisson distributed independently of each other, so that there is no bunching. For a thermal source, the fields are gaussian random variables and Wick's theorem tells us that $\langle\psi^\dagger\psi^\dagger\psi\psi\rangle = 2!\ \langle\psi^\dagger\psi\rangle^2$, so that there are very large intensity fluctuations. The quantum interpretation is that there is excess "noise" or "bunching" in the photon distribution. In the atom trap, the potential energy of the short-range interacting bosons is given by the probability of finding two bosons at the same place at the same time, $\langle V\rangle \propto \langle\psi^\dagger\psi^\dagger\psi\psi\rangle$. Experiments comparing the normal and condensed states at the same density have shown that, just as expected, $\langle V\rangle_{\mathrm{normal}}/\langle V\rangle_{\mathrm{condensed}} = 2!$. The atomic vapor is unstable to decay into bound atom pairs, which fall (literally) out of the trap. Because a third body is needed to carry off the binding energy, the decay rate obeys $\langle\psi^\dagger\psi^\dagger\psi^\dagger\psi\psi\psi\rangle_{\mathrm{normal}}/\langle\psi^\dagger\psi^\dagger\psi^\dagger\psi\psi\psi\rangle_{\mathrm{condensed}} = 3!$. Remarkably, this third-order coherence effect has also been observed experimentally. Thus an atom trap emitting a coherent beam of alkali atoms (a "boser") has all the coherence properties of a laser.

not been possible in traditional condensed-matter systems. In addition, some of the characteristic timescales are extremely slow (fractions of a second) opening up the possibility of studying macroscopic quantum coherence effects in a totally new time domain.

The other aspect that is new is the unprecedented ability to dynamically change the trap parameters and optically probe the detailed response of the condensate. It is now possible, for example, to measure such quantities as the potential energy stored in the condensate and the momentum distribution function. In addition, it is possible to suddenly remove a barrier between two condensates and directly see the quantum interference fringes that result from their overlap. In the future it will likely be possible to extend these results to multicomponent condensates and study the separate response of each component.

Atomic condensates have played a truly useful role in promoting cross-disciplinary communication and productive interactions among the condensed-matter, atomic-physics, and quantum-optics communities. The condensed-matter theory community is supplying expertise in two areas: many-body calculation techniques and experience with the study of collective effects. It turns out that in this low-density regime, straightforward and standard mean-field theory calculation methods appear to be quite accurate for low temperatures, so there appear to be few theoretical challenges in this regard (except for questions of metastability for systems with negative scattering lengths that have not yet been fully settled). On the other hand, there remain quite a few challenges in understanding collective effects. These include the mechanism for damping of collective modes at finite temperatures, two-fluid hydrodynamics, effects of the spin degrees of freedom, the details of the different time regimes in the dynamics of condensate formation, and how the systems carry angular momentum via vortices and multipole shape distortions. There are connections with models of nuclei here.

Fermi systems are also of considerable interest. Because Pauli exclusion limits the phase space available for scattering, the two-body collision rate drops rapidly as the temperature is lowered, and it is difficult to cool and equilibrate a Fermi system in isolation. (In a degenerate Fermi system the Fermi energy is much larger than the temperature. Evaporation carries away highly energetic particles, but this mostly results in lowering the Fermi energy, rather than cooling the system.) However, sympathetic cooling in Bose-Fermi mixtures is possible because the Bose-Fermi collision rate is not limited by the Pauli principle. The fermions cool by losing energy to the bosons, which are in turn cooled by the usual evaporative means. This opens up new possibilities similar to those studied in superfluid ^{3}He-^{4}He mixtures but now in a very different regime.

A profound physical problem that the atomic BECs seem well suited to address is that of the dynamics of a macroscopic quantum system approaching an equilibrium state with long-range phase coherence. This physics is important both in connection with the dynamics of the cooling process and for the development of atom lasers ("bosers," see Figure 3.3). Similar questions have been addressed in the condensed-matter literature—for example, in connection with the development of nematic order in liquid crystals quenched from high temperatures. The theory of such ordering kinetics is reasonably well developed. However, the experimental systems studied so far are all well described by a classical, overdamped dynamics; that is, their time evolution consists merely of a frictional descent into the nearest local energy minimum. The theoretical analyses have also all been for purely relaxational models. The dynamics of the atomic BEC systems is clearly not in this regime, as the unitary time evolution of Schrödinger's equation is surely important in the development of macroscopic phase coherence. Experimental and theoretical studies examining these fundamental issues are beginning, and their rapid advancement offers exciting prospects for the future.

QUANTUM SPIN CHAINS AND LADDERS

The study of quantum magnetism has experienced a resurgence in the last decade with the synthesis of whole new families of organic and inorganic compounds containing spin-1/2 and spin-1 degrees of freedom. One very important consequence of the discovery of high-temperature superconductors has been progress in the synthesis of oxide compounds having spins arranged in two-dimensional planes, one-dimensional chains, quasi-one-dimensional ladders with both even and odd numbers of legs, and even two-dimensional arrays of intersecting one-dimensional chains. These new systems have fascinating properties and provide an excellent testing ground for theories of strongly correlated electronic systems. The fact that some of these ladders can be doped provides a testing ground for ideas about high-temperature superconductors in a simplified quasi-one-dimensional setting. There may also be even more direct connections because of the existence of quasi-one-dimensional domain wall structures in the so-called "striped phase" of nominally two-dimensional high-T_c materials. Spin-1/2 is the most quantum-mechanical spin length and had previously been available for study only in complicated organometallic compounds. In those materials, the very weak coupling between spins makes the interesting quantum effects only occur at extremely low temperatures.

Paradoxically, magnetism arises from electrostatic not magnetic forces. Magnetic dipole forces are weak and generally detrimental to the creation of magnetic alignment. Instead, it is the Coulomb interaction, combined with the Pauli principle, that can lead to spin alignment and, hence, to magnetism. The overall antisymmetry of the quantum wave function for a pair of electrons requires that the spatial wave function be symmetric for the spin singlet and antisymmetric for the spin triplet. If the spatial wave function is antisymmetric, it vanishes when two electrons are at the same location; thus the probability of close approach is reduced. This in turn reduces the mean Coulomb repulsion. However, this reduction in the potential energy comes at the expense of increased kinetic energy associated with the extra nodes in the spatial wave function. Roughly speaking, if the potential energy term dominates, the system will have a ferromagnetic ground state. If the kinetic energy dominates, the system will have an antiferromagnetic ground state.

Classically, it makes no formal difference to the statistical mechanics whether a local moment system is ferromagnetic or antiferromagnetic (as long as we consider a bipartite lattice so that the antiferromagnetism is not frustrated) or has a large or a small spin angular momentum (say, $S = 5/2$ or $S = 1/2$). A classical spin can point in any direction, and its dynamics can be understood in terms of a simple gyroscope precessing under the influence of torques as a result of interactions with its neighbors. This mutual precession leads to the existence of a continuum of spin waves with precession frequencies proportional to the square of the wave vector in ferromagnets and linear in the wave vector for antiferromagnets.

A quantum spin is a very strange and different beast. If its component in a particular direction is measured, it will always be found to be one of a discrete set of only $2S + 1$ values: $-S, -S + 1, \ldots, S - 1, S$. This discreteness suggests that quantum magnets could be very different, perhaps having a discrete excitation gap or other properties that depend in detail on the particular value of S.

At the same time, quantum mechanics has another novel feature that opposes this discreteness. Quantum mechanics allows for the possibility of uncertainty as to which state a spin is in. In fact, there exists a continuum of "coherent states" that are linear superpositions of the $2S + 1$ discrete basis states. This continuum of possibilities allows the quantum spin to act somewhat classically. Thus, for example, a spin-1/2 can be chosen to be in a linear combination of "up" and "down" that is pointing in some arbitrary transverse direction. This competition between discrete and continuum pictures, which in a sense is analogous to the usual wave-particle duality of quantum mechanics, leads to very rich physics in quantum magnets.

Many spin systems can, to a good approximation, be represented by a quantum Heisenberg model. In spatial dimensions greater than 1, it turns out that for this model, the particular length and discrete nature of the spin is not very crucial, but there is still a major distinction between ferromagnets and antiferromagnets. The ferromagnet has a purely "classical" ground state, in which the spins are fully aligned in some arbitrary direction. The low-lying excited states are spin waves carrying $S = 1$ and behaving approximately like free bosons whose energy increases quadratically with momentum. The antiferromagnet (whose order parameter is the staggered magnetization) is somewhat different: quantum fluctuations reduce the order parameter below the classical value even at zero temperature; hence, the exact ground state is not known. The low-lying excitations are nevertheless spin waves carrying $S = 1$ and having an energy linear in the momentum, just as the classical precession frequency is linear rather than quadratic in the wave vector.

These simple pictures break down completely for the case of one-dimensional antiferromagnets, for which quantum fluctuations are especially severe. In particular, there is no long-range order in the staggered magnetization even at zero temperature (the spin-spin correlation functions fall off algebraically at zero temperature and exponentially at finite temperature). For $S = 1/2$, it turns out that the naive spin wave picture fails because the $S = 1$ bosonic spin wave "fractionalizes" into

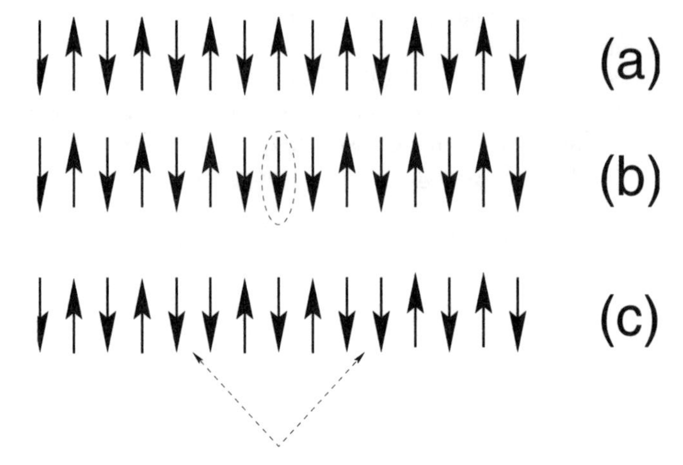

FIGURE 3.4 Spin fractionalization process in which a $\Delta S = 1$ spin flip in a spin chain breaks up into two domain walls each carrying spin-1/2. (a) Perfectly ordered antiferromagnetic configuration for a spin-1/2 chain. (b) The circled spin has been flipped. (c) Adjacent pairs of spins mutually flip producing a pair of domain walls (indicated by the dashed lines) separating perfectly ordered antiferromagnetic configurations. These domain walls ("spinons") are unconfined and free to move independently.

two independent $S = 1/2$ objects known as "spinons." A graphic depiction of the fractionalization process is shown in Figure 3.4. The first row shows a perfect antiferromagnetically ordered array of spins. In the second row, angular momentum $S = 1$ has been added to the array by flipping a single spin in the center. The third row shows the time-evolved state obtained by an angular momentum conserving mutual spin flip of the pair of spins just to the left of the center spin and the pair just to the right. We now see that the original flipped spin has broken up into a pair of "domain walls," which can separate indefinitely because the interior region between the walls is still perfectly ordered antiferromagnetically. It is possible to show that each of these domain-wall defects acts like a quantum particle carrying half of the $S = 1$ angular momentum of the original spin flip.

An immediate experimental consequence of the spin-wave fractionalization is that there is much more phase space available for spin excitations. An ordinary spin wave has a definite energy associated with each momentum. A pair of spinons, however, can be in many different individual momentum states for a given total momentum. This results in a band of possible energies, rather than a single energy, for a given momentum (see Figure 3.5).

The picture we have just described for one-dimensional spin-1/2 chains applies *mutatis mutandis* to the case of all odd-half-integer spins (1/2, 3/2, 5/2, . . .). However, the case of integer spin values is radically and surprisingly different. It turns out that integer spin chains described by the Heisenberg model have an

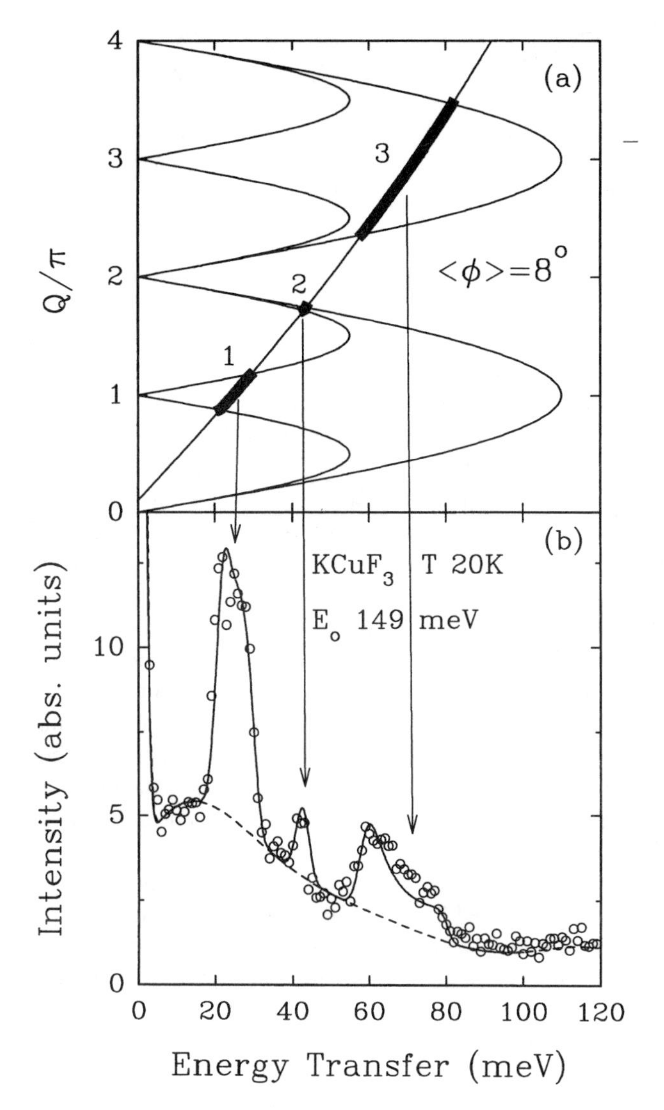

FIGURE 3.5 Spinon production observed by spin-flip inelastic neutron scattering. There is much more phase space available for spinons than for ordinary spin waves. The curve with period π in the upper panel is the single spinon dispersion curve. The curve with larger amplitude and period 2π is the upper bound on the range of allowed energies for a pair of spinons of total momentum Q. The parabolic curve in the upper panel is the kinematically allowed momentum and energy transfer to the scattered neutrons. The lower panel shows large peaks in the cross section at energies correctly predicted by the picture in which a single flipped spin decays into a pair of independent spinon excitations. Ordinary spin wave theory would predict zero intensity in the first and third peaks. [*Physical Review Letters* **70**, 4003 (1993).]

$+ 0 0 - 0 + 0 0 0 - + 0 -$

FIGURE 3.6 Typical spin configuration in the ground state of the Affleck-Kennedy-Lieb-Tasaki (AKLT) model. The symbols refer to the z component of the spin at each site. Notice that, if the zeros are ignored, the state has perfect antiferromagnetic order.

apparently featureless "spin liquid" ground state with an excitation gap and spin correlations that decay exponentially with distance even at zero temperature. The spin degrees of freedom seem to have disappeared because of some "confinement" mechanism.

The origin of the excitation gap is a novel "hidden" topological order not visible in the ordinary spin-spin correlation function. This order is best understood by studying the Affleck-Kennedy-Lieb-Tasaki (AKLT) model, which is closely related to the Heisenberg model but whose ground state is more readily soluble. Figure 3.6 shows a typical configuration (in the S^z basis) of the spins in the ground state of the AKLT model. We see that an up spin can be followed by an arbitrary number of zeros ("sideways spins") but then *must* be followed by a down spin. That is, if we removed all the zeros, there would be perfect antiferromagnetic order. This novel order is completely invisible to the ordinary, experimentally measured, spin correlation function and can only be detected theoretically using a nonlocal "string order" correlation function that includes a factor of –1 for each of the nonzero spins within the string of spins connecting two sites. The most obvious experimental manifestation of this hidden topological order is that it costs a finite amount of energy to break it; hence, the system has a spin excitation gap.

The origin of the excitation gap can also be understood by examining a different graphic of the AKLT ground state shown in Figure 3.7a. Each $S = 1$ spin on a site is visualized as being made up of a pair of spin-1/2 particles. These spins are formed into singlet bonds with their neighbors to create a "valence bond solid" as shown. Enforcing the rule that the state be symmetric under interchange of the pair of spins on a site guarantees that they will form a triplet and correctly represent the $S = 1$ on that site (see Figure 3.8).

Just as in the case of the spin-1/2 chain, it is possible to split a single spin flip with $\Delta S = 1$ into a pair of spinon excitations each carrying $S = 1/2$. This is illustrated in Figure 3.7b. Notice now, however, that in order to avoid generating more unpaired spins, the two sites containing the unpaired $S = 1/2$ spins are connected by a "string" of alternating double bonds and missing bonds. There is a finite "string tension," meaning there is a finite energy cost per unit length to produce this string. Thus, in contrast to the $S = 1/2$ chains, the spinons are confined (much as quarks are) and the resulting excitation ("meson") has a finite minimum energy cost. It can be shown that this excitation breaks the topological order discussed above. The situation for larger integer spins is similar, but the

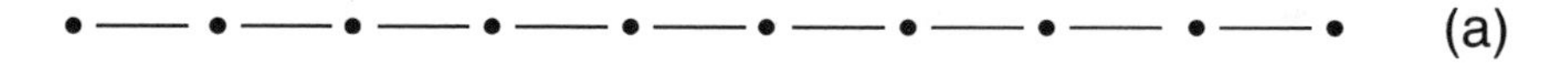

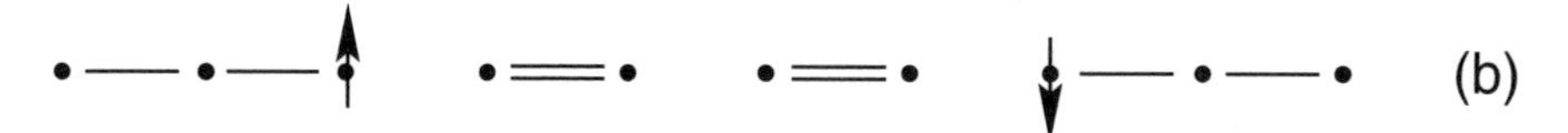

FIGURE 3.7 (a) Valence bond solid picture of the same Affleck-Kennedy-Lieb-Tasaki (AKLT) model ground state shown in Figure 3.6. The dots represent a site that contains a single spin $S = 1$. This is viewed as being made up of a pair of spin-1/2 particles, each of which forms a singlet bond (solid line) with one of its neighbors. (b) An excited state of the AKLT model. Because sites are viewed as containing two spin-1/2 particles, there must be either two bonds or one bond and an unpaired spin on each site. Thus there are two spin-1/2 particles liberated here. The string of alternating double and zero bonds costs a finite energy per unit length. This "string tension" confines the spin-1/2 particles together much as quarks are confined.

gap must become exponentially small in order to match onto the gapless classical limit at $S = \infty$.

A great deal of theoretical progress has been made in understanding the role of disorder in one-dimensional quantum spin chains for both $S = 1/2$ and $S = 1$. The $S = 1/2$ system develops into a random singlet phase in which there are strong singlet bonds over short distances and arbitrarily weak bonds over arbitrarily long distances. The $S = 1$ chain is quite different because it is initially gapped and, therefore, stable against weak disorder. For moderately strong disorder, the gap is destroyed, but the topological order remains in a "Griffiths" phase. Thus, paradoxically, the spins that disappeared at low energies in the clean system can be made to reappear by the addition of strong *nonmagnetic* impurity disorder. This is illustrated in Figure 3.9, which shows a segment of an $S = 1$ chain cut off from the rest of the chain by a nonmagnetic impurity at each end. The disruption of the valence bond solid ground state liberates a nearly free spin-1/2 at each end of the segment. This is manifested experimentally in the magnetic susceptibility, which becomes algebraic rather than exponential at low temperatures.

There are many open questions still to be addressed. In two and three dimensions there are a rich variety of highly frustrated lattices such as the kagome for which we still lack a complete understanding of the low-energy physics. Debate continues as to whether high-temperature superconductivity occurs because of, or in spite of, antiferromagnetism in the insulating parent compounds. Mixtures of itinerant electrons and local moments occur in heavy fermion, Kondo lattice, and disordered systems near the metal-insulator transition. These continue to be dauntingly complex theoretical and experimental challenges. In addition, the entirely new classes of oxide systems now being synthesized will produce fascinating new realizations of ladders, chains, and planes of spins, which will doubtless raise new theoretical and experimental challenges.

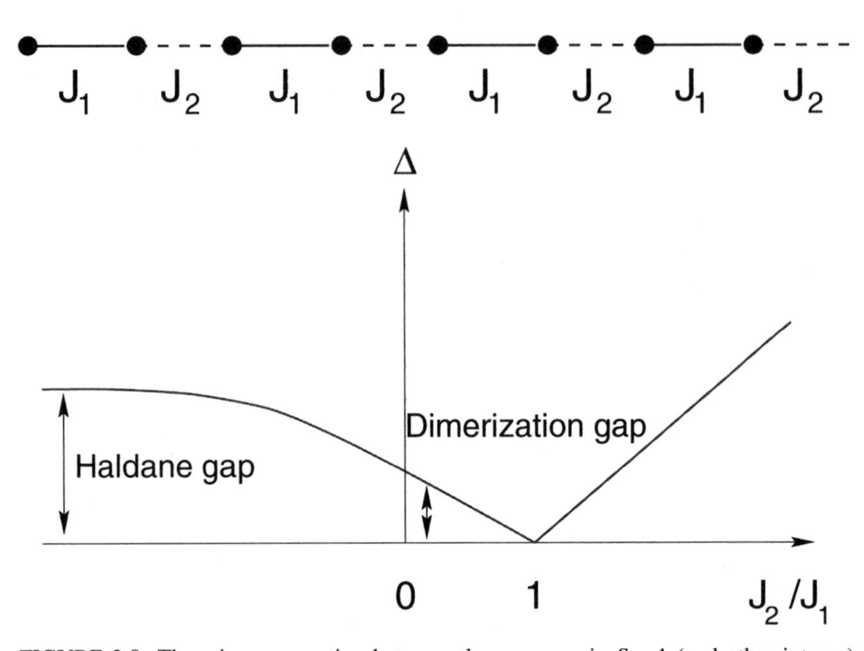

FIGURE 3.8 There is a connection between the gap seen in S = 1 (and other integer) chains and the gap seen in dimerized systems. (Upper) Consider the S = 1/2 chain with alternating weak and strong bonds. In the limit $J_2 = 0$, the exact ground state is a "valence bond solid" (VBS) of singlets on the J_1 bonds. For J_2 nonzero, the bonds begin to fluctuate, but the VBS state still captures the essential physics, and the gap survives throughout the region $J_2 \neq J_1$. (Lower) The gap persists even as the strength of the weak bond passes through zero and changes sign to become ferromagnetic. As these bonds become infinitely ferromagnetic, the associated pairs of spins become locked into triplet states, and one recovers the S = 1 spin chain description. Because the gap never closes during this process, the system does not undergo any phase transition as the dimerization is varied adiabatically. It follows that the dimer system has the same type of topological order, measured by the "string" correlation function, as the S = 1 system and that the Haldane gap in an S = 1 system is a special limit of the dimerization gap in an S = 1/2 system. [Reprinted with permission from S.M. Girvin, "Exotic quantum order in low-dimensional systems," *Solid State Communications* **107**, 623 (1998). Copyright © 1998 Elsevier Science.]

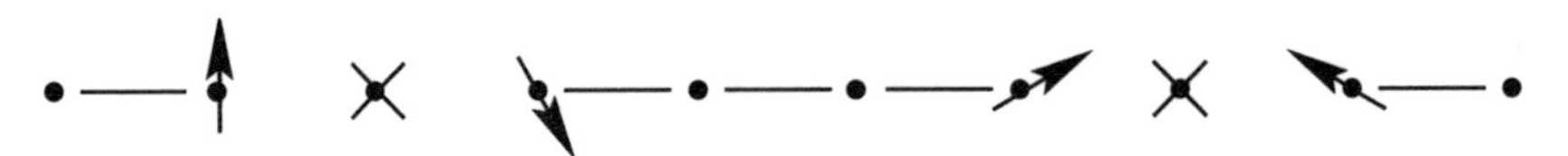

FIGURE 3.9 Nonmagnetic impurities (denoted by crosses) liberate spin-1/2 degrees of freedom in a gapped spin-1 chain. This follows from the fact that every site must have either two bonds or one bond and a free spin. These liberated spins are only weakly coupled to each other and hence produce low-energy excitations inside the Haldane gap. They are experimentally observable via their Curie-like contribution to the susceptibility.

THE QUANTUM HALL EFFECT

The various quantum Hall effects are arguably among the most remarkable many-body phenomena discovered in the second half of the twentieth century, comparable in intellectual import to superconductivity and superfluidity. They are an extremely rich set of phenomena with deep and truly fundamental theoretical implications. The integer effect, whose discovery was awarded the 1985 Nobel Prize (see Table O.1), has revolutionized our understanding of localization and transport in high magnetic fields. The fractional effect, whose discovery and theoretical explanation was awarded the 1998 Nobel Prize (see Table O.1), has yielded new paradigms of fractional charge, spin, and statistics as well as unprecedented order parameters. There are beautiful connections with a variety of different topological and conformal field theories studied as formal models in particle theory, each here made manifest by the twist of an experimental knob. Where else but in condensed-matter physics can an experimentalist change the number of flavors of relativistic chiral fermions, or set the Chern-Simons angle by hand?

Some of the themes discussed above for quantum spin chains reappear here. The gapped-spin liquid with hidden topological order is replaced by a gapped-charge liquid, which also has a novel form of hidden topological order. The concepts of quantum number fractionalization and confinement and deconfinement of quasiparticle excitations also reappear.

Because of the recent technological advances in molecular beam epitaxy and the fabrication of artificial structures, the field continues to advance with new discoveries, even well into the second decade of its existence. Experiments in the field were limited for many years to simple transport measurements that indirectly determine charge gaps. However, recent advances have led to many successful optical, acoustic, microwave, specific heat, and nuclear magnetic resonance (NMR) probes that continue to advance our knowledge as well as raise intriguing new questions.

The quantum Hall effect takes place in a two-dimensional electron gas subjected to a high magnetic field. In essence, it is a result of commensuration between the number of electrons, N, and the number of flux quanta in the applied magnetic field, N_Φ. The electrons undergo a series of condensations into new states with highly nontrivial properties whenever the filling factor $\nu = N/N_\Phi$ takes on simple rational values. The original experimental manifestation of this effect was the observation of an energy gap yielding dissipationless transport (at zero temperature) much like in a superconductor. The Hall conductivity in this dissipationless state is universal, given by $\sigma_{xy} = \nu e^2/h$ independent of microscopic details. As a result of this, it is possible to make a high-precision determination of the fine-structure constant and to realize a highly reproducible quantum-mechanical unit of electrical resistance, now used by standards laboratories around the world to maintain the ohm.

The integer quantum Hall effect (IQHE) owes its origin to an excitation gap associated with the discrete kinetic energy levels (Landau levels) in a magnetic field. The fractional quantum Hall effect (FQHE) has its origins in very different physics of strong Coulomb correlations that produce a Mott-insulator-like excitation gap. In some ways, however, this gap is more like that in a superconductor because it is not tied to a periodic lattice potential. This permits uniform charge flow of the incompressible electron liquid and, hence, a quantized Hall conductivity.

The microscopic correlations leading to the excitation gap are captured in a revolutionary wave function, developed by R.B. Laughlin, that describes an incompressible quantum liquid. The charged quasiparticle excitations in this system are "anyons" carrying fractional statistics, intermediate between bosons and fermions, and fractional charge. This sharp fractional charge, which despite its bizarre nature has always been on solid theoretical ground, has recently been observed directly in two different ways. The first is an equilibrium thermodynamic measurement using an ultrasensitive electrometer built from quantum dots (see Figure 3.10). The second is a dynamical measurement using exquisitely sensitive detection of the shot noise for quasiparticles tunneling across a quantum Hall device.

Quantum mechanics allows for the possibility of fractional average charge in both a trivial way and a highly nontrivial way. As an example of the former, consider a system of three protons, arranged in an equilateral triangle, and one electron tunneling among their $1S$ atomic bound states. The electronic ground state is a symmetric linear superposition of the quantum amplitudes for being in each of the three different $1S$ orbitals. In this trivial case, the mean electron number for a given orbital is 1/3. This is a result of statistical fluctuations, however, because a measurement will yield electron number 0 two-thirds of the time and electron number 1 one-third of the time. These fluctuations occur on a very slow timescale and are associated with the fact that the electronic spectrum consists of three very nearly degenerate states corresponding to the different orthogonal combinations of the three atomic orbitals.

The $\nu = 1/3$ quantum Hall effect has charge-1/3 quasiparticles, but it is profoundly different from the trivial scenario just described. An electron added to a $\nu = 1/3$ system breaks up into three charge-1/3 quasiparticles. If the locations of the quasiparticles are pinned by, say, an impurity potential, the excitation gap still remains robust and *the resulting ground state is nondegenerate*. This means that a quasiparticle is not a place (like the proton above), where an extra electron spends one-third of its time. The lack of degeneracy implies that the location of the quasiparticle completely specifies the state of the system; that is, it implies that these are fundamental elementary particles with charge 1/3. Because there is a finite gap, this charge is a sharp quantum observable that does not fluctuate (for frequencies below the gap scale).

To understand this better, imagine that you are citizen of Flatland, living in

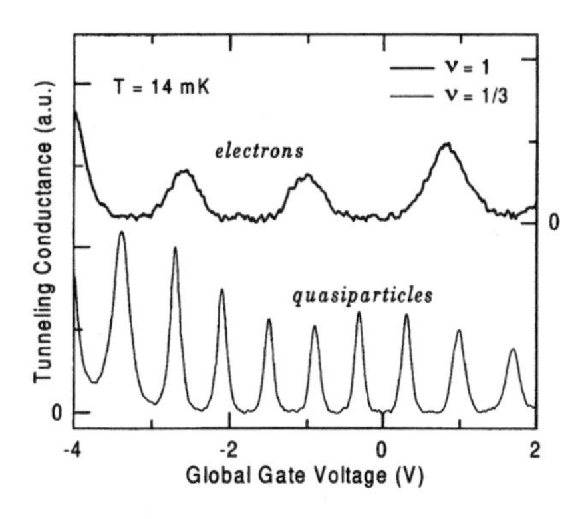

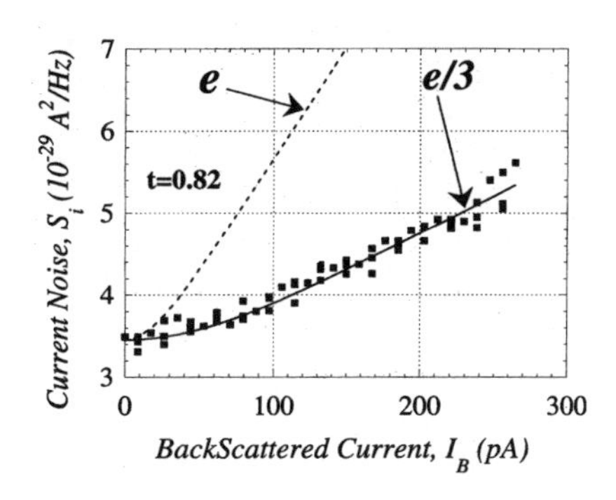

FIGURE 3.10 (Left) Variation of the conductance for tunneling through a quantum dot as a function of bias voltage on the dot. The oscillations indicate the discrete charging of the dot under quasi-thermodynamic equilibrium conditions. The period of the oscillations for the $\nu = 1/3$ quantum Hall plateau is three times smaller than for the $\nu = 1$ plateau, indicating that the quasiparticle charge is one-third that of an electron. (Right) Absolute shot noise for currents tunneling through a $\nu = 1/3$ Hall fluid. The deviation from linearity at low currents is the crossover from shot noise to thermal Nyquist noise. Both the intensity of the noise and the location of the crossover from shot noise to Nyquist noise are in quantitative agreement with the quasiparticle charge being $e^* = 1/3$. The dashed line shows the predicted shot noise if the current were carried by objects with charge *e* instead of *e*/3. [Right: Reprinted with permission from R. De-Picciotto, M. Reznikov, M. Heiblum, V. Umansky, G. Bunin, and D. Mahalu, "Direct observation of a fractional charge," *Nature* **389**, 163 (1997). Copyright © 1997 Nature.]

the cold world of a two-dimensional electron gas in a high magnetic field. Flatland cosmologists have theorized that the charged particles seen drifting around are topological defects left over from the Big Chill at the beginning of the universe. Flatland particle theorists decide that the apparently featureless vacuum in which everyone lives each day is actually a roiling sea, filled with strange but invisible objects that have precisely three times the charge of one of these quasiparticles. To study this possibility, a Flatland high-energy particle accelerator is constructed that can reach the unprecedented energy scale of 10 K. On smashing together three charged particles, it is found that they do indeed temporarily coalesce into an object with the bizarre property of having integer charge. It is decided to name this short-lived object "the electron."

The message here is that the charge of the quasiparticles is sharp to the observers so long as the gap energy scale is considered large. If the gap were 10 GeV instead of 10 K, we (living at room temperature) would have no trouble accepting the concept of fractional charge.

Composite Particles

It is a peculiarity of two dimensions that the $\nu = 1/3$ vacuum represented by the Laughlin wave function can be viewed in more than one way in terms of composite particles. One way is to make a singular gauge transformation that attaches three quanta of magnetic flux to each electron. This induces an Aharonov-Bohm phase of 3π when two particles are interchanged. The physics will therefore remain invariant if we change the particle statistics from fermion to boson to cancel this phase change. At $\nu = 1/3$ there are three flux quanta from the externally applied uniform magnetic field for each particle. Thus, if we make a mean-field approximation in which the flux quanta attached to the particles are smeared out into a uniform field, they will precisely cancel the external field, leaving a theory of composite bosons in zero (mean) magnetic field, as illustrated schematically in Figure 3.11. The condensate wave function of these bosons defines a hidden off-diagonal long-range order not visible in the ordinary correla-

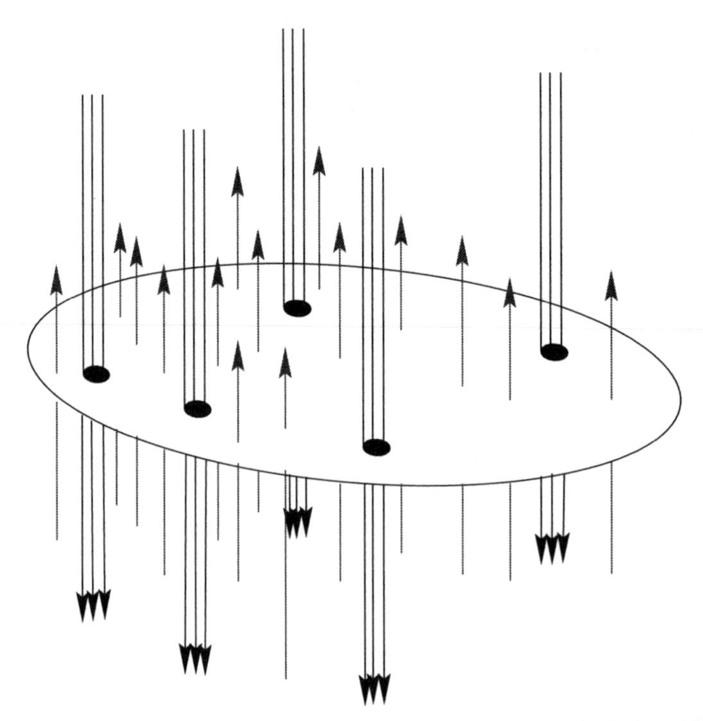

FIGURE 3.11 Illustration of the condensation of composite bosons in the $\nu = 1/3$ fractional quantum Hall effect. The three flux quanta attached to the electrons convert them into bosons moving in zero average field. Vortices in the condensate are the Laughlin quasiparticles carrying charge $\pm 1/3$.

tion functions of the original electron variables. (There are deep analogies here with the hidden string order in quantum spin chains discussed previously.)

The natural excitations in a bosonic condensate are (Goldstone mode) phonons and vortices. The analogs here ("magnetophonons") have recently been observed directly in Raman scattering. Vortices in two dimensions normally cost a logarithmically divergent amount of energy and are confined in neutral pairs at low temperatures. In the FQHE, however, something peculiar happens. Because each composite boson carries three flux quanta, binding one-third of a charge to the vortex is equivalent to binding one quantum of flux to the vortex. Just as in a type-II superconductor, this quantized flux screens out the currents at large distances and removes the divergence in the energy. Thus in this picture, the Laughlin quasiparticles are topological defects, and the same mechanism that gives them fractional charge also deconfines them. By analogy with type-II superconductors, the magnetophonon acquires a mass gap, which was predicted theoretically and has been observed experimentally.

An alternative picture of the $\nu = 1/3$ vacuum can be developed by attaching two rather than three flux quanta to each particle. These composite objects remain fermions and see a mean magnetic field of one flux quantum per particle, as illustrated schematically in Figure 3.12. Thus the $\nu = 1/3$ FQHE is mapped onto the IQHE at $\nu_{\text{eff}} = 1$. The Laughlin quasiparticles become additional composite fermions added to the next Landau level. In this formulation the off-diagonal long-range order remains hidden, but there are two significant advantages. First, accurate variational wave functions for various hierarchical quantum Hall states at different rational filling fractions can be written down explicitly and studied numerically with relative ease. Second, the special case of $\nu = 1/2$ is naturally described as composite fermions in zero mean magnetic field. The characteristic Fermi surface wave vector $2k_F$ of these composite fermions has been observed in surface acoustic wave attenuation experiments. If the mean field picture is taken literally, then moving slightly away from $\nu = 1/2$ puts the composite fermions in a weak magnetic field that should cause the quasiparticles to follow curved trajectories. Remarkably, this too has been observed experimentally.

It should be emphasized that experiments to date have all dealt with the kinematics of the composite fermions and the associated length scales, not their dynamics and the associated frequency scales. Hence, there is no unambiguous evidence for long-lived Fermi liquid-like quasiparticles above a sharply defined Fermi surface, as opposed to well-defined length scales at the Fermi surface. There is great theoretical interest currently in trying to understand the nature of fluctuations around the mean-field solution and their effect on the composite fermions. Considerable progress has been made, but many questions still remain to be definitively settled.

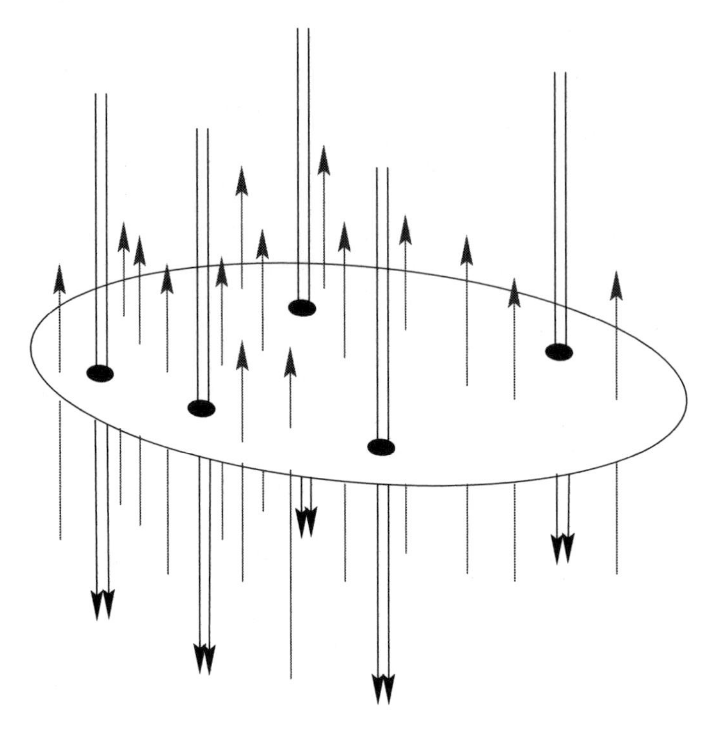

FIGURE 3.12 Illustration of the formation of composite fermions in the $\nu = 1/3$ fractional quantum Hall effect. The two flux quanta attached to the electrons turn them into fermions moving in an average field corresponding to the $\nu = 1$ integer quantum Hall effect. A Laughlin quasiparticle is represented by a composite fermion in the next Landau level.

Edge States

At low energies, the bulk of an FQHE system appears as a featureless vacuum with an excitation gap; however, very unusual gapless modes exist at the edges. These are shape distortions that preserve the area of the incompressible fluid. In a certain sense, the quantized-edge density fluctuations can be viewed as a gas of Laughlin quasiparticles liberated from the bulk gap.

Because these objects carry fractional charge and statistics in the bulk, they do not form an ordinary Fermi liquid at the edge. Instead, they constitute a nearly ideal realization of a chiral Luttinger liquid. The edge modes are chiral because they propagate in only a single direction, controlled by the direction of $E \times B$ drift in the edge-confinement potential. The density of states for tunneling an ordinary electron into a Luttinger liquid vanishes with a power-law singularity at low energies because of an orthogonality catastrophe that results from the fact that the electron must break up into fractionally charged quasiparticles. Recent progress

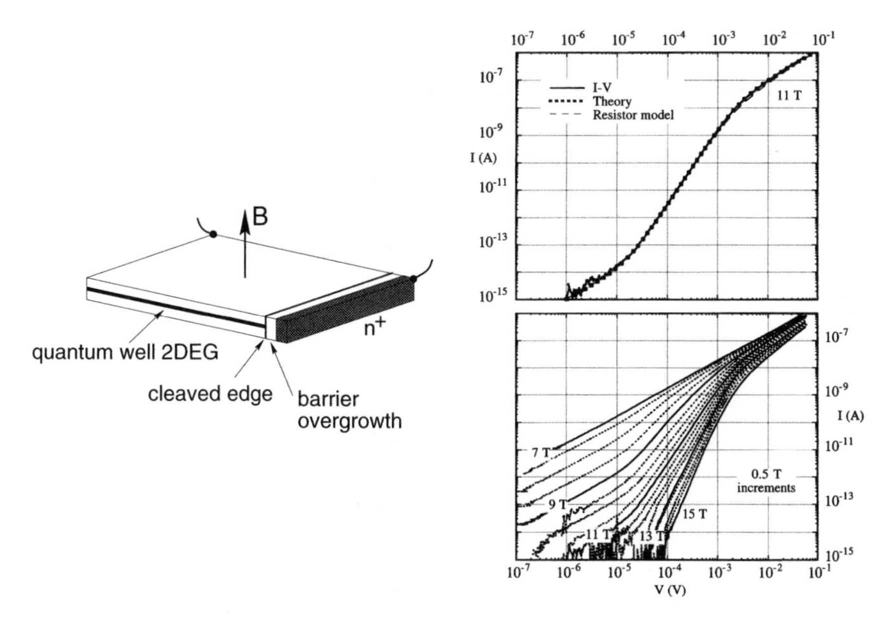

FIGURE 3.13 (Left) Schematic of a device produced by cleaved-edge overgrowth methods to measure tunneling into the edge of a two-dimensional electron gas. (Right) Nonlinear current-voltage response for tunneling an electron into a fractional quantum Hall effect edge state. Because the electron must break up into three fractionally charged quasiparticles, there is an orthogonality catastrophe that leads to a power-law density of states. The flattening at low currents is caused by the finite temperature. [*Physical Review Letters* **77**, 2538 (1996).]

in sample fabrication using cleaved-edge overgrowth techniques has made it possible to observe the resulting singularity experimentally in the tunneling current-voltage characteristic, as shown in Figure 3.13.

The various power-law correlations in a Luttinger liquid are characterized by a critical exponent g, which normally is continuously variable, depending on the details of the particle interactions. The value of g is therefore nonuniversal and can only be roughly estimated theoretically. According to current theory, one of the most important features of the chiral Luttinger liquids realized at the edge of FQHE systems is that g is universal, dependent only on the quantized value of the Hall conductivity in the bulk and independent of all details of the electron interactions. In particular, this makes the temperature and voltage dependence of the tunneling current have a power-law form that is universal and independent of all microscopic details. The theory works extremely well on the 1/3-quantized Hall plateau, but the unexpectedly smooth variation of the exponent with magnetic field away from the plateau shown in Figure 3.13 is not yet understood.

Magnetic Order of Spins and Pseudospins

At certain filling factors ($\nu = 1$ in particular), quantum Hall systems exhibit spontaneous magnetic order. For reasons peculiar to the band structure of the gallium arsenide host semiconductor, the external magnetic field couples exceptionally strongly to the orbital motion, giving a large Landau level splitting, and exceptionally weakly to the spin degrees of freedom, giving a very small Zeeman gap. The resulting low-energy spin degrees of freedom of this ferromagnet have some rather novel properties, which have recently begun to be probed by NMR, specific heat, and other measurements.

Because the lowest spin state of the lowest Landau is completely filled at $\nu = 1$, the only way to add charge is with reversed spin. However, because the exchange energy is large and prefers locally parallel spins, and because the Zeeman energy is small, it is cheaper to partially turn over several spins, forming the topological spin texture shown in Figure 3.14. Because this is an itinerant magnet with a quantized Hall conductivity, it turns out that this texture (called a skyrmion by analogy with the corresponding object in the Skyrme model of nuclear physics) accommodates precisely one extra unit of charge. NMR Knight shift and other measurements have confirmed the prediction that each charge added or removed from the $\nu = 1$ state flips over several spins—from 4 to 30, depending on the pressure. In the presence of skyrmions, the ferromagnetic order

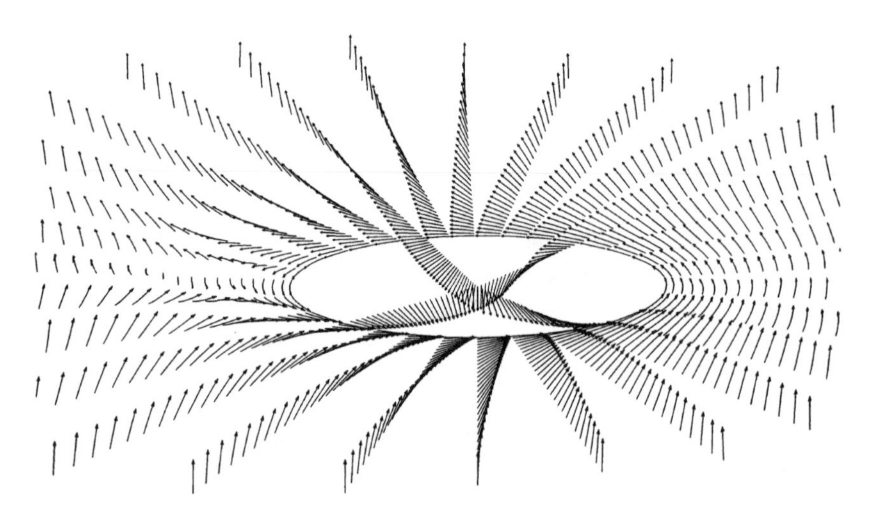

FIGURE 3.14 Skyrmion spin texture in a quantum Hall ferromagnet. Note that the spins are all up at infinity but down at the origin. At intermediate distances they have a vortex-like configuration. Because of the quantized Hall conductivity, these objects carry quantized charge. [Reprinted with permission from S.M. Girvin, "Exotic quantum order in low-dimensional systems," *Solid State Communications* **107**, 623 (1998). Copyright © 1998 Elsevier Science.]

is no longer collinear, leading to the possibility of additional low-energy spin wave modes that remain gapless even in the presence of the Zeeman field (somewhat analogous to an antiferromagnet). These low-frequency spin fluctuations have been indirectly observed through a dramatic enhancement of the nuclear spin relaxation rate $1/T_1$. In fact, under some conditions T_1 becomes so short that the nuclei come into thermal equilibrium with the lattice via interactions with the inversion layer electrons. This has recently been observed experimentally through an enormous enhancement of the specific heat by more than five orders of magnitude (see Figure 3.15).

Spin is not the only internal degree of freedom that can spontaneously order. There has been considerable recent progress experimentally in overcoming technical difficulties in the molecular beam epitaxy fabrication of high-quality multiple-well systems. It is now possible, for example, to make a pair of identical electron gases in quantum wells separated by a distance of about 100 Å—comparable to the electron spacing within a single quantum well. Under these conditions strong interlayer correlations can be expected. One of the peculiarities of quantum mechanics is that it is possible, even in the absence of tunneling between the layers, for the electrons to be in a coherent state in which their layer index is uncertain. To understand the implications of this, we can define a pseudospin that is up if the electron is in the first layer and down if it is in the second. Spontaneous interlayer coherence corresponds to spontaneous pseudospin magnetization lying in the x-y plane (corresponding to a coherent mixture of pseudospin up and down). If the total filling factor for the two layers is $\nu = 1$, then the Coulomb exchange energy will strongly favor this magnetic order, just as it does for real spins. This long-range transverse order has been observed experimentally through the strong response of the system to a weak magnetic field, applied in the plane of the electron gases, in the presence of weak tunneling between the layers. Very recent work indicates that a two-layer quantum Hall system at filling factor $\nu = 2$ may even allow for an antiferromagnetic or a canted spin phase, further demonstrating the complexity and richness of the magnetic phase diagrams of quantum Hall systems.

Another interesting aspect of two-layer systems is that, despite their extreme proximity, it is possible to make separate electrical contact to each layer and perform drag experiments in which current in one layer induces a voltage in the other as a result of Coulomb or phonon-mediated interactions.

The many-body physics of two-layer systems can also be found in wide single-well systems with the two (nearly degenerate) lowest electric sub-band states playing the role of the pseudospin degrees of freedom.

Stacking together many quantum wells gives an artificial three-dimensional structure analogous to certain organic Bechgaard salts in which the quantum Hall effect has been observed. Interest has been growing recently in the bulk and edge ("surface") states of such three-dimensional systems and in the nature of possible Anderson localization transitions.

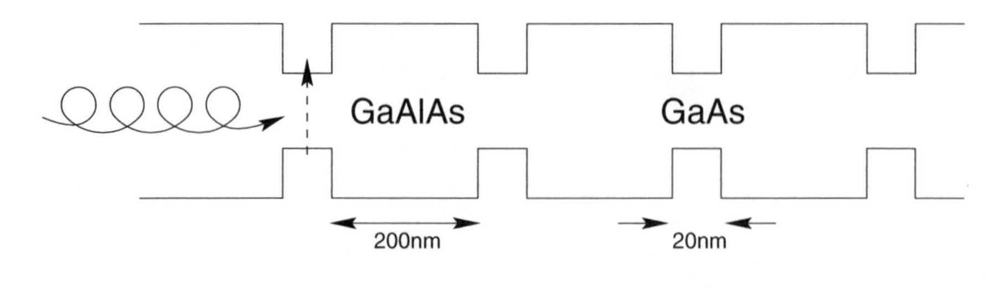

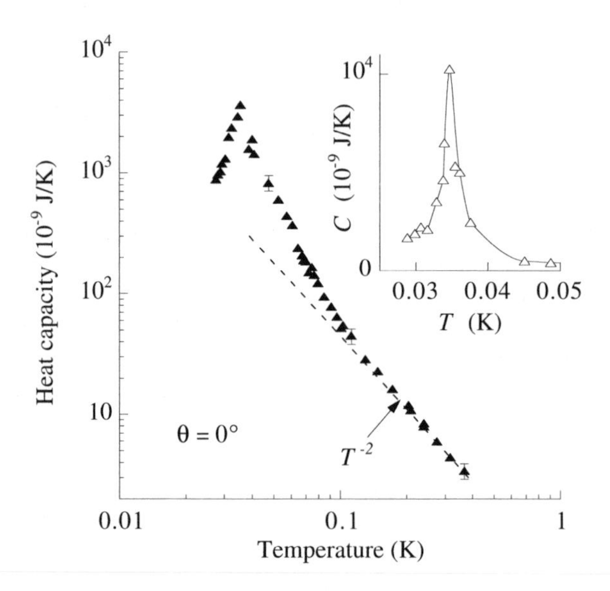

FIGURE 3.15 Even in a sample with many quantum wells, there are not enough nuclei in the 20-nm-thick wells to do an ordinary nuclear magnetic resonance (NMR) experiment. The technique of optically pumped NMR therefore had to be developed [Barrett et al., *Physical Review Letters* **74**, 5112 (1995); Tycko et al., *Science* **268**, 1460 (1995)]. (Upper) Circularly polarized light above the gallium arsenide (GaAs) band gap but below the band gap of gallium aluminum arsenide (GaAlAs) is absorbed only in the quantum wells. The angular momentum of the photons is transferred to the orbital motion of excited electrons and then, via the spin-orbit interaction, to the electron spins. Finally, the hyperfine interaction transfers the polarization to the nuclei in the quantum wells. This polarization enhancement allows the NMR signal to reach detectable levels for samples with as few as 10 wells. NMR experiments clearly demonstrate the existence of skyrmions—collective excitations with charge $\pm e$ and large-spin S. These objects enhance the nuclear relaxation rate, bringing T_1 down from many hours to about 20 seconds. (Lower) With the nuclei in thermal equilibrium, the specific heat is enhanced by many orders of magnitude, rising from picojoules to microjoules per kelvin. [*Physical Review Letters* **76**, 4584 (1996) and **79**, 1718 (1997).]

Although the quantum Hall effect is extremely well understood at this point, there remain some important mysteries yet to be resolved. In the IQHE, the main question continues to be the nature of the delocalization transition in which the Hall conductivity jumps from one quantized value to the next. The general scenario by which this happens is well understood; it is almost certain that there exists a critical point near the center of each Landau level at which the localization length diverges, and numerical estimates of the critical exponent agree well with experiment—in selected samples, at least. It is not understood at present, however, why generically there are often deviations from the expected scaling behavior (although the answer probably has to do with macroscopic inhomogeneities). One problem only just beginning to be addressed is the possible relevance of Coulomb interactions to the transition. In addition, despite valiant efforts, there does not yet exist a simple quantum field theory for this transition from which we can analytically compute the critical exponent. Finally, there remains an interesting set of puzzles about what happens at weak magnetic fields as Landau level mixing becomes strong and direct transitions apparently occur from quantum Hall effect states to insulating states.

In general, the ordering that produces the hierarchy of fractionally quantized states is very well understood. The most interesting remaining problem is to understand the physics of the $\nu = 1/2$ state, which in the composite fermion picture is a Fermi liquid-like state with zero mean magnetic field. The nature and effect of fluctuations around the mean field still need to be better understood.

The theory of quantum Hall edge states has successfully made detailed predictions of the observed temperature and voltage dependence of the tunneling current-voltage characteristics of the $\nu = 1/3$ fractional plateau edge. One unexpected experimental discovery, however, has been that the edge-tunneling density of states has a power law-form over a wide range of magnetic fields, not just on the plateaus. Furthermore, as shown in Figure 3.15, the exponent of the power law varies continuously and linearly with magnetic field and seems quite insensitive to whether the bulk of the sample is in a quantized Hall state or not. Current theory predicts that the exponent should be quantized, just as the Hall conductivity is, and should vary discontinuously with magnetic field.

Another significant question involves the sharp peak in the specific heat shown in Figure 3.15. The large linear region in the plot is explained quantitatively by the Schottky anomaly of the nuclei in the quantum wells. The extra peak is known to involve the additional nuclei in the barriers between the quantum wells, but the mechanism that gives rise to a sharp feature is not understood at present. It may be related in some way to the freezing of the skyrmions.

SUMMARY

The committee has detailed here just a small sampling of the wide variety of novel quantum effects that have been explored successfully in the past decade. Some of the major accomplishments are listed below:

- The discovery of high-temperature superconductivity and identification of the symmetry of the order parameter and elucidation of novel vortex fluctuations.
- Discovery of quantum phase transitions in two dimensional systems such as antiferromagnets in high-temperature superconductors, the superconductor-insulator phase transition in metal films, the metal-insulator transition in two-dimensional electron gases, and a wide variety of novel quantum Hall transitions.
- Observation of Bose-Einstein condensation and macroscopic coherence effects in atom traps.
- Synthesis of new families of spin-1 and spin-1/2 quantum magnets with a variety of novel geometries, progress in understanding dynamics in disordered one-dimensional magnets, and the development of giant and colossal magnetoresistance materials.
- Numerous new discoveries in quantum Hall physics including direct observation of the fractional charge of quantum Hall effect quasiparticles, composite particles, edge-state modes, Raman observation of the fractional quantum Hall effect neutral excitation gap, discovery of drag and spontaneous coherence in double-layer systems, and NMR and nuclear specific heat anomalies due to skyrmions.
- Mesoscopic device physics observations of conductance quantization in electron wave guides, parity effects in superconducting grains, and Andreev reflection in disordered systems.

FUTURE DIRECTIONS AND RESEARCH PRIORITIES

The novel discoveries of the last decade have raised many new questions and issues that require further exploration both theoretically and experimentally:

- New tools and paradigms for studying the interplay between interactions and disorder in quantum systems would shed light on phenomena like the recently discovered metal-insulator transition in two-dimensional electron gases.
- Carbon nanotubes are likely to present a great opportunity for study of novel electronic properties.
- Many of the remarkable quantum effects discovered in the last decade have been observable only at relatively low temperatures. Can quantum energy scales be boosted so that, for example, room-temperature mesoscopic and single-electron devices can be constructed?

- Can we develop practical theoretical tools to describe real-time dynamics and nonlinear response of interacting quantum systems as well as their electronic structure?
- Can we develop general principles to guide us in the study of modern materials that are vastly more complex than materials of the past?
- What is the physics of the normal and condensed states of high-temperature superconductors? Can we synthesize room-temperature superconductors and overcome practical materials difficulties?
- Finally, if the past is any guide, we will be faced with completely unexpected and surprising quantum phenomena as new materials are synthesized. What new techniques will have to be developed to deal with these surprises?

4

Nonequilibrium Physics

The term "nonequilibrium physics" means "the study of physical systems that are not in mechanical and thermal equilibrium with their surroundings." In some cases, these are systems spontaneously approaching equilibrium, such as a molten metallic alloy being allowed to cool and solidify. In other cases, they are systems changing their shapes or properties as forces are exerted on them, heat applied to them, or their states of repose are otherwise disturbed. Examples include flowing fluids driven by thermal or pressure gradients, solid materials deforming or breaking under the influence of external stresses, or quantum systems—atomic spins, perhaps—being driven by oscillating electromagnetic fields.

As a field of research, nonequilibrium physics is simultaneously very new and very old. The natural origins of complex patterns such as dendritic snowflakes or chaotic motions such as those of turbulent fluids have puzzled human beings since the dawn of history, but it is only in the last decade or so that we have begun to understand these phenomena in any depth. This field is also simultaneously very pure and very applied. The processes used to produce industrial materials—for example, casting alloys for jet engines or fabricating microscopically small features of computer chips—are all applications of nonequilibrium physics. In many cases, however, the fundamental research needed to predict and control these processes with the precision necessary for advanced technologies has yet to be done.

It is in this area of research, also, that condensed-matter and materials physics interact especially strongly with other scientific disciplines. Nonequilibrium physics is inherently interdisciplinary. It has involved engineers, mathematicians, and chemists in addition to physicists since it emerged as a recognized specialty about a generation ago. In recent years, the same nonequilibrium concepts being tested in the design of alloys are also being applied to galaxy formation in the cosmos, climatic changes on the Earth, and the growth of forms in biological systems.

The Brinkman report[1] was remarkably prescient in its discussion of nonequilibrium physics. Its authors recognized the growing importance of topics such as pattern formation, chaotic behavior, turbulence, and fractal geometries. Understandably, they missed today's emerging interests in friction, fracture, and granular materials and the speculations that some of these spatially extended chaotic phenomena might be exhibiting previously unanticipated collective behaviors. They also could not have predicted the invention of the scanning-probe microscopes and optical tweezers that are only just now beginning to open the world of biological phenomena to first-principles physical investigations.

In the last decade or so—the period since the Brinkman report—important progress has been made in many of these areas. We now understand in much more systematic ways how complex patterns emerge from simple ingredients in hydrodynamic, metallurgical, and chemical systems. Notable progress has been made in sorting out the mechanisms that control pattern formation, for example, in convecting liquid crystals, on the surfaces of vibrating fluids, in chemical reaction-diffusion systems, and in some biological phenomena such as cellular aggregation and membrane morphology. New understanding of spiral waves in active media has found application in the analysis of cardiac arrhythmia. We are beginning to understand how complex systems—for example, those in which fluid flow and chemical reactions are occurring simultaneously—sometimes become intrinsically self-organized, sometimes exhibit large critical fluctuations, sometimes become chaotic, and sometimes do all three of those things at the same time.

Nonequilibrium physics has grown into a major enterprise, one that cannot be described fully in this report. The committee has therefore selected a special set of topics as illustrative examples of the themes and issues to emphasize. The first of these topics is pattern formation and turbulence in fluid flow. The next two are in the areas of processing and performance of structural materials, specifically, microstructural pattern formation in solidification and a group of topics in solid mechanics: friction, fracture, granular materials, and polymers and adhesives. The final section includes some brief remarks about nonequilibrium phenomena in biology and in the quantum domain. Each of these topics, in different ways, illustrates the four themes listed below:

1. Much of the most important progress in recent years has consisted simply of recognizing that fundamental questions remain unanswered in the physics of

---

[1]National Research Council [W.F. Brinkman, study chair], *Physics Through the 1990s*, National Academy Press, Washington, D.C. (1986).

familiar situations. We regularly see turbulent fluids in the atmosphere and in our kitchen sinks, as well as in a wide variety of less obvious but equally important engineering applications. Snowflake-like solidification patterns are familiar to us and are even more common than most of us realize because they occur on microscopic scales within most structural metals and alloys. We deal with friction, fracture, and the mechanical performance of solid materials in essentially every action that we take in our daily lives. Yet, as we shall point out in this chapter, our basic scientific understanding of these phenomena remains incomplete and is inadequate for many practical purposes.

2. *Nonequilibrium systems that form patterns or exhibit complex, perhaps chaotic, behaviors are especially sensitive to small perturbations; thus their detailed behavior may be extremely difficult to predict or control.* As a general rule, complex patterns form in simple systems when they undergo instabilities. For example, smoothly flowing fluids become turbulent, or smooth solidification fronts sprout branches. In a technical sense, the term "instability" is synonymous with extreme sensitivity; a system is said to be "unstable" when it requires only an infinitesimally small perturbation to cause a qualitative change in its behavior. Turbulent eddies and dendritic branches are triggered by tiny fluctuations in their surroundings, and their subsequent development is also controlled sensitively by those fluctuations. It is in this sense that the properties of these systems are often extremely difficult to predict. One of the biggest challenges to condensed-matter and materials physics in the next decade may be to understand the intrinsic limitations, if any, on the quantitative scientific analysis of such phenomena.

3. *Progress in nonequilibrium physics will depend in many cases on our ability to bridge length scales and timescales.* For example, solidification is controlled both by mechanisms of molecular attachment at crystal surfaces and by heat diffusion across large ingots. Fracture is controlled both by breaking of molecular bonds at crack tips and by the macroscopic flow of elastic energy in solids. In general, we need to understand the interrelations between phenomena at scales ranging from atomic and molecular all the way up to fully macroscopic.

4. *Progress in nonequilibrium physics, as in all of condensed-matter and materials physics, will depend on our ability to bridge cultures.* We shall have to understand the importance of, and the impediments facing, our efforts to bring new science to bear on technology and to take advantage of new technologies to advance basic science.

PATTERN FORMATION AND TURBULENCE IN FLUID DYNAMICS

Nonequilibrium Phenomena in Fluids

When a fluid system is driven far from mechanical or thermodynamic equilibrium, it has a spontaneous tendency to form patterns and defects. Some examples are the spiral waves and ordered oscillations in reaction-diffusion systems; rolls, hexagons, and plumes in thermal convection of fluid layers; turbulent "spots" in the transition from the laminar state to turbulence in a boundary layer; and large-scale circulation patterns in the atmosphere. It is likely that, at least to some extent, living organisms are the result of this tendency for nonequilibrium systems to self-organize. This pattern-forming tendency owes its existence to nonlinearities that dominate the dynamics under conditions of strong forcing.

Although patterns found in nonequilibrium systems are varied in character and combine an astonishing labyrinth of order and disorder, they do share some common features. For example, the details of pattern formation are generally sensitive to small perturbations. In small systems, boundary conditions determine the positions and orientations of patterns. The nonlinearities in pattern-forming systems often produce intermittency; that is, such systems may undergo irregular, large excursions away from their most probable states.

It is only natural, then, to think that these common features might imply a deeper layer of truth, and that there might exist a general theory of nonequilibrium phenomena. Such a theory, if it exists, is still outside our reach; but we have made substantial progress in developing special theories for some particularly simple cases. Examples include liquid-crystal hydrodynamics, Rayleigh-Benard convection, Taylor-Couette flow, and fully developed turbulence in boundary layers. The main advantage of studying simple-fluid systems, as opposed to more complex-fluid systems such as those used in industrial processes, is that the laws of motion for the simple fluids are well known. If there exist common underlying principles, they will be most easily discovered in simpler systems. Specific examples will continue to provide useful insights, and the methods of analysis that they generate will find broad utility. What is unclear is whether a deep general theory will emerge from the knowledge acquired by studying special systems; whether nonequilibrium phenomena, like thermodynamic critical phenomena, fall into a small number of universality classes; and whether a broad-based understanding will eventually enable us to predict and control complex, technologically important processes.

Pattern Formation

Consider the simple case of Rayleigh-Benard convection. When a fluid, initially at rest between two horizontal plates, is heated from below, it experiences a temperature gradient. For small gradients, the heat transfer from the bottom to the top occurs purely by conduction—that is, by molecular collisions. When the gradient exceeds a certain threshold, however, the conductive state becomes unstable and yields to convective states involving bulk motion of the fluid. If the system is confined so that the fluctuations are correlated across the entire system, or if the system is modulated externally, the convective dynamics is largely temporal rather than spatial in character; one then observes a variety of universal properties associated with temporal chaos of low-dimensional systems.

Universality here pertains to quantitatively identical dynamics in apparently dissimilar systems.

Most systems are loosely confined, or unconfined for all practical purposes, and so the dynamics has a strong spatial complexion. This appears in the form of cellular structures such as squares, hexagons, rolls, and traveling waves whose properties depend on features such as boundary conditions or the strength of the forcing. In principle, these features of "collective behavior" are fully described by the Navier-Stokes equations. Unfortunately, the latter are rather complex and cannot be used directly for studying the conformation and dynamics of patterns. Much recent progress has come via the study of relatively simple, so-called "amplitude equations" that accurately approximate the full Navier-Stokes equations near the onset of instability where the nonlinearity is weak.

The amplitude equations are closely related to the Ginzburg-Landau equations that describe the behavior of order parameters in transitions from one state to another. They have been used successfully to describe major features of pattern formation, including defects, that occur in weakly nonlinear regimes. For systems where the initial instability is a traveling wave, as in binary mixtures, the dynamics can be described by means of the complex Ginzburg-Landau equation. This equation is rich in character; features such as "defect mediated turbulence" also appear to be contained in it. However, not all of its facets have yet been explored. Other complexities arise in spatially extended systems where, for example, the phases of the patterns may vary slowly in time and space. Equations for this kind of dynamics have been written and their outcome verified for special cases where the basic patterns are known and their shapes do not change. Again, much work remains to be done.

One of the grand unifying principles in equilibrium statistical mechanics is the minimization of free energy. Much thought has been expended on possible analogies in nonequilibrium phenomena. Although a suitable integral of the amplitude function and its gradient serves as an approximate free energy in some cases, this is not a viable procedure in general. A class of pattern formation problems that has not been fully explored is the nucleation and growth of turbulent "spots" in boundary layers (or "slugs" and "puffs" in pipe and channel flows). A spot is a compact object of well-defined shape; it moves in the flow direction at a specific speed; the flow within the spot is turbulent; and the distinction between its inside and outside is generally unambiguous. The spot preserves its characteristic shape as it grows, and the growth rate increases as the Reynolds number (a measure of the flow strength) grows beyond a critical value. A suitably defined length scale seems to diverge at the critical Reynolds number. Although there is a random element to the initiation of the spot in space and time, morphological characteristics such as its shape and growth rates seem universal. These and some other facets of turbulent spots bear superficial resemblance to critical phenomena and to growth processes like diffusion-limited aggregation or directed percolation. How to bring these similarities to bear quantitatively on the formation of spots has eluded us so far. Although there is a large body of engineering literature on this problem, very little of it is motivated by the more general considerations of nonequilibrium physics.

Turbulence

More complexities develop as one increases the stress applied to a fluid system at its boundaries, for example, by increasing the heat flux or the shear rate. ("Increasing the stress" means increasing the Reynolds number.) Among the complexities are the decay of the long-range order of the patterns, the development of new length scales, and the appearance of a strong flux of energy across the range of length scales (the "inertial range") on which turbulent motion is occurring. The scale range increases with the Reynolds number and is bounded, on the one hand, by the characteristic size of the system (the "large scale") and, on the other, by the small "dissipation scale" at which viscous effects become dominant. The flow is said to be fully turbulent when the scale range is large.

Well-developed turbulence has some interesting and important features. A tracer substance such as a dye, when injected into a turbulent flow, is mixed efficiently and diffused at unusually high rates; isosurfaces of the dye concentration are fractal; the small scales are spatially intermittent and amenable to multifractal description and modeling; correlations and fluctuations are anomalously large; and externally imposed perturbations decay slowly. These features are characteristic of phenomena far from equilibrium.

A quantitative theory of turbulence is likely to be valuable in the study of other nonequilibrium phenomena. This is why turbulence merits some attention and discussion here; indeed, until the 1960s, fluid turbulence was the clearest example of a phenomenon in which a large range of length scales are simultaneously important (see Box 4.1). The successful application of scaling, universality, and renormalization group theory to thermodynamic critical phenomena has altered this situation, but turbulence still offers one of the cleanest examples of scaling behavior in nonequilibrium physics.

Physicists generally like to focus on "universal" aspects of the phenomena they are studying. The conventional wisdom in turbulence theory is that small-scale turbulence possesses universal properties that are independent of specific large-scale flows. However, the notion of absolute universality, initiated brilliantly by Kolmogorov and others, is not strictly valid for turbulence, let alone for all nonequilibrium systems. Universality may pertain, at best, only to certain scaling exponents. The universality of scaling exponents is a compelling notion—one that clearly invites comparisons with other nonequilibrium problems—and principal questions regarding them are just beginning to be resolved. There are lingering impediments. For example, at present there is no theory in turbulence for effects of finite Reynolds number or finite shear. Despite advances in modern experimental methods, properties of turbulence continue to be probed

BOX 4.1 Length Scales in a Turbulent Jet

Figure 4.1.1 illustrates one essential feature of turbulent flows—namely, the coexistence of many superimposed length scales. The present example is a turbulent jet in which a small amount of fluorescent dye has been mixed. The dye concentration has been mapped by the so-called "laser-induced fluorescence" technique. Different scales of the dye concentration field have been extracted by convolving the data with wavelets of different sizes. That is, from top left to bottom right, we see pictures of the same flow pattern in which the size of the features being resolved has been reduced by a factor of 2 from frame to frame. Note that, the smaller the structure, the more string-like it becomes. There is no apparent self-similarity and, in contrast to earlier ideas, individual structures do not become more isotropic at smaller length scales. Note, however, that the anisotropy of individual structures does not necessarily preclude statistical isotropy.

The jet shown here has a Reynolds number of about 4000. This is rather low, and so the scale range is not large. In Earth's atmosphere at a height of 30 m, the scale range is about 10^5, yielding 10^{15} degrees of freedom in three dimensions. If the equations of motion are to be solved explicitly on a computer, the memory requirements grow roughly as the cube of the Reynolds number. This feature limits the Reynolds numbers at which turbulent solutions can be obtained by numerical techniques. There are also inherent complications such as sensitivity to initial conditions, but the statistical averages are believed to be independent of them.

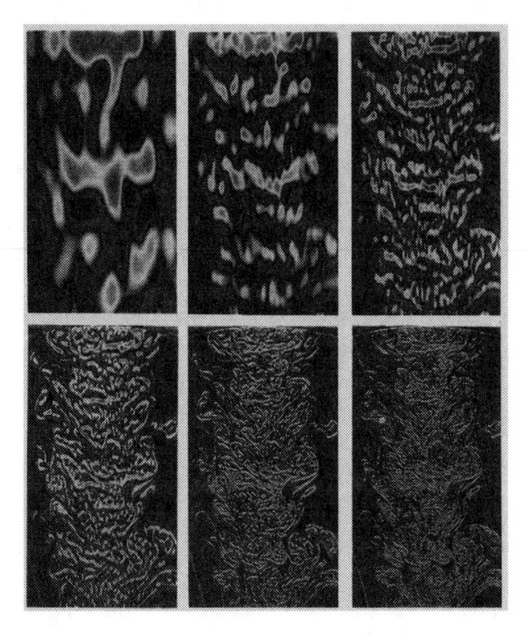

FIGURE 4.1.1 Depiction of the coexistence of many superimposed length scales. [Reprinted with permission from R. Everson, L. Sirovich, and K.R. Sreenivasan, "Wavelet analysis of the turbulent jet," *Physics Letters A* **145**, 314 (1990). Copyright © 1990 Elsevier Science.]

only partially at high Reynolds numbers, and quantities of theoretical interest can be measured only approximately. A major advance in this regard is the use of powerful computers to solve the equations of motion explicitly and thus elucidate spatio-temporal details of turbulent solutions. The Reynolds numbers of the numerical solutions are approaching the range of interest for addressing important issues.

An interesting question is whether the coherence of the small-scale motion, in the form of elongated and anisotropic vortex structures, is consistent with the universal scaling presumed to exist in fully developed turbulence. In an anisotropic ferromagnet near its critical point, for example, the critical indices do not depend on the magnitude of the anisotropy (although they are different for isotropic and anisotropic cases). In turbulence, however, the critical indices may, in some instances, depend on the magnitude of the anisotropy. The relation between scaling, which emphasizes the sameness of various scales, and structure, which becomes better defined and topologically more anisotropic for larger amplitudes, is at present quite obscure.

In summary, the issues considered here are the changes occurring in a fluid flow that is increasingly stressed at its boundary. The stresses may be applied by mechanical, thermal, or other means. The changes include instabilities, bifurcations, temporal chaos, pattern formation, phase modulations, defects, growth of localized structures, interactions among dissimilar length scales and timescales, universal and anomalous scaling, intermittency, anomalous transport, and the like. These phenomena have strong similarities to those that are seen in other nonequilibrium systems. If these similarities can be exploited intelligently, there will be many new opportunities for understanding turbulence better. Conversely, turbulence poses a rich variety of problems and has an array of tools of analysis that should be useful to other branches of nonequilibrium physics.

Fluid turbulence is a difficult problem with a long history, but the pace of progress has accelerated in recent years. Much of the recent progress is the result of a powerful combination of modern experimental methods, computer simulations, and analytical advances. The present picture of turbulence is generally self-consistent despite lingering uncertainties, and recent advances have further improved our qualitative and quantitative understanding. That the qualitative understanding should impact practical and industrial problems is substantially an article of faith; much remains to be bridged between the fundamental developments of recent years and practical problems of industrial relevance. To some extent, this is a problem of bridging cultures. To a larger extent, however, this is a reflection of the difficulties of strongly nonlinear problems that occur far from equilibrium.

What matters in turbulence is the ability to quantify properly the mix of the universal and system-specific aspects and to describe that mix economically. Such an understanding will propel forward not merely the study of fluid turbulence but the entire subject of nonequilibrium physics.

PROCESSING AND PERFORMANCE OF STRUCTURAL MATERIALS: METALLURGICAL MICROSTRUCTURES

The nature of nonequilibrium physics in condensed-matter and materials physics—its intellectual vitality, its technological potential, and some of the difficulties it faces—is concisely illustrated by the history of our understanding of dendritic solidification patterns and their relevance to the microstructural properties of metallic alloys. In more familiar terms, this is the "snowflake" problem. Only in the past few years have we finally learned how these elegant dendritic crystals emerge literally out of thin air, and why they occur with such diversity that no two seem to be exactly alike.

Much of the research on dendritic crystal growth has been driven not only by our natural curiosity about such phenomena, but also by the need to understand and control metallurgical microstructures. The interior of a grain of a freshly solidified alloy, when viewed under a microscope, often looks like a collection of overly ambitious snowflakes. Each grain is formed by a dendritic (tree-like) process in which a crystal of the primary composition grows out rapidly in a cascade of branches and sidebranches, leaving solute-rich melt to solidify more slowly in the interstices. The speed at which the dendrites grow and the regularity and spacing of their sidebranches determine the observed microstructure which, in turn, governs many of the properties of the solidified material such as its mechanical strength and its response to heating and deformation.

One logical and possibly realistic goal for research in this area might be the development of quantitative methods for predicting the metallurgical microstructures that emerge in the processing of industrial materials. An especially important class of examples includes the materials used for high-temperature applications such as gas turbines and jet engines. It would be useful to be able to develop predictive models for forging turbine disks, or for casting complex shapes such as turbine blades, or for the vapor deposition processes used to apply thermal barrier coatings to those blades. In principle, industrial laboratories could benefit greatly if they were able to design such processes cheaply and quickly by computer rather than by expensive, time-consuming trial-and-error methods. We now have in hand some of the conceptual bases for such calculations, and our computational capabilities continue to grow exponentially. Let us ask, therefore, what will be needed—what new experiments, new theory, new mathematics—in order to write usefully predictive computer codes to design and control such manufacturing processes.

Because many of these processes involve dendritic crystal growth, we know that one starting point must be an understanding of the dynamics of isolated, freely growing dendrites. Box 4.2 describes some of the remarkable progress made in the last few years. The free-dendrite problem is most easily defined by reference to the xenon dendrite shown in the figure in Box 4.2, a pure single crystal growing into its liquid phase. The speed at which the tip advances, the radius of curvature of the tip, and the way in which the sidebranches emerge behind the tip are all determined uniquely by the degree of undercooling—that is, by the degree to which the liquid is colder than its freezing temperature. The question is, How? (An equivalent problem is one in which the dendritic behavior is controlled not by the temperature but by the degree of chemical supersaturation.)

As described in Box 4.2, a rich understanding of the behavior of isolated dendrites has been found in theories of morphological instability and the discovery that very weak forces—crystalline anisotropy of surface energies, for example, or even atomic-scale thermal fluctuations in some cases—can completely control the patterns that emerge from these instabilities. These new conceptual developments, however, still leave us very far from being able to predict metallurgically relevant microstructures. Current simulations of casting, for example, include heat flow and fluid convection in complex geometries but succeed in only very rudimentary ways in coupling those effects to the formation of dendritic microstructures.

Perhaps the most important theoretical challenge is a quantitative understanding of what is called the "mushy zone"—the region between the fully formed solid and the molten fluid where the dendrites are forming and interacting among themselves. Within this region, the thermal, chemical, and hydrodynamic degrees of freedom of the system are all active. Even if each dendrite is behaving according to the rules already discovered, it is doing so in an environment where the local growth conditions are determined by its neighboring dendrites and their associated diffusion and flow fields. This behavior is almost certainly chaotic, and therefore most likely will have to be described in probabilistic rather than deterministic terms. We know that the mushy zone has its own collective instabilities that can produce fatal structural defects in the solidified materials.

The situation in the real world is even more complicated. In many casting processes, new dendrites nucleate at impurities throughout the molten fluid as it cools. Thus these processes are highly sensitive to the purity of the materials. Moreover, heterogeneous nucleation of this kind is extremely difficult to predict or control, even under ideal conditions. Other complications arise from the fact that, in welding, for example, the molten fluid itself is turbulent.

Can such behavior be modeled in a usefully predictive way? Can the relevant dynamics be described with sufficient accuracy by some coarse-grained, many-dendrite theory; or will there be such sensitivity to details and such a huge variety of possibilities that this problem will forever be beyond our reach? And even if we can make substantial progress, will we be able to translate our theoretical understanding into decision-making tools that will be applicable to real-life manufacturing?

These questions regarding intrinsic limits of predictability are unavoidable. Nevertheless, we should be able to do better than we can at present. We already know enough about these systems to recognize that a coordinated experimental

BOX 4.2 The Free Dendrite Problem

Under most metallurgical processing situations, dendritic growth is controlled by diffusion—either the diffusion of latent heat away from the growing solidification front or the diffusion of chemical constituents toward and away from that front. These diffusion effects very often lead to shape instabilities; small bumps grow out into fingers because, like lightning rods, they concentrate the diffusive fluxes ahead of them and therefore grow out more rapidly than a flat surface. The key to understanding pattern formation in such situations is understanding the nature of these instabilities.

Today's prevailing theory of free dendrites is generally known as the "solvability theory" because it relates the determination of dendritic behavior to the question of whether there exists a sensible solution for a certain diffusion-related equation that contains a singular perturbation. The term "singular" means that the perturbation, in this case the surface tension at the solidification front, completely changes the mathematical nature of the problem whenever it appears, no matter how infinitesimally weak it might be.

This theory has been worked out in detail for many relevant situations, such as the xenon dendrite shown here (Figure 4.2.1). It predicts how pattern selection is determined, not just by the surface tension (itself a very small correction in the diffusion equations), but by the crystalline anisotropy of the surface tension—an even weaker perturbation in this case. It further predicts that the sidebranches are produced by secondary instabilities near the tip that are triggered by thermal noise and amplified in special ways as they grow out along the sides of the primary dendrite. The latter prediction is especially remarkable because it relates macroscopic features—sidebranches with spacings on the order of tens of microns—to molecular fluctuations whose characteristic sizes are on the order of nanometers.

Each of those predictions has been tested in the xenon experiment, quantitatively and with no adjustable fitting parameters. They have also been checked in less detail in experiments using other metallurgical analog materials. In addition,

and theoretical investigation of mushy-zone dynamics in relatively simple situations would yield useful new information.

Yet here is another example of today's widening gap between fundamental research and applied technology in the United States. Few if any materials manufacturing companies in this country continue to support research in this area, and most no longer even maintain technical staffs that could take advantage of new developments. Some serious efforts are being made at government-supported laboratories to remedy this situation, but the university-industry-government consortia that exist in Europe and Asia do not exist here at present.

PROCESSING AND PERFORMANCE OF STRUCTURAL MATERIALS: SOLID MECHANICS

Solid mechanics, by necessity, has been a very well studied part of engineering and the applied sciences. We depend on experts in this field to design the theory has been checked in numerical studies that have probed its nontrivial mathematical aspects. As a result, although we know that there must be other cases (competing thermal and chemical effects, for example, or cases where the anisotropy is large enough that it induces faceting), we now have reason to feel confident that we understand at least some of the basic principles correctly.

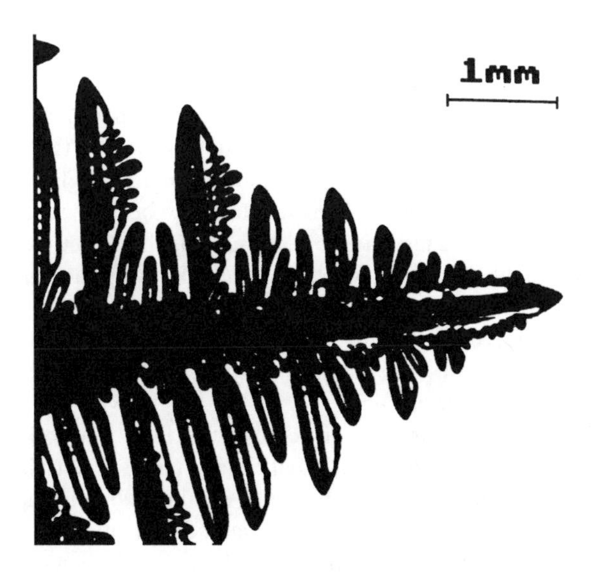

FIGURE 4.2.1 Xenon dendrite (Courtesy of ETH, Switzerland.)

structures—buildings, automobiles, airplanes, connections in electronic devices, and so on—that will be strong, safe, and reliable.

Although some of the basic principles of modern solid mechanics were established by physicists (for example, the dislocation theory of plastic deformation in crystals), the field has developed largely outside of mainstream physics throughout the last half of the twentieth century. The experts in solid mechanics have been engineers, applied mathematicians, materials scientists, geologists, and so on, but seldom physicists.

That situation is changing. In just the last few years, there has been a healthy revival of interest in this area within the physics community, and interdisciplinary activities have been expanding. Perhaps the most important reason for this change is that many of the most challenging modern problems in solid mechanics involve nonequilibrium phenomena and therefore pose novel fundamental questions. Another important reason is that, for the first time in history, we have the experimental and computational tools needed to answer those questions.

Brittle and Ductile Solids

As an introduction to some of these issues, consider the distinction between brittle and ductile solids. A brittle solid breaks when enough stress is applied to it, whereas a ductile material deforms plastically. Remarkably, we do not yet have a truly fundamental understanding of the distinctions between these two behaviors; we do not even have a sharp definition of "brittleness" and "ductility." The conventional theory says that dislocations form and move more easily through ductile materials than brittle ones, thus allowing deformations to occur in one case and fracture in the other. But this picture is incomplete and far too simple.

One difficulty in the standard picture emerges when we realize that all solids deform plastically under very high stresses, such as those in the neighborhood of a crack tip. Those stresses, by definition, must be high enough to break the bonds between neighboring molecules; thus they ought to be high enough to rearrange the molecules and thus cause plastic deformations that blunt the tip or limit the stresses in its neighborhood. How, then, can cracks propagate through any material? Do some materials spontaneously harden in regions of concentrated stress? If so, why? Or might they sometimes fail to deform plastically if the stresses change too rapidly? These are fundamental nonequilibrium questions that lie at the very heart of solid mechanics.

Another equally fundamental difficulty is that the concept of "dislocation" is meaningful only for crystalline solids; but noncrystalline materials such as glasses and amorphous metals can also be brittle or ductile. What is happening in such systems? Materials scientists have speculated for many years about an amorphous analog of dislocation motion, but we do not yet have a clear idea of what that might be, and we are very far from being able to describe such mechanisms quantitatively.

In fact, we still have no satisfactory understanding of the equilibrium properties of the glassy state; we do not even know whether it makes sense to use the term "thermodynamic equilibrium" for such systems (see Box 4.3). Perhaps glassy materials exist only as long-lived metastable or unstable states. If so, where do they fit into our classification of solid-like materials? What language do we use to describe their properties? Questions of this kind will become even more urgent and difficult when, later in this chapter, we talk about granular materials, biomaterials, materials in constrained geometries, or a variety of other states of matter whose unconventional mechanical properties have yet to be explored and understood.

Instabilities in Dynamic Fracture

The field of fracture mechanics is among the most elegant and important in all of the engineering sciences. In the last several decades, its practitioners have developed powerful mathematical theorems and numerical algorithms that permit

BOX 4.3 Glassy States of Matter

It is quite remarkable that, despite literally centuries of scientific study, we still do not have a fundamental understanding of glassy states of matter. This gap in our knowledge looks even more remarkable when we realize that the category of glassy materials—i.e., amorphous, noncrystalline solids, includes a very large fraction of the substances that we encounter in everyday life. It includes window glass, of course, and metallic glasses; and it includes many polymeric materials. Indeed, very many simple substances—even ice—can exist with great stability in glassy states. Condensed-matter physicists accurately use the term "glass" to describe yet other strongly disordered systems. For example, "spin glasses" are magnetic systems in which the interactions between the atomic spins are randomly distributed in sign, thus frustrating any tendency toward long-range order. The term "vortex glass" refers to the tangle of field-carrying vortices that occurs in certain kinds of superconductors.

For ordinary practical purposes, glassy materials become solids when cooled below their so-called "glass temperatures." They acquire immeasurably large viscosities and they support shear stresses. The uncertainties are whether these materials might continue to behave like liquids if only we could measure their properties over sufficiently long times or whether, on the other hand, this solid-like behavior is characteristic of some new, intrinsically nonliquid state. This is not an academic question. If we need to predict the structural or thermal properties of amorphous materials, then we need to know what fundamental concepts are relevant.

A great deal of progress has been made in the study of glassy materials during the last decade or so. That progress has consisted of many new phenomenological insights about a wide variety of different kinds of glasses and also some careful new experiments that seem to point to the need for unconventional descriptions of the ways in which molecules move and interact in glassy systems. In addition to new understanding of the behavior of glasses themselves, the concepts that have emerged in these investigations have generated novel ideas in fields as far from condensed-matter physics as numerical mathematics and psychology. "Simulated annealing," for example, is a powerful numerical strategy for solving optimization problems. It is based on a model of how a glass-forming material finds its state of (near) equilibrium when it is cooled at a controlled rate through its glass temperature. "Neural network" models of associative memory originally were developed in conjunction with studies of the properties of spin glasses with different kinds of couplings between the spins. It seems likely that this deep and subtle field of research will continue to generate new concepts of far-reaching impact.

them to compute stress distributions and failure criteria for highly complex solid objects. With the help of modern computers, they are now able to predict with confidence the mechanical properties of structural materials in a wide variety of engineering applications.

Almost all of this progress, however, pertains to static—or very nearly static—phenomena. Roughly speaking, conventional fracture mechanics has been concerned primarily with predicting when materials will break and much less with understanding what happens after failure occurs. The latter topic, "fracture dynamics," remains largely unexplored. It is fundamentally more challenging than its static counterpart because it involves very deep issues in nonequilibrium physics. Some of the most important outstanding questions are, How fast do cracks move? What mechanisms limit their growth? How is energy dissipated in fracture? What determines the various kinds of fracture patterns that we see in nature?

One of the most important recent developments in fracture dynamics has been the experimental demonstration that fast brittle cracks undergo material-specific instabilities (see Figure 4.1). Fracture surfaces frequently are rough; they may even be fractal. We now know that this roughness often occurs because fast cracks are unstable with respect to bending away from their directions of propagation or dissipating energy in the form of tip-splittings or sidebranches. High-speed fracture is frequently a complex, chaotic, pattern forming process. But instability is not a universal phenomenon. Since our ancestors first made stone tools and later learned how to "cut" diamonds, it has been clear that sharp, smooth fracture surfaces can be made by producing cleavage cracks in glassy or crystalline solids. Apparently the trajectories of those cracks are stable.

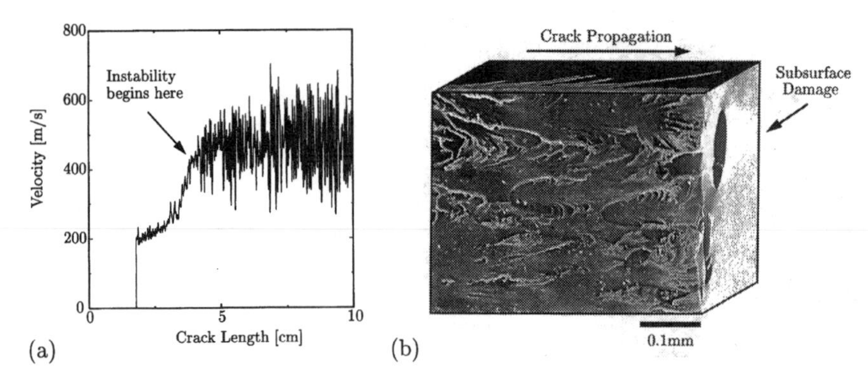

FIGURE 4.1 Unstable fracture in a polymeric glass. This figure illustrates an experiment in which a crack was observed with high precision as it moved along the center line of a long strip of the polymer. (a) The graph of crack speed versus crack length. We see that the initial crack in the unstressed sample was about 2 cm long. The stress applied to the sample was increased until the crack suddenly started moving at a speed of about 200 m/s. The crack then accelerated smoothly up to a critical speed of about 400 m/s, at which point an instability occurred that shows up on the graph as a rapid and irregular oscillation of the crack speed. (b) Photograph of the fracture surface left by the unstable crack. The front face is the fracture surface itself, with visible roughness on the scale of 0.1 mm. The top and right faces show that the instability also generated sidebranching cracks and subsurface damage as the main crack moved through the system. (Courtesy of University of Texas, Austin.)

Despite literally thousands of years of familiarity with these phenomena, we still do not know what mechanisms control fracture stability. There exist some candidate ideas, a few of which have yet to be tested; but there is as yet no theory that is sufficiently plausible and well developed that it can serve as a guide to experiment. One hint, known for about 30 years, is that the stresses in the neighborhood of a crack tip moving at an appreciable fraction of the speed of sound transform in a way that might cause bending. But this observation does not address crucial questions pertaining to the dynamics of deformation and decohesion near the crack tip, which must surely be relevant. The situation is further complicated by the lack of a good theory of ductile yielding in situations where stresses and strains are varying rapidly in both space and time.

In some respects, the present state of the theory of dynamic fracture resembles that of dendrite theory almost half a century ago, before scientists had identified the diffusive instability that underlies pattern formation in crystal growth. To be sure, elastodynamics, plasticity, and decohesion in solids are much more complex phenomena than diffusion. Also, it is quite possible that there are many qualitatively different mechanisms that cause instability in fracture. Perhaps this additional complexity explains why progress has been so slow in this field. With modern tools and interdisciplinary modes of research, we should be able to do better in the future.

Polymers and Adhesives

As discussed elsewhere in this report, an increasing fraction of the structural materials used in modern technological applications are polymers or polymeric composites. Nonequilibrium phenomena are involved both in the chemical and thermal processing of these materials and in the way they respond to stresses during use. These materials occur both as structural elements by themselves and as the adhesives that bind different kinds of structural components in complex applications.

Many structural polymeric materials are alloys of two immiscible polymers, consisting typically of micron-sized droplets of one polymer dispersed in a matrix of another. The mechanical behavior of such a composite is not some simple average of the properties of its constituents. Rather, it depends crucially on the shapes of the droplets and the properties of the interfaces between the droplets and the matrix. Common examples are rubber-toughened glassy polymers used to produce materials that resist high-speed impact damage and, in recent applications, three-phase blends consisting of one rubbery, one glassy, and one semicrystalline phase. Control of both the particle morphology and the fracture toughness of the internal interfaces depends on so-called "compatibilizers," copolymers usually produced in situ by reaction during a mixing flow in the molten state. Besides promoting the breakup of particles into a fine dispersion by their action as polymeric surfactants and inhibiting particle coarsening by preventing coalescence, these compatibilizers help strengthen the interfaces between the phases.

Some progress has been made in understanding how these connectors work, in large part because we now have instrumental probes that allow us to see what is happening on submicron length scales or smaller. Techniques for using copolymers to strengthen interfaces between immiscible glassy polymers are now fairly well established. However, it is still far from clear how to ensure entanglement between polymers, and thus achieve fracture toughness, at interfaces without connectors. The rules for strengthening interfaces involving semicrystalline polymers are especially important for practical purposes but as yet remain undiscovered. Consider polyethylene, the most common semicrystalline polymer, as an example. Outstanding questions include, How long do polyethylene connectors at an interface have to be to allow a zone of plastic deformation to form ahead of a crack tip? If parts of such connectors are incorporated into crystals, is that sufficient to lock them into place, or do they have to be both entangled and run through the adjacent crystals for effective anchoring? What role does crystallization play in the strengthening or weakening of such interfaces? Such evidence as exists suggests that the influence of crystallization is significant. Do the answers to the above questions depend on crack growth rate?

Most adhesives are polymeric, and these play increasingly significant roles in a wide variety of technologies. Adhesives span the range from the very rigid materials used in structural applications to very soft solids such as the pressure-sensitive adhesives used in adhesive tape. In all cases, the performance of the adhesive joint is determined by a complex interplay between the bulk mechanical properties of the participating materials, including the adhesive itself, and the detailed microstructural features of the interface. As in the case of the polymer alloys, these molecular-scale features are becoming accessible to physical probes such as secondary-ion-mass spectrometry, ion-beam analysis, and neutron reflectometry. Thus, while adhesives traditionally have been developed by time-consuming, empirical approaches, the prospect now exists for rational design based on molecular-scale information.

Friction

Another classic part of materials research that is enjoying a resurgence of interest among physicists is the science of friction. This topic has much in common with dynamic fracture and adhesion. Two interacting solid surfaces sliding past each other look in many ways like a dynamic shear crack. Mechanisms such as cohesion and decohesion, energy dissipation, elastic deformation, and so on are all relevant. But friction is an even larger and more complex topic than fracture because it occurs in such a wide variety of circumstances and, apparently, with an equally wide variety of underlying physical mechanisms.

The conventional goal of research on this topic is to determine frictional forces as functions of the relative state of motion of two solid surfaces and the stress holding the surfaces in contact with each other. Real friction, however, is far more interesting and complex than this conventional statement would make it seem.

Much of the recent progress in this area has been based on novel techniques for visualizing the microscopic processes that take place during friction-controlled sliding. Several atomic-scale probe microscopies have been used, as well as some relatively simple and direct methods for following the motions of larger features such as contact points and asperities. Numerical simulations, especially via molecular dynamics, are now beginning to provide very valuable insights; and good use also is being made of analog systems for making accurate observations of slipping events. A recent example of the latter technique involves layers of carefully characterized granular substances confined between sliding plates.

Friction problems fall very roughly into three different categories: friction between molecularly flat crystalline surfaces, friction between deformable rough surfaces, and, in a very general sense, lubricated friction—that is, friction controlled by the dynamic behavior of substances constrained to move between the surfaces that are sliding across one another. In the first of these categories, the clean crystalline surfaces, it is possible to make plausible models that involve only atomic-scale degrees of freedom. Although such models still must include assumptions about irreversible behavior, they are relatively well posed and, in some cases, they are now beginning to produce credible agreement between theory and experiment.

The other two categories of friction problems are fundamentally more challenging because they involve two or more widely separated length scales and timescales. They may also be of broader practical importance.

In dry friction between polycrystalline, noncrystalline, or otherwise imperfect surfaces, the actual area of contact is much smaller than the nominal area of the surfaces. The behavior of the small contact regions is crucial in determining frictional forces and dissipation rates, but there is as yet no clear understanding of the physical mechanisms that occur there. The problem seems to have issues in common with fracture; the behavior is governed by cohesion and decohesion at atomic-scale contacts that are strongly coupled to larger-scale elastic and plastic modes of deformation. One useful way of dealing with systems of this kind is to describe them not just by the relative positions and speeds of the sliding surfaces but also by "state variables" that might represent, for example, the density and strength of the contacts, and that obey equations of motion of their own. Such "rate and state dependent" friction laws have been developed especially by seismologists.

The ostensibly most complex problems in this field are those in which a "lubricant"—that is, some extraneous substance—is present in the space separating the sliding surfaces and transmits the frictional forces from one surface to the other. In some of the most interesting recent experiments, the lubricant is confined to a very small region, just a few molecular diameters across, and thus its properties—especially under shear—may be quite different from those of the same substance in bulk. Now the use of state variables is absolutely essential. The lubricant may respond to changes in the shear rate by changing its state, perhaps from liquid-like to solid-like, and such variations may occur on many different space scales and timescales. The challenge is to identify the essential degrees of freedom for these complex systems and to understand the interrelations between the relevant microscopic and macroscopic phenomena.

One of the most interesting and characteristic kinds of behavior seen in friction experiments is stick-slip motion. In many circumstances, surfaces in contact with one another will stick together until the applied shear stress reaches some threshold, and then will slip past each other in accord with a rate-dependent friction law until, under the influence of external forces perhaps, they come to rest and restick. Familiar examples include squeaky door hinges and the motion of a violin string driven by a bow.

It is easy to imagine how stick-slip motion can occur at a localized asperity, that is, at a point where irregularities on opposite surfaces are attached to each other via contact forces or molecular bonds. Slipping begins when the bond breaks and stops when a new bond is established. On macroscopic scales, friction-limited slipping may be the average of very many uncorrelated microscopic stick-slip events. Macroscopic motions also may have a stick-slip character, as in the case of the squeaky hinge. Such behavior occurs when the combined action of dynamic friction and external loading induces some kind of mechanical instability.

One interdisciplinary research topic that combines many of these ingredients—stick-slip friction plus fracture—is earthquake dynamics. Earthquakes, by definition, are stick-slip events. They are triggered when some piece of a fault is brought to its slipping threshold by the tectonic forces in the Earth's crust. They have the additional features that they occur on large length scales and have an extremely broad range of sizes, even on single fault segments. Both physicists and seismologists have been interested recently in the discovery that models of earthquake faults consisting simply of elastically coupled stick-slip slider blocks are deterministically chaotic systems that exhibit some of the characteristic behavior of real faults. Of course, these models do not account for the geometric complexity of real seismic phenomena; but the qualitative picture that they provide, in which large events occur intermittently as cascades of small events, is at the least an intriguing caricature of many kinds of self-organized phenomena. It might even prove to be useful in seismology.

Granular Materials

Granular substances such as sand provide an especially clear example of a familiar class of materials whose properties have yet to be understood from a fundamental scientific point of view. These materials have been studied empirically for centuries in civil engineering, geology, soil mechanics, etc., because they are essential ingredients in a wide variety of natural phenomena and have many practical applications. But we do not know how to answer some of the most basic questions about their behavior.

There are several clear distinctions between granular materials and the other, superficially comparable, many-body systems that are more familiar to physicists. Because they have huge numbers of degrees of freedom, they clearly need to be understood in statistical terms. However, individual grains of sand are enormously more massive than atoms or even macromolecules; thus thermal kinetic energy is irrelevant to them. On the other hand, these grains also have infinitely many internal degrees of freedom; thus they may—or may not—be highly inelastic in their interactions with each other or with other objects. They also may—or may not—have irregular shapes; arrays of many grains may achieve mechanical equilibrium in a wide variety of configurations and packings. It seems, therefore, that the concept of entropy must be relevant. We shall need some way of deciding which are the statistically most probable states under various constraints. But is there any analog of temperature or internal energy? What other quantities might be necessary for describing the states of these substances?

The questions become even more interesting when we consider the analogs of nonequilibrium properties for granular materials. What happens to sand when it is made to vibrate? Or when it is exposed to shear stresses? In some circumstances it behaves like a solid; close-packed sand can support limited shear stresses. In other circumstances—strong shaking in an earthquake, for example—it flows like a liquid. In yet other circumstances, granular materials behave in ways that we do not yet know how to characterize (see Figure 4.2). Their free surfaces spontaneously form regular patterns when shaken in special ways; their internal stresses organize themselves into chain-like structures under certain kinds of loading; flow patterns sometimes look roughly like localized shear bands.

Granular materials are only the simplest examples of states of matter that are unfamiliar and relatively unexplored from a fundamental point of view, yet appear in many ordinary circumstances. To change the granular system just a little, we might consider cases in which the grains cohere to each other. If the coherence is weak, such substances may behave like viscous fluids—wet sand or clay, for example. If it is strong, then we have materials like concrete or sandstone which, for the most part, behave like ordinary solids. They support shear stresses, and they can be brittle or ductile in their failure modes. In both cases, however,

FIGURE 4.2 Localized standing wave in a vertically vibrated layer of 0.2 mm diameter bronze balls. (Courtesy of Center for Nonlinear Dynamics, University of Texas, Austin.)

when we try to understand the nonequilibrium properties of these materials, we find ourselves in uncharted territory.

We find ourselves even further afield when we consider the nonequilibrium physics of yet more dynamically complex materials such as foams, or the colloidal suspensions, gels, and so on discussed in Chapter 5. Foams, for example, are close-packed collections of bubbles separated by fluid films. They may be similar in some respects to granular materials; their fluid-like properties may be controlled by local rearrangements of the bubbles.

Such considerations lead us inevitably to think about the nonequilibrium behavior of biological materials. Cellular membranes, for example, are double layers of large phospholipid molecules (see Chapter 5). These membranes obviously have a high degree of structural integrity; they do not pull apart easily and they resist bending. Yet the molecules rearrange themselves easily within the layers, which behave in many ways like two-dimensional fluids. There has been substantial progress in the last decade in understanding the properties of these basic biological materials. But it is only very recently, with the advent of new experimental techniques, that the outstanding questions in this immense field are beginning to become well-posed problems in nonequilibrium materials physics.

Length Scales, Complexity, and Predictability

There is a growing consensus among seismologists that it is impossible, even in principle, to predict earthquakes. By this, they mean that they may never be able to tell us, say a month or a year in advance, that an earthquake will occur at a certain time and place. They have good reasons to be optimistic that, with new scientific techniques, they will be able to make increasingly accurate estimates of seismic hazards; and they may even be able to identify some precursory phenomena that, in special circumstances, are warnings of imminent seismic events. But earthquake faults are extremely complex, unstable, dynamical systems. When near their slipping thresholds, they are highly sensitive to everything from large-scale motions of the Earth's crust down to the detailed friction laws governing the motions of rocks at the points where failure might occur. Even if we had an arbitrarily large computer and a complete understanding of the physics at each of the relevant length scales and timescales, it seems highly unlikely that we could ever determine the state of such an immensely complex system with enough accuracy to be able to make event-specific predictions.

Are there similar in-principle limitations in other areas of solid mechanics? In the foreseeable future, we ought to be able to determine from first principles the elastic-plastic response and fracture toughness for single-crystal copper or silicon, or perhaps even for some amorphous materials. As we have seen, these are challenging but potentially solvable problems. Is there any hope, however, of being able to make such determinations for multicomponent, polycrystalline, structural alloys, or for multiphase polymeric composites, and then go on to predict how those materials will perform in service over extended periods of time? Or are such predictions as difficult as predicting earthquakes?

As seen from the perspective of the last decade of the twentieth century, it appears that the next decades of research in nonequilibrium solid mechanics should be characterized by a diversity of goals and modes of inquiry. It seems clear that we must continue to move in certain well-established and productive directions—ab initio calculations of dislocation dynamics in crystals, for example, or experimental tests of such theories using synchrotron radiation. The results of those investigations should provide data and new insights that will be essential for a wide range of related research and applications, and they should point the way to new research in directions that we cannot now anticipate.

In pursuing these investigations, however, we must recognize that making the transition from atomistic theories of single-crystal plasticity to engineering design of large-scale, complex, structural materials is not just a matter of building bigger computers. We are almost certain to find new physics problems that need to be solved at each of the many stages of this process. If the analogy of earthquake prediction is accurate, then many of these problems will have to be solved in ways that are not now familiar to us.

FURTHER PROSPECTS FOR THE FUTURE

Nonequilibrium Phenomena in the Quantum Domain

Modern technological advances in mesoscopic and atomic systems, described elsewhere in this report, have made it imperative that we extend the study of nonequilibrium phenomena to the quantum domain. The inclusion of quantum mechanics makes an already difficult task even more difficult. In addition to having to understand which microscopic details are relevant to observed macroscopic effects, quantum mechanics forces us to consider the idea of "coherence," that is, the degree to which quantum mechanical wavefunctions preserve not just their amplitudes, but also their phases during dynamical processes. We have to think carefully about the physical meanings of both classical statistical fluctuations and quantum uncertainties. Not only do we have only partial information about the relative probabilities of various atomic or electronic configurations, but now we must take into account that we have only partial information about the quantum phases.

Such considerations become yet more challenging when we consider the problem of "quantum chaos"—the quantum mechanics of systems whose classical behavior is chaotic. In classically chaotic systems, we lose predictability as the system evolves in time. What happens to such systems in the quantum domain? Until just a few years ago, such a question seemed to be more important philosophically than practically; but that situation has changed dramatically with recent developments in mesoscopic physics. The electronic orbits in artificially structured nanosystems are often chaotic, and the control of these systems is now so precise that the electronic states can remain coherent long enough for the chaotic behavior to be relevant. Thus, what had previously been an abstract problem in quantum measurement theory is now becoming a practical problem in the design of advanced electronic devices.

An even more advanced example of the growing importance of quantum dynamics in nonequilibrium phenomena is the current effort to develop quantum computers. Here, as in the mesoscopic devices, we are dealing with systems far from equilibrium coupled to an external environment and a "classical" measurement apparatus. But, in this case, the whole concept depends on the possibility of maintaining systems with many degrees of freedom in quantum mechanically coherent states for times long enough actually to perform useful computations. Related questions have been raised by the production of Bose-Einstein condensates by evaporative cooling in atom traps. We are only beginning to understand how to describe the dynamics of the formation of a coherent condensate in a quasi-isolated system driven out of equilibrium by evaporative cooling.

Yet another example of the importance of dissipation and decoherence can be found in the study of quantum critical phenomena such as the metal-insulator transition. The dynamics of critical phenomena are often best studied using both linear and nonlinear response to applied external probe fields. It is often the case that "external" dissipation (due to phonons, for example) is irrelevant to the equilibrium-critical behavior because the effective electron-phonon coupling vanishes at low temperatures. However, this coupling can be "dangerously irrelevant" to the nonlinear transport because the external dissipation is the only mechanism available to carry away the Joule heat. Hence, the less relevant the phonons are to the equilibrium properties, the larger the bottleneck they represent for energy dissipation in the nonequilibrium case. This can lead in some cases to a singular temperature rise that obscures the underlying singular response of the quantum critical fluctuations.

In short, there continues to be a host of deep and largely unresolved philosophical questions in quantum measurement theory, forced on us by the development of systems at the interface between the quantum and classical domains. These questions may well be of growing practical importance in the next decade.

Nonequilibrium Phenomena in Biology

Living systems are, by definition, in states that are far from thermodynamic equilibrium. They are very different from any of the examples that we have discussed so far, however. The most obvious difference is that they are highly complex in their basic ingredients. Even the simplest biological materials are composed of large multicomponent molecules that, individually, perform specific chemical and mechanical functions. More important, the selection of these molecular constituents, and the ways in which they are assembled to make living organisms, has taken place not according to some global optimizing principle of the kind cherished by physicists, but rather according to the incremental and perhaps chaotic processes of evolution. In dealing with systems of this kind, physicists are learning to make basic changes in their research strategies.

Biology always has been a far more empirical and phenomenological science than physics. Until very recently, there has been little room in biology for what physicists call "theory." The complex phenomena being observed and interpreted by biologists are taking place in systems whose fundamental properties are not understood in the way we understand, for example, the physics of solid xenon or the mechanical properties of grains of sand. Physicists usually have not had the information they need for developing quantitative theories of biological phenomena or the tools they need for testing those theories.

As described in more detail elsewhere in this report, that situation is now beginning to change. Laser tweezers, atomic-force microscopes, and the like are permitting us to see what individual molecules are actually doing during biological processes. It is now possible, for example, to measure forces between cellular membranes, to watch those membranes change their shapes in response to various kinds of stimuli, or to see how proteins are formed and transported from one place to another within cells. From the wealth of information just now becoming available, we are beginning to understand that large biological molecules often function as machines, absorbing energy from their chemical environments, dissipating energy, and doing biologically useful work—all in accord with the basic principles of nonequilibrium physics.

There seems little doubt that, so far, we are seeing only a very small part of the huge world of biological materials and biophysical phenomena. The near-term challenges for physicists working in these areas will be to identify those biological systems that are ripe for quantitative investigation, to develop the instruments and techniques for data analysis that will be needed to characterize those systems, and to induce quantitative and predictive theories that can serve as guides for further experimentation. Ultimately, the goal is to acquire a deep, detailed understanding of the most extraordinary of all nonequilibrium phenomena: life itself.

FUTURE DIRECTIONS AND RESEARCH PRIORITIES

In summary, the committee draws the following conclusions regarding recent developments and future directions for research in the nonequilibrium physics of condensed matter and materials.

1. Fluid dynamics, in addition to being relevant to a wide range of topics in science and technology, remains a uniquely valuable laboratory for the study of complex, nonequilibrium phenomena that emerge in relatively simple, easily characterized systems.

2. Very significant progress has been made in the last decade in understanding dendritic pattern formation in crystal growth. That progress, however, has yet to have a major impact on efforts to predict and control solidification microstructures in industrially important materials. In part, the difficulty is that there remain some challenging scientific problems to be solved, such as the problem of

the "mushy zone." Another part of the difficulty is that there is relatively little effort in this area in the United States, especially in industrial laboratories.

3. Recent developments in scientific instrumentation, especially atomic-scale resolution in probe microscopy, plus extraordinary advances in computing power, mean that long-standing problems in solid mechanics should now be solvable. These are fundamentally challenging problems that involve non-equilibrium statistical physics, nonlinear dynamics, and the like. They are also, essentially without exception, directly relevant to modern technology. Among those problems are the following:

a. The origin of dynamic instabilities in brittle fracture;

b. The fundamental distinction between brittleness and ductility in both crystalline and amorphous solids;

c. The relation between molecular and mesoscopic structure and mechanical properties, especially fracture toughness, in composite materials containing, for example, varieties of polymeric constituents;

d. The relation between molecular and mesoscopic structure and the dynamics of friction in an extremely wide variety of situations, ranging from atomically flat surfaces interacting across molecularly thin layers of lubricants, to tectonic plates interacting across earthquake faults; and

e. The relation between elementary interactions between grains and the macroscopic mechanical behavior of granular materials.

4. In all probability, the next major frontier for research in nonequilibrium physics will be in the area of biological materials and phenomena.

5. The same recent advances in scientific instrumentation and computing power that portend both major advances and major surprises in nonequilibrium materials research also force us to face fundamental issues in the physics of complex systems. The problem of understanding the limits of predictability in these systems must be addressed with every bit as much skill and objectivity as the more familiar problems of understanding specific properties of specific systems. These issues lie, not just at the interface between different scientific disciplines, but also at the interface between science and public affairs.

5
Soft Condensed Matter: Complex Fluids, Macromolecular Systems, and Biological Systems

The vast territory of "soft" materials extends to paints, surfactants, porous media, plastics, pharmaceuticals, coatings, ceramic precursors, minerals in suspension, foodstuffs, textiles, proteins, fats, blood, and guitar strings. These materials are composed of colloids, gels, emulsions, micelles, foams, liquid crystals, polymer melts, and other regimes of organization.

Biological physics, widely regarded as a major growth area for physics in the next decades, is already closely allied with the study of macromolecules and complex fluids—both active subfields of physics today. Labeling these as separate fields is already arbitrary, and with good luck, separation will become impossible. Chemical synthesis using biological methods will create huge classes of materials amenable to physical experiments. Physical techniques and principles will come to be applied to biological materials in unexpected ways.

This new domain of physics includes phenomena ranging from the precise agility of liquid crystals to the lurches and shudders of sandpiles, from the resilience of rubber to the self-organizing structure of soft surfaces, and from neuromuscular twitch to single-molecule mechanics.

A major difficulty in preparing this report was how to organize it into topics. The word "interdisciplinary" inadequately denotes the new paths of learning, distinct from those identified in earlier surveys and in university curricula. This chapter discusses the following:

- The necessary combination of traditionally distinct fields of physics, chemistry, and biology;
- The relentlessly increasing complexity of the materials examined;

- The imaginative and exciting use of these materials; and
- The birth of new physics based on the theme of softness, malleability, and fragility.

It is only through basic research, rather than goal-driven manipulations of materials, that the essential connections between structure and function will emerge. Practical applications, whose development is often easier to support, will flow more easily once basic connections between structure and function have been elaborated. This chapter argues for research that, at its best, celebrates the complexities of the source material while taxing scientists' ingenuity to knead those complexities into tractable and relevant form.

There are abundant new industrial and medical products based on the softness and fluidity of materials. In fact, there have been uses from ancient times. The softness of rubber was enjoyed in pre-Columbian American ballgames. Emulsions and foams such as mousse and hollandaise sauce have long been used in food preparation. Today's new products are often discovered by ingenious trial and error rather than by systematic theory and physical understanding (see Box 5.1). The experience of practical innovation becomes a source of information to learn the general features of these soft materials.

Understanding inevitably leads to practical use, but it can take decades to progress from initial scientific query and curiosity to practical application. From the invention of the transistor to the first useful integrated circuits took about 20 years. From the realization that DNA stores the genetic code to the beginnings of a viable biotechnology industry took even longer.

With soft materials there are several unusual challenges to be recognized. Our effort to see what is common and mutually constructive in working with diverse soft materials immediately encounters cultural differences between physicists, engineers, and biologists. Particularly at the intersection between biology and physics, barriers to learning from each other are daunting. More than in any other chapter of this report, the primary emphasis here is on educational needs and opportunities for students and professionals. Following are a few examples of learning opportunities that would cost little compared to the large potential rewards:

- Summer schools,
- "Bilingual" survey texts and tutorials,
- Continuing education,
- Industry-academic visitation and collaboration,
- Grant programs to encourage truly basic research, and
- Graduate training in chemistry, physics, and biology.

Education must be deep as well as broad; it must obviate departmental boundaries. Nothing is more challenging than the creation of optimal modes of training

BOX 5.1 The Use of Milk Proteins

The world produces 5×10^{11} (500,000,000,000) liters of milk each day. Some of this is directly consumed. A major part is processed or used to supply industrial components. The milk protein casein, for example, 170 million pounds of it per year in the United States, is put into bakery products, medicines, adhesives, paper, low-fat coffee whiteners, and synthetic whipped dessert toppings.

Milk is a fragile mixture of fat and proteins in water. The structure and composition of fat globules, casein micelles, globular proteins, and lipoprotein particles have the malleability to allow them to be made into hundreds of butter, cream, yogurt, and cheese products. Their natural "complex fluid" properties are the kinds that physicists are now beginning to recognize.

In the dairy industry, processing is often guided by ingenious trial and error. The condensed-matter physics of soft matter can now have a chance to contribute here. The gelation of fat globules and proteins, the distribution of gel networks, and the size of nano-droplets of dispersions—which change the texture, taste, and feel of food—are in fact physical properties amenable to systematic physical investigation.

in fields that traditionally disdain each other. The complexities of materials are themselves challenging enough to require no elaboration. New materials and properties are now studied in physics departments; at the same time there is increasing need for good physics in biology and engineering departments.

It is said that one of Isaac Newton's greatest achievements was to extract from Johannes Kepler's notebooks the two Kepler Laws that showed Newton the way to the discovery of gravity and the explanation of planetary motion. The notebooks and their calculations were themselves inspired by Tycho Brahe's astronomical observations. There is an analogy here to modern soft-materials research.

Biological systems present a set of successful molecular mechanisms that create the living state. The path of trial and error to industrial success leaves a valuable though diffuse trail of information. To pick out tractable essentials from these data is a challenge that might lead to the discovery of what makes a system live, today's equivalent of Newton's realization of gravity. The very mass of new data creates its own challenges. Entire genomes of species, including our own, are being mapped. Already one hears biologists speak of a "post-genomic era" when new thinking will be needed to work with the new information and new materials. Much of that thinking is expected to come from physicists.

Condensed-matter and materials physicists are used to thinking in terms of emergent phenomena in large complex systems (see Chapter 3) and understand that the simple paradigm that "structure determines function" can easily fail because of collective phenomena. However, much of our experience is in relatively simple systems with a great deal of symmetry (translation symmetry in a ferromagnet, for example). We need to be looking for new deep concepts in biology. These will necessarily require thorough knowledge of the details of structure, but will, we hope, capture fundamental and general principles of function.

The committee foresees possibilities for a shift to biologically based industry, for advances in medicine, for insight to understand and simulate the thinking process itself. The study of soft materials—such as complex fluids, macromolecules, colloidal suspensions, and biological preparations—is already creating materials and processes with useful medical, industrial, and domestic applications. So successful has physics been in medicine that the view of physics is distorted by its success. Virtually all the sophisticated hardware used in medicine is based on physical techniques and innovation of the kind described in this report. Physical gadgets are saving so many lives that nonphysicists often believe that gadgetry is the right place for physics in biology and medicine. Practical byproducts should not deflect attention from the central message that basic research will ultimately bring much greater rewards.

COMPLEX FLUIDS

As though condensed fluids were not already sufficiently complex (see Box 5.2), condensed-matter physics has defined "complex fluids" in an effort to investigate the suspensions and solutions of large molecules. Here "large" begins with the nanometer size of proteins and high polymers and extends to the micron-plus dimensions of colloids, liquid crystals, and grains of sand. Particles of this size organize themselves by steric collisions that create unexpected symmetries and sensitivity to boundary surfaces. Their interactions are governed by electrostatic and solvation forces in forms not seen between smaller particles.

Liquid Crystals and Microemulsions

Most large asymmetric molecules, viruses, and lipids assemble spontaneously into ordered structures whose dimensions and macroscopic properties vary dramatically with small changes in the conditions under which they are formed. The statistical mechanics of these assemblies challenges the best theorists, and measurements of macroscopic behavior and microscopic structure are a primary activity in materials research. Since the early 1960s it has been known that what underlies the phase diagrams of lipid-water mixtures is a set of lamellar, hexagonal, cubic lattices whose dimensions and symmetry change with temperature, concentration, and the salts dissolved in the water phase.

Force measurements on lamellar stacks of sheet-like membranes reveal steric repulsion resulting from the loss of thermal motion. The collisions between these sheets can be hard bumps or soft encounters through spatially varying electrostatic, van der Waals, and hydration forces. The stability of a lamellar array reflects the interplay of all these factors, together with the layer flexibility that allows thermal undulation in the first place.

Similar reasoning holds for the packing of rod-like particles such as slightly flexible linear polymers and some viral particles.

Pursuing a sudden opportunity to examine the huge number of new liquid-crystal phases with natural and artificial materials in solution, theorists are

BOX 5.2 An Enduring Challenge: The Structure of Liquids

Formulation of a paradigm for the structure and transport properties of liquids that is as compelling as periodicity is for crystals or "sparseness" is for gases remains one of the grand challenges of science. The periodicity of the crystal allows precise definitions of lattice excitations (phonons) and crystal defects (vacancies, dislocations), which let us understand in detail its thermodynamic properties, atomic transport, and deformation. Kinetic theory plays a similar role for gases.

The complexity of the liquid structure arises from liquids being almost as dense as crystals without the organizing principle of periodicity. The phase space of a liquid contains many closely spaced energy levels with low barriers between them. Computer simulations have produced many models of liquids, in excellent agreement with the data from diffraction experiments. Nevertheless, a simple, unifying picture of what gives rise to these structures and their associated transport behavior is still missing.

One approach to achieving such a picture is the introduction of large numbers of defects into a crystalline lattice. These models fail to make a specific link to the experimental and computational features of the liquid structure and are thermodynamically unsatisfactory. The alternative approach starts from the gas and gradually densifies it. Keeping track of the atoms, however, is a difficult task, and the formalism required to account for the thermodynamic and transport properties becomes quite complex.

The closest we have come to formulating a simple paradigm for the liquid is the "polytetrahedrality" of monatomic liquids, which goes back to a suggestion by F.C. Frank. The densest local configuration of equal spheres is a tetrahedron. Close packing of these tetrahedra around a common center leads to the formation of (imperfect) icosahedra, which, through their fivefold symmetry, explain the lack of periodicity. Although it is not possible to fill three-dimensional (3D) Euclidean space with perfect tetrahedra, it is possible to make a perfect packing of tetrahedra as the curved 3D surface of a four-dimensional polytope (see Figure 5.2.1). This surface can be mapped onto 3D Euclidean space by the introduction of line defects (disclinations), and the pair distribution function of the resulting structure is in good agreement with experiment.

Although this approach comes closer than any to defining a structural paradigm, more work is clearly needed, such as an extension to larger clusters and the detailed study of the defect lines—their identification in computer models, their systematic enumeration and statistical mechanics, and a demonstration of their role in atomic diffusion and viscous flow.

The study of glasses is particularly useful here because they allow us to narrow the energy range and hence the number of configurations of the liquid that need to be considered. Metallic glasses are the prime experimental systems to test the above paradigm because, unlike network-forming or organic glasses, the building blocks of metallic glasses are single, spherically symmetrical atoms. All glasses are intrinsically unstable; they are formed when the liquid goes out of equilibrium when cooled below the glass transition temperature. As a result they can continuously lower their free energy by a process of structural relaxation. Study of these relaxation phenomena yields unique additional information about the structure and defects of these glasses.

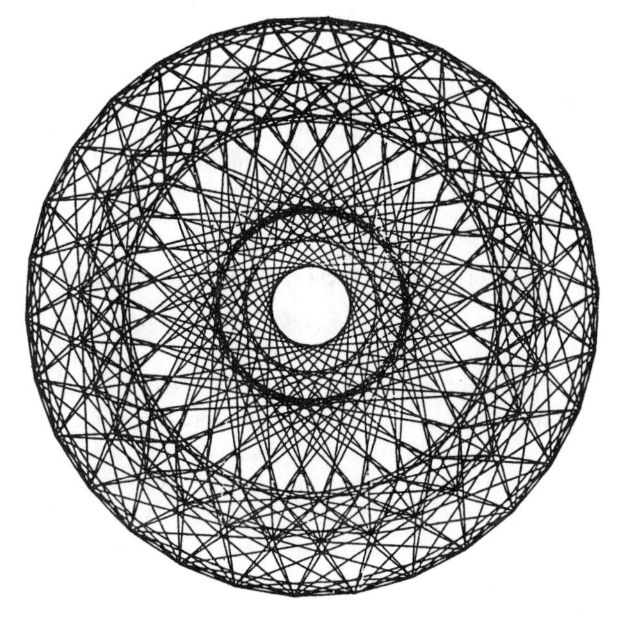

FIGURE 5.2.1 Two-dimensional projection of the 120 vertices of 720 nearest-neighbor bonds of a four-dimensional polytetrahedral polytope. The perfect tetrahedral packing in this structure is the organizing principle of the structure of simple liquids.

creating a new language of structure and symmetry. From the observed structure of these phases, experimentalists construct materials of controlled microscopic structure, symmetry, density, and thermal conductivity. For example, the fragile cubic lattice of a lipid-water microemulsion can be perfused by water-soluble monomers that are then polymerized to create a hardened tortuous network. These materials have extraordinarily high surface areas for their volume and can be engineered to host chemical reactions that must progress on surfaces.

Polymeric microemulsions—i.e., polymer equivalents of microemulsions—have been identified. Thus it may now be possible to make inexpensive blends out of immiscible polymers and create particles whose sizes are tens of nanometers. To control particle size precisely, we must understand the kinetics of formation of these microemulsions out of their highly viscous polymer components.

Liquid crystals of lipid and water can hold DNA or drugs to deliver them through the lipids of the membranes that protect biological cells. Strategies are being developed to transfect cells by delivering alien DNA across this protective barrier, while protecting the DNA so as to allow it to become part of the target cell's genome.

Specialized domains in cell membranes that surround cells confine proteins to force them into two-dimensional order. Physical theory and measurement of lateral organization is already an important part of the search to explain the origin and function of these functional regions.

Colloidal and Macromolecular Interactions

During the past 10 years there has been a burst of measurements of the forces between colloids and macromolecules: polystyrene lattices, membrane bilayers, and biological macromolecules such as DNA and collagen in solution. Optical tweezers (see Figure 5.1), atomic-force microscopes, osmotic stress, electric and magnetic fields, and immobilizing surfaces have been used to position nanometer- to micrometer-size particles and measure the forces between them. At more than nanometer separations, there are clearly defined electrostatic double layer interactions; at smaller distances, solvation and the properties of the restructured solvent often dominate intermolecular forces. Taken together, this body of measurements introduces us to a new physics of the forces that act to organize biological macromolecules and colloidal suspensions.

Having entered the realm of solvation forces, one is immediately confronted by the embarrassing realization that many of them would not have been predicted beforehand. Measured forces frequently look nothing like those predicted by the computer simulations of biomolecules generated by programs used to predict structure and design drugs. Double-layer electrostatic forces have been known for decades and behave much as expected, even though qualitative discrepancies between observation and expectation occasionally occur. Short-range, exponentially varying forces at the nanometer range, usually dubbed "hydration forces," are still only formally described. A rigorous theory needs to recognize not only the restructuring of the solvent in the face of the intrusive interface, but also the way that the macromolecular surface has been engineered by nature to work into

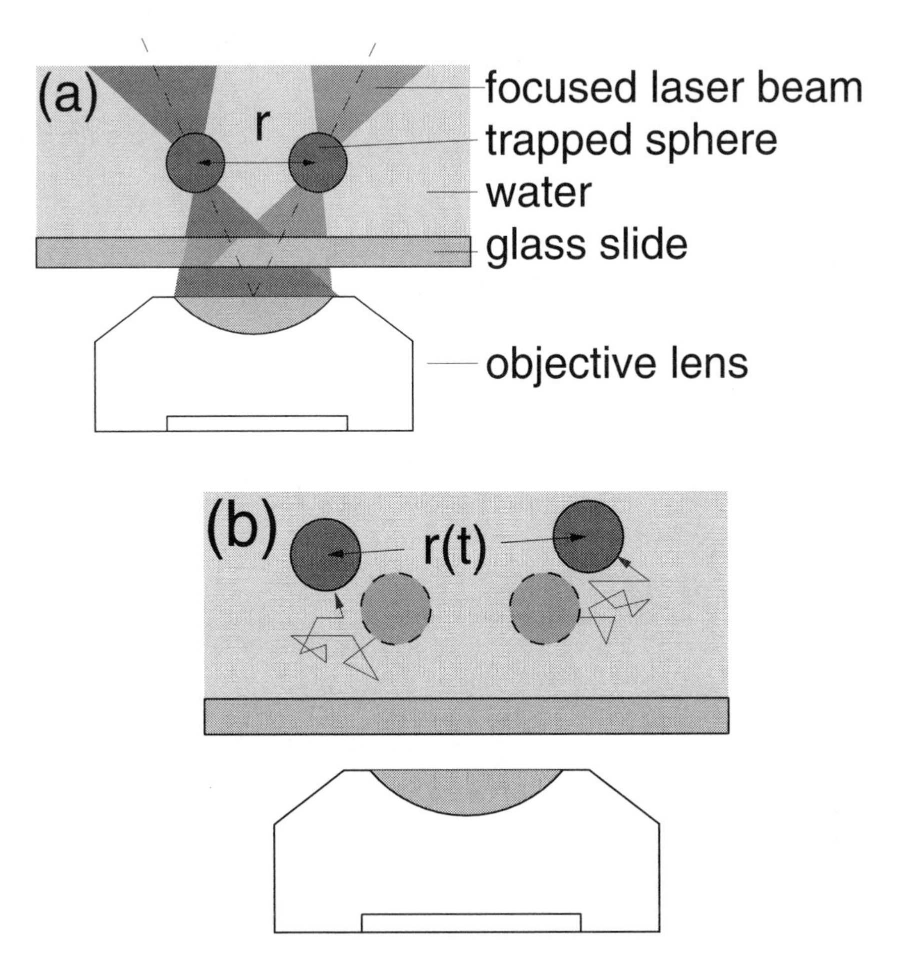

FIGURE 5.1 The laser grip of optical tweezers makes it possible to position two micron-sized spheres at a specified separation. When the laser is turned off, the spherical colloids begin to diffuse. This diffusion is followed through computerized image analysis of the spheres' motions as seen through a light microscope. From the paths of the two colloids after release, it is possible to infer the forces between the spheres. If they repel, the spheres drift apart. Micron-sized spheres can be positioned at various locations to test the consequence of proximity to surfaces. (Courtesy of the University of Chicago.)

the structure of the solvent. All the usual difficulties that impede understanding of highly structured liquids are amplified by the minute details of macromolecular structure.

Still, the empirical facts speak for themselves, telling us how to think more logically about molecular organization. New forms of colloidal crystals and suspensions can be designed using measured forces. Computer algorithms can be reconstructed to incorporate observed forces. Phase diagrams of assembling molecules and colloids can goad far more careful thinking about the statistical mechanics of liquids and liquid crystals.

For colloids, practical applications will probably come sooner than fundamental theories. New ways to position materials to effect force measurements are also new ways to organize those materials for practical use. Colloidal systems are particularly promising, in that there is already extensive experience in their synthesis and commerce. The peculiarities of unexpected forces in response to manipulation become new ways to make industrial materials such as paints, slurries, and hitherto unimagined suspensions or colloidal gels.

We can expect that basic research will be increasingly able to point the way to practical design. For example, it has been known for years that polymers attached to colloidal particles can act as "bumpers" to prevent aggregation by steric repulsion. Now we have measurements of forces between polymer layers attached to solid surfaces or to bilayer membranes. Theories of polymer steric repulsion can be critically tested. Application of polymers to surfaces can be designed for desired properties.

Conversely, polymers that are in solution but unable to associate with colloids can act to create depletion forces that act to drive the colloids together. These forces can be created, measured, and controlled by connecting known properties of polymers to observations of how those polymers change colloidal forces. Micron-size colloids can be concentrated into lattices of such good order that they are beginning to be used to create optical devices that work by diffraction and absorption of light.

Polyelectrolytes

Polymers whose properties are dominated by their electrostatic charge are instructive because of their solution properties, their ability to control ion activities, and their propensity to form liquid crystals. Among biopolymers, DNA has been the most intensely studied from the viewpoints of liquid-crystal physics as well as its "solution" properties in the cell. Intermolecular forces have been measured in detail, both for molecules in simple salt solution and for those organized by natural condensing agents. There has been extensive work aimed at modeling the electrostatic potential around DNA.

Among artificial materials, there are several electrically charged or polarizable polymers, both natural and synthetic, that form networks controlled by applied electric fields. Some of these materials are block copolymers, neutral as well as charged, whose long-range order can be seen to emerge from mesogenic organizing centers in the molecules themselves. Enhanced stiffness in polyelectrolytes can be achieved by neutralization with oppositely charged aliphatic surfactant molecules. Modest electric fields can be particularly effective in creating organization in liquid-crystalline polymers. Large-scale organization can also be induced by shear processing these materials because, like most mesomorphic materials, they readily align in shear flow.

The interplay between the various degrees of flexibility and freedom and the electrostatic forces within and between molecules creates many modes of packing and stimulates theories of molecular organization. The need is for complete theories that include real force potentials rather than analytically convenient approximations. Precise determination of phase structure from x-ray and neutron diffraction, combined with direct measurement of the work of assembly, will be essential.

At the moment, DNA is probably the friendliest polyelectrolyte, natural or synthetic, used to create well-defined liquid crystals in water. Soon there should be other made-to-order polymers in practically any single length. Polydispersity, a major nuisance when testing theories of polymer assembly, is not an impediment with uniformly engineered DNA. It may even be that the size regulation so useful for fundamental studies will also make DNA and other precisely prepared synthetic model compounds similarly useful for practical applications. Because they are highly soluble in water, polyelectrolytes might come to replace organic polymers, which must be dissolved in environmentally unsuitable solvents.

Polysaccharides

Although the substance of enormous industries (see Box 5.3), polysaccharides have been relatively unappreciated by most polymer chemists and physicists. Cellulose, whose biomass exceeds that of any other natural polymer, was not even mentioned in the National Research Council's 1994 survey of polymers.[1] At present, modifications in chemistry and physical processing are creating new research questions and many practical applications, from the design of new paper currencies to the creation of industrial fibers to cosmetics to artificial food and blood thickeners.

It has recently become possible to measure equations of state of several polysaccharide systems, a development that should demand better theories of polymer assembly. Considering the mass of polysaccharides in the world, their economic and practical importance, and the excellent chemical and biochemical work already done, it is surprising that physical research has been so limited. Because polysaccharides are often polymers of repeating units, they would seem to be an ideal material for physicists to study. There is a splendid opportunity here for instructive physics on materials far less complicated than the more popular proteins. Their swelling properties, their viscous and elastic capabilities, and their stability over a wide range of solution conditions and temperatures are theoretically intriguing and technically enticing. There is already the expectation of creating bacterial polyesters and biodegradable ther-

---

[1] *Polymer Science and Engineering: The Shifting Research Frontiers*, National Academy Press, Washington, D.C. (1994).

BOX 5.3 Growth of the Use of Polysaccharides: Xanthan, Guar, and Cellulose

The label on a bottle of salad dressing, a box of ice cream, a coffee whitener reminds us of the various polysaccharides that we consume. Among these many polymers of sugar are guar from seeds, carrageenan from seaweed, pectin from fruits, and xanthan from the coats of microbes. These compounds are used as thickeners and preservatives, often playing a role parallel to that played in natural circumstances. Xanthan is so versatile, so stable in its physical properties in the face of heating and mixing with salts that it is pumped into the ground to stimulate the recovery of oil wells and into your stomach after giving food the right "mouth feel." The animal polysaccharide hyaluronic acid is a significant component in cartilage and connective tissue; commercially extracted from animals, it is increasingly used medically for the repair of joints and cartilage.

It is no surprise then that natural polymers, or slightly modified natural polymers, are industrially popular. Several billion pounds of starches from plants are used in the United States alone for processing paper and in sizing, binding, and adhesive applications. Hundreds of millions of pounds of modified cellulose find their way into foods as well as paper and construction materials.

Viscous and elastic properties make these polymers industrially valuable. These are physical properties. Yet physics has paid surprisingly little attention to polysaccharides and related polymers. With its new capabilities, soft-matter physics can be expected to recognize and to modify the behavior of these materials that have traditionally enjoyed the attention mainly of chemists, colloid scientists, and chemical engineers.

moplastics. The natural polysaccharides that coat some bacteria are able to direct the precipitation of minerals dissolved in the surrounding solutions; heroic hopes for deep-sea mining might be coupled to learning how tiny bacteria collect minerals.

Taken in the context of polymer studies in general, there is the possibility to study materials that have already been selected by nature for their physical properties. Most industrial use has been guided by trial and error rather than by combination with systematic physical theory and experiment. In the food industry particularly there are huge potential benefits.

MACROMOLECULES AND MACROMOLECULAR FILMS

Phase Separation and Ordering in Thin Polymer Films

Because diffusion is so slow in molten polymer systems, physicists have been able to observe the very early stages of phase separation—stages in which change is so slight that simplified, linear theories can be applied. It is possible to see the importance of polymer chain length on the kinetics and evolution of chain morphologies, the different classes of behavior approaching critical points, and the effects of surfaces and interfaces on phase separation. Phenomena that can only be observed in atomic or molecular films when they are, at most, 5-nm thick can be seen in 500 nm thick polymer films. Because of this extra range, it has been possible to see phase separations as a function of time and position from an organizing surface. Patterned surfaces can be templates to create three-dimensionally patterned films (see Box 5.4).

Structured films of synthetic and of natural polymers can be used as substrates to host and guide the growth of biological tissues. New processing treatments of surfaces for medical implants might follow from fundamental discoveries on thin polymer films.

New Macromolecular Materials

Dendritic and hyperbranched polymers have quite dense structures but nearly monodisperse size distributions (from 2 to 50 nm). These macromolecules can be tethered to surfaces to act as "bumps" to prevent the "stiction" of hard-disk drive heads; they can also be used to separate active ions such as erbium in optical amplifiers. The nature of the surface of these dense macromolecules, their behavior in solutions and melts, and their internal structure all have physical consequences. Their compact, well-defined shape stops mutual interpenetration or strong entanglement, which creates low-viscosity melts and solutions. Potential applications include their use as highly reactive, low-viscosity components for construction of thermosetting polymer networks. Spherical and wedge-shaped "monodendrons" can be produced (see Box 5.5). Attached hydrophobic groups create a variety of cylindrical crystal structures with ion-conducting channels. The possibilities of design are so encouraging that we have begun to speak of biomimetic polymers. The scale of the opportunities afforded can perhaps be gauged by the degree to which planar macromolecular assemblies (i.e., self-assembled monolayers) are currently being investigated by polymer physicists.

Flexible polymerized or "tethered" membranes are natural generalizations of linear polymers. Examples include graphite oxide sheets, the "rag" structure of MoS, and the spectrin skeleton of red blood cells. Theory predicts a remarkable flat phase, with anomalous roughness and singular elastic constants, caused by a delicate interplay between thermal fluctuations and the Gaussian curvature. The behavior of defects such as dislocations in unpolymerized flexible membranes (such as lipid bilayers) with local crystalline order is also remarkable. Unlike crystalline films forced to be flat by a surface tension, it is energetically favorable for these defects to buckle into the third dimension (see Figure 5.2). The macroscopic result is a "hexatic" membrane with a finite dislocation density at any low, but nonzero temperature, i.e., a membrane with zero shear modulus but with extended bond orientational order and a continuously variable fractal dimension.

BOX 5.4 Control of Phase Morphology

Films of phase-separated polymers can be forced to take on a microscopic lamellar structure when the film is "capped" by stiff coating.

A mixture of the immiscible polymers polystyrene (PS) and polymethylmethacrylate (PMMA) in a bulk melt forms a disordered arrangement of large (>1 μm), almost pure PS and PMMA phase domains. Remarkably, if a thin film of a mixture of PS and PMMA on a silicon oxide (SiOx) substrate is quenched so that phase separation is prevented, then capped with an additional layer of SiOx, and finally heated to allow demixing to occur, a very regular pattern of the PS and PMMA domains can result. A schematic of the domain structure through the film thickness is shown (Figure 5.4.1).

The figure shows that there are many long, parallel phase-separated domains with a well-defined periodicity within the plane of the film, necessarily accompanied by a periodic deformation of the upper film surface. The periodicity scales with the thickness of the polymer film and disappears if the SiOx capping layer is very thick; in this case the film consists of a uniform, five-layer sandwich, SIOx/PMMA/PS/PMMA/SiOx. The structure comes from a balance between the free energy associated with forming the interfaces between PS- and PMMA-rich domains and the free energy increase associated with the elastic bending of the silicon oxide capping layer. These factors create both the lateral morphology observed for small capping layer thickness and the transition to only five layers (SiOx/PMMA/PS/PMMA/SiOx) as the capping layer thickness is increased.

A notable feature of many liquid crystals and macromolecular films is their extraordinary sensitivity to their bounding surfaces. A change in surface tension or surface stiffness can be felt even millimeters away from the surface. This kind of sensitivity to outside influence is an unexpected tool for arranging molecules into desired microscopic form.

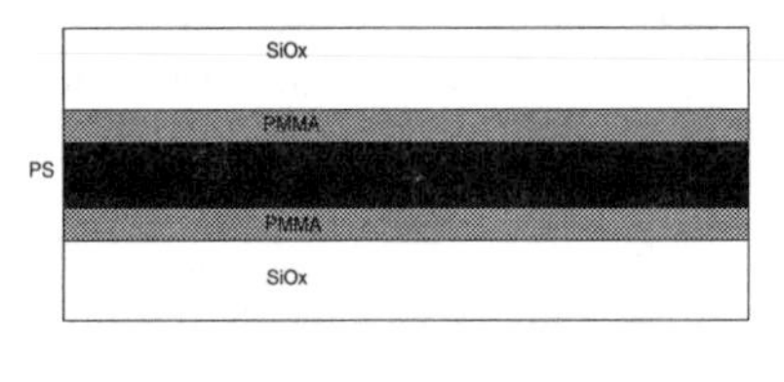

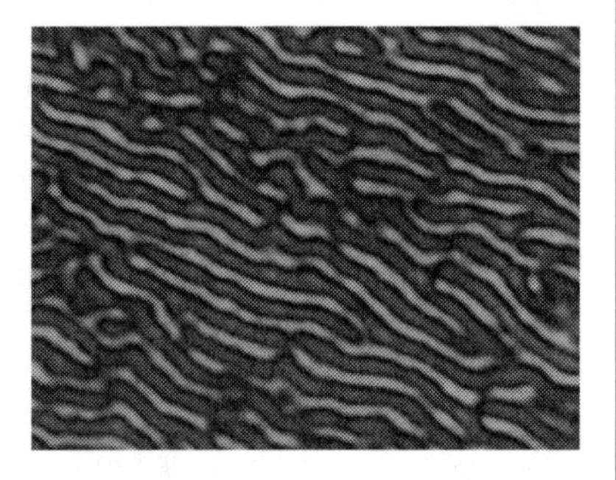

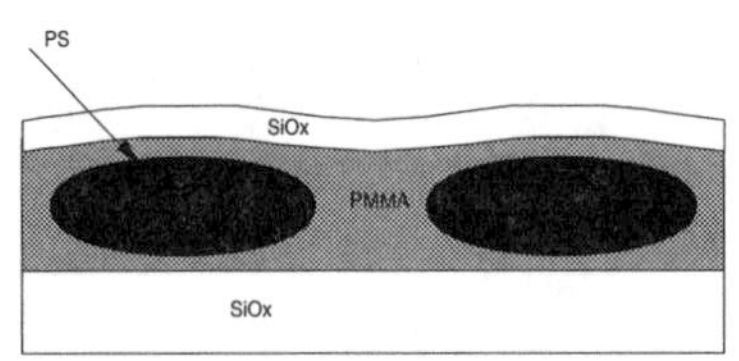

FIGURE 5.4.1 Schematic of PS and PMMA domain structures. (Courtesy of the University of Guelph.)

BOX 5.5 Dendrimers, Strictly Spherical Polymers

We usually think of polymers as a string of flexibly linked beads. It is possible, though, to link together Y-shaped units. The pattern of regular branching creates spherical dendrimers, whose precisely determined diameters are ~100 Å (Figure 5.5.1). These spheres can be coated with a variety of chemical groups so that their microscopic surfaces can have different qualities. Surface chemistry controls the way these tiny spheres move and pack under concentrated-solution conditions.

The viscosity of such dendrimer fluids is much lower than that of normal polymers, leading to novel applications. In dilute suspensions, chemically labeled dendrimers can be directed to react specifically with desired sites. One possibility is that single dendrimers, loaded with drugs, will release those drugs at the places to which dendrimers have been directed to bind (Figure 5.5.2). Another is that partial dendrimers, dendrons, with liquid-crystal elements will self-assemble into arrays of cylinders with precisely tailored inner and outer surfaces. These cylinders may find uses as engineered channels through cell membranes.

Because they can be engineered so reliably and because of their rigidly simple spherical shape, dendrimers are a powerful research tool. The idealized models used in many physical theories resemble the well-defined spherical structure of dendrimers. In addition the structures of dendrimers can be systematically modified to test predictions of physical theories.

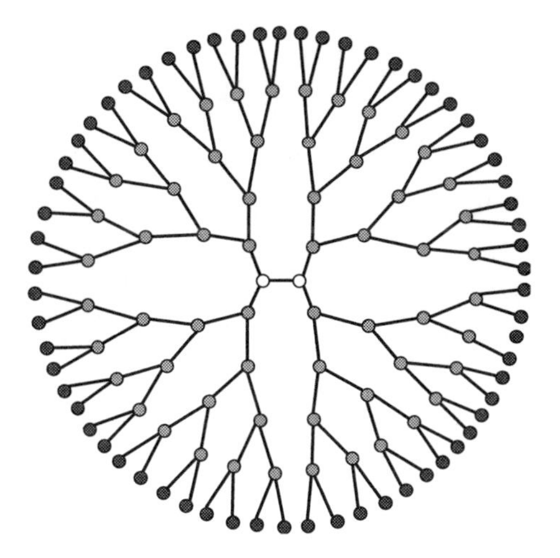

FIGURE 5.5.1 Single dendrimer. (Courtesy of the National Institute of Standards and Technology.)

BOX 5.5 Continued

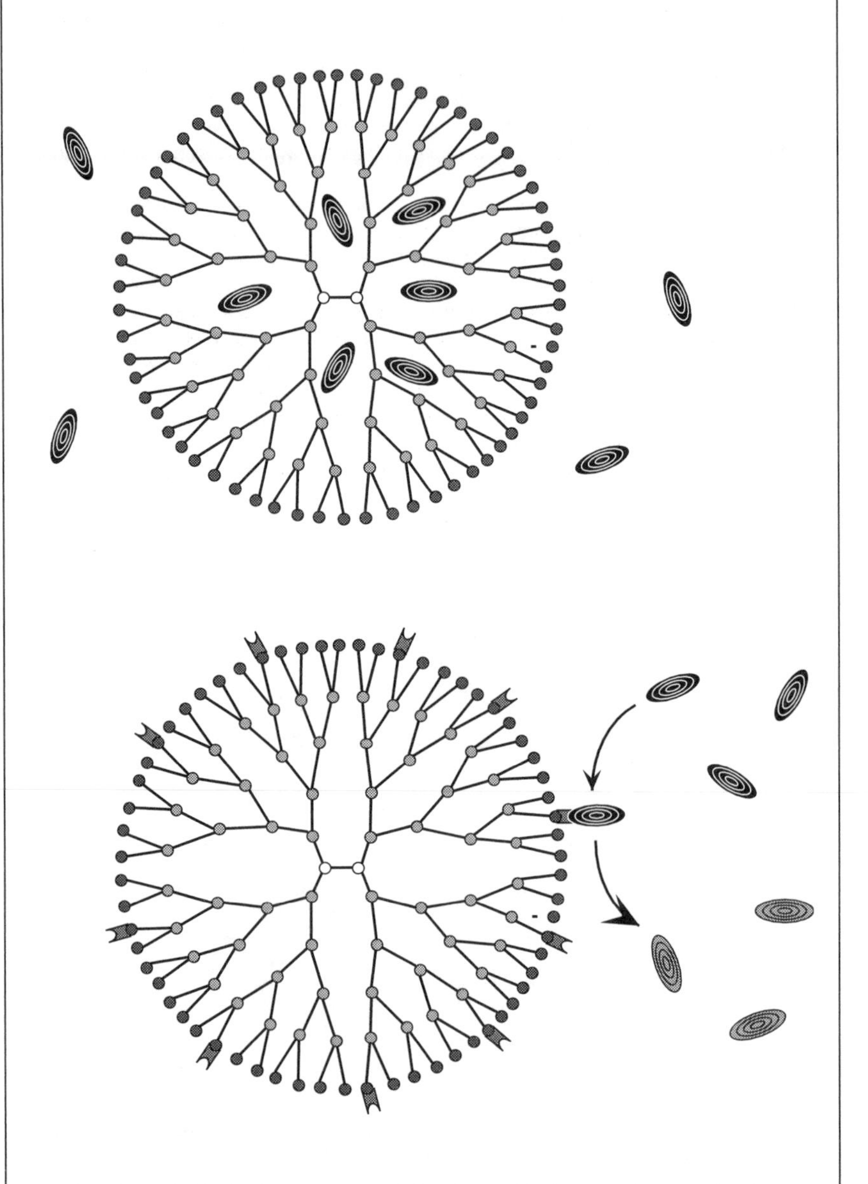

FIGURE 5.5.2 Single dendrimers (top) "packed" with drugs and (bottom) releasing the drugs at specified binding sites. (Courtesy of the National Institute of Standards and Technology.)

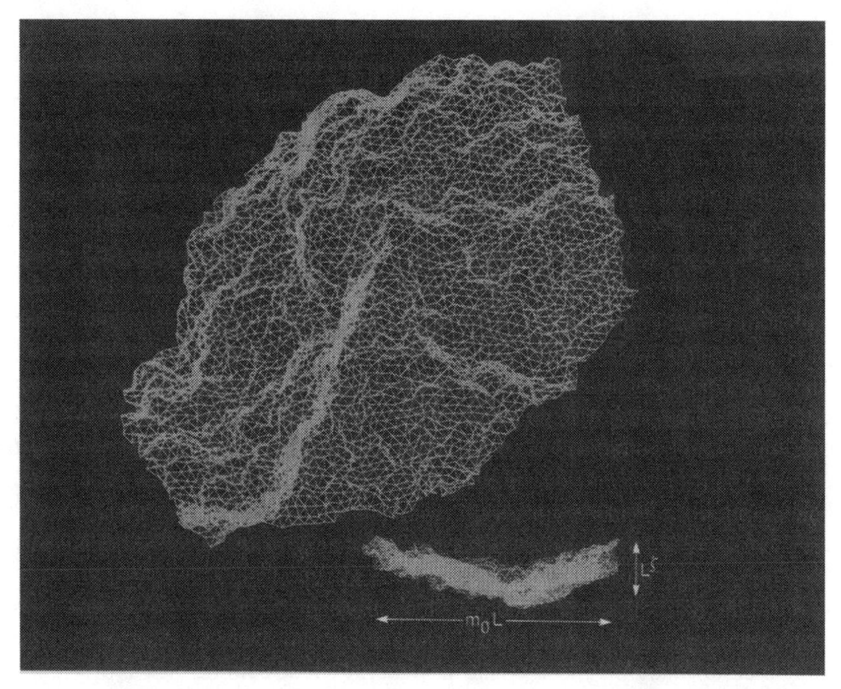

FIGURE 5.2 The crumpling of a membrane sheet. Perspective and side view of an instantaneous configuration of a large tethered membrane composed of 4219 monomers. [Reprinted with permission from F.F. Abraham and D.R. Nelson, "Diffraction from polymerized membranes," *Science* **249**, 394 (1990). Copyright © 1990 American Association for the Advancement of Science.]

Associating polymers are molecules with hydrophilic backbones and two or more associating hydrophobic groups ("stickers") per chain. These polymers have very unusual rheological properties because of the tendency of the associating groups to cluster and form transient networks at sufficiently high polymer concentrations. Their dramatically altered solution elasticity and viscosity allows applications as thickening agents and rheology modifiers.

Unfortunately, most systems have stickers that are randomly placed along the chain; such associating polymers are poorly understood. New synthetic methods now allow us to prepare systematically varied molecules with specified number and location of stickers. Polyelectrolytes and polyampholytes, with similarly daunting structural heterogeneity can now be more systematically produced. The possibility of learning from the properties of these molecules leads naturally to related phenomena, such as protein folding and the organization of polymer systems of increasing but controlled "randomness."

Polymers associated into gels will hold drugs whose diffusive release into the body can be controlled by gel compactness. These gels can be placed in a particular part of the body for local, directed, sustained drug release. Stable substrates of synthetic and natural polymers create substrates for directed cell culture.

Structural Polymers: Controlling Properties of New Polymers from Old Monomers

Polyolefins, so-called "commodity polymers" such as polyethylene and polypropylene, are produced in huge quantities by the petrochemical industry. Polymer production for U.S. consumption alone is about 10^{11} pounds per year. This is roughly equal to steel production in weight, but it is about seven times steel production in volume. Polyolefins are nearly 60 percent of the total. These polyolefins are cheap. Their chain architecture can be manipulated by relatively crude methods that offer some control over branching and stereoregularity of the chain. In the early 1990s, polymer chemists developed new transition metal catalysts. These catalysts offered unprecedented freedom to copolymerize the olefins with other monomers; they also allowed much tighter control over branch content and stereoregularity. It is now possible to achieve properties for ethylene-containing polymers that range from those of good elastomers (soft, with only a very small fraction of tiny crystals to act as crosslinks) to highly crystalline (hard) polymers. This wide range of properties can be achieved using a very cheap monomer to make up most of the polymer. Great commercial and scientific opportunities exist for polymer physicists to develop the relation between chain architecture and the properties of these new materials.

For example, high-density (unbranched) polyethylene of high molecular weight is used routinely for the socket of artificial hip joints. For this use it has an elastic modulus of 10^3 GPa and a failure stress of only 50 MPa. The same polyethylene, processed with solvent into a dilute gel, dried, and then drawn into fibers, has an elastic modulus of 170 GPa and a failure stress of 3000 MPa along the fiber direction. Processing methods such as fiber spinning, blow molding, injection molding, extrusion, film blowing, solution casting (for adhesives and coatings), and reactive blending subject the molten polymer or polymer solution to a flow field that strongly distorts or aligns the molecular conformations. Such processing is able to effect the final material properties after cooling, cross-linking, or evaporation of solvent. Trial-and-error improvement in commercial processing will be more efficient with better theories of polymer organization.

The most serious lack of knowledge concerns what determines the mechanical properties of semicrystalline polymers and how they depend on flow history and polymer architecture. Because ease of processing and final properties are equally important attributes of a polymer, both melt rheology and final properties need basic study. Studies in the 1970s did not emphasize the strong effects of flow, which can reorganize the orientation of crystals and change crystallization rates by orders of magnitude. Modern physical probes (e.g., small-angle neutron and synchrotron x-ray scattering, infrared dichroism, and Raman spectroscopy) could provide the tools to understand the transient effects of flow on the molecular conformation, crystallization, plastic deformation, and fracture.

Biological Connections

In contrast to synthetic polymers, which are composed of at most three monomer types, normally arranged in random order, proteins are synthesized according to a programmed sequence involving 20 kinds of monomer. They are also produced to a precise length. Other classes of biopolymers, such as saccharides and nucleic acids, are also very precisely specified compared with artificial synthetics; even heterogeneity appears to be intentionally created. Polymer chemists and molecular biologists have begun to collaborate to synthesize copolymers with the same precise control of monomer sequence and chain length. The challenge for polymer physicists, as precision-architecture polymers become available, is to understand the link between architecture and polymer system properties. Lured by the wide range of properties conveyed by biological macromolecules, from spider silk to the elastin of blood vessel walls, we expect to see polymers with similarly remarkable properties emerge from these new syntheses. Given the vast range of possibilities and the small initial quantities of each polymer, it will be necessary to develop methods for rapid screening. Scanning-probe microscopies have begun to be used to determine mechanical properties. The aim is to be able to screen polymers for desired properties with the same efficiency that is achieved for developing and producing biological polymers by the natural checks of evolution and growth.

BIOLOGICAL SYSTEMS

Biological molecules are substances that have evolved to do highly specific jobs on highly specific timescales. Physicists have had impressive success studying the dynamics of macromolecules, particularly proteins. The mechanics of single molecules can be measured and used to test theories of molecular conformation. The measured energies of packing biopolymers inform us of the work needed to package them into cells and viruses and challenge us to explain and manipulate macroscopic properties.

It helps to distinguish incidental and essential physical properties of biomolecules. For some materials, such as DNA, the cell works with physical properties because it must. For others, such as RNAs, proteins, lipids, and polysaccharides, molecular physical properties are themselves useful to the organism. There are bulk materials that provide structural stability to the cell or organism. Happily, the language and concepts of lyotropic liquid-crystal physics often meet the need to examine biological materials.

Two Traditions of Learning Must Merge to Allow Systematic Progress

For there to be the necessary crossover between biology and physics, there will have to be a significant change in the way we teach ourselves and our students to handle new materials and to respect the natural ways designed by patient evolution. In the ideal scenario, physics students will learn the facts and methods of biochemistry and molecular biology, and biology students will lose their fear of and skepticism about talking with physicists (and perhaps even learn some physics). In this area more than in any other addressed by this report, progress will have to follow education. It will be built on a developed respect for the very different thinking and experimental methods used by the two "sides." This point is far more pertinent than any review of particular systems or prediction of future events. A 1996 study on self-assembling biological materials[2] nicely summarizes the current state of learning about molecular organization and provides many examples of doing physics with biomaterials, including:

- New paths for synthesis of peptide (proteinaceous) chains with practical properties as well as properties that will be available to test physical models of assembly;
- Spontaneously assembled surfactants, both natural and synthetic, with their possibilities for structuring surfaces, hosting proteins or DNA or molecular sensors, binding stabilizing polymers, and (again) creating structures amenable to systematically developed theories of lipid assembly;
- Layered polymer systems that can act as sieves, scaffolds for growth and hardening of nonorganic materials, and mechanically stabilizing films; and
- Gene therapy, in which physical methods might be used to design the delivery and incorporation of new genes to be reproduced within the genome of target cells.

The committee enthusiastically agrees with that study and will not repeat its advice here. Rather, this section will provide more examples for both basic and applied research. It will suggest possibilities for a physics of biological materials in which physical thinking will be essential to the understanding of how biological substances are designed to work in their native habitat. There are many biological phenomena that traditionally or currently have been studied productively through physical thinking. Some examples are as follows:

- Protein-DNA and protein-protein interactions;
- Molecular motors and locomotion;
- Protein interaction, folding, and dynamics;

[2]National Research Council, *Biomolecular Self-Assembling Materials: Scientific and Technological Frontiers,* National Academy Press, Washington, D.C. (1996).

- Photosynthesis and energy transfer; and
- Nerve action.

Physics and Structural Biology

If allotted text were in proportion to progress, the discussion of molecular structure determination and structural biology would nearly fill this chapter. Because of advances in physics, we now expect to know the structure of large functional biological molecules in crystals and in solution to angstrom resolution. Any thinking about functional mechanisms is expected to build on known structures. Although structure determination of proteins in solution is usually by nuclear magnetic resonance (NMR) and is considered the province of chemists, most structures are delineated by x-ray diffraction, whose procedures are predominantly the work of physicists. New sources—synchrotrons, free-electron lasers, and high-power x-ray laboratory generators—together with better lenses (e.g., optical fibers, multilayer optics, and single-capillary microbeams) and a new generation of fluorescence and charge-coupled device detectors, offer the possibility not only to determine structure but to observe several classes of molecular motion (see Figure 5.3).

Synchrotrons, neutron sources, laboratory x-ray generators, and NMR machines, with their ancillary detectors and computation algorithms, are practical microscopes whose success can be measured in the number of publications involving their use, number of students studying and using them, and numbers of dollars expended on them. Increasingly precise molecular structures, frequently with better than 1 Å resolution, are the product of strong light sources as well as better handling of delicate materials. Bright synchrotron sources are especially useful for time-resolved studies, in which snapshots of the structure can be taken in small time increments to make it possible to study intermediate states during kinetic processes such as protein folding, allosteric changes, and association or dissociation. Observable time resolution can be as short as nanoseconds, with about 0.1 ns expected soon. Here, as elsewhere in this report, we note the need to nurture these facilities, improve them, and expand them so that they can continue to meet and encourage the growing demand for revealing molecular structure, assembly, and dynamics.

More urgent is that, while the world grows x-ray and NMR rich, it remains neutron poor. Neutron and x-ray scattering, as well as NMR, allow us to study the solution structure and dynamics of proteins and lipids and, perhaps, devise industrial applications. Neutrons are sensitive to the positions of biologically abundant light elements such as hydrogen, carbon, nitrogen, and oxygen. Because of this, with hydrogen-deuterium isotope substitution for contrast, small-angle neutron scattering can provide unique information about the structure and function of biological macromolecules, protein-nucleic acid or protein-lipid complexes, and multi-subunit proteins. Neutron reflectivity has come to reveal fea-

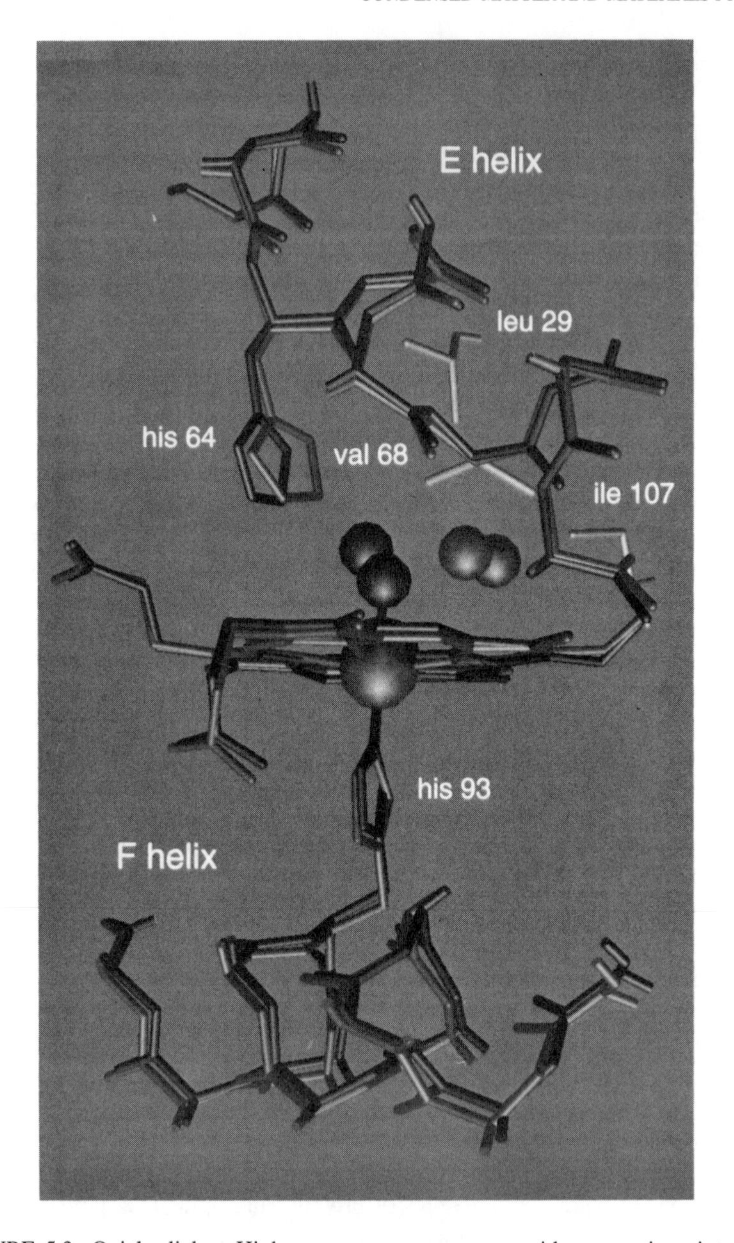

FIGURE 5.3 Quick clicks. High-power x-ray sources provide successive pictures of a protein process. [Reprinted with permission from W.A. Eaton, E.R. Henry, and J. Hofrichter, "Nanosecond crystallographic snapshots of protein structural changes," *Science* **274**, 1631 (1996). Copyright © 1996 American Association for the Advancement of Science.]

tures of lipid monolayer and bilayer systems in aqueous environments, especially for preparations that are difficult to crystallize. Reflectivity could similarly be used to aid the biosensor industry by measuring the interactions between biological components and organic or inorganic surfaces and to assist in the understanding of nonspecific binding of proteins to surfaces. Dynamic processes in macromolecules can also be probed directly by neutron scattering because scattering cross sections relate directly to oscillation frequencies and amplitudes of motion. Optical microscopic techniques like fluorescent energy transfer are very useful for obtaining information about proteins in vivo.

The need for good physics goes beyond mere observation. Today, about 35 years after the first hemoglobin structure, we still do not understand how that protein changes its binding affinity for oxygen by assuming distinct allosteric forms. Energy storage and energy flow through this nanometer-sized machine still deserve the attention of molecular biologists and biophysicists, who still have difficulty conversing.

Molecular Conformation and Protein Folding

The protein-folding problem attracts some of the best minds in biological physics and statistical mechanics, but it will require much more effort to reach satisfactory results. The question of how a linearly synthesized protein curls up to its three-dimensional functional form continues to challenge and excite theorists. The committee concurs with the eloquent review found in a recent National Science Foundation study.[3] Decades of witty measurements and imaginative theories suggest that there can be some progress on this vexing problem, though success in prediction is still elusive.

At this stage, it is clear that physicists have defined the language in which folding is being examined. Cellular chaperone proteins, which wrap a protein during its transformation from a linear chain to a three-dimensional structure, are now believed to act by reversing random errors in folding. Their action is discussed by biologists and physicists alike in terms of their kicking the protein out of snarls that might otherwise trap it on its way to correct conformation. In this way, cellular activity is thought to facilitate what is essentially a statistical physical process. The tools to follow folding are increasingly based on physical techniques, particularly high-power, high-speed x-ray scattering and spectroscopy.

Single-Molecule Motions and Mechanics

Laser tweezers allow one to hold the ends of linear molecules, particularly double-helical DNA, to measure the force of extension in different solutions and

[3]Ken Dill, NSF Workshop on Interdisciplinary Macromolecular Science and Engineering (unpublished), chaired by S.I. Stupp, University of Illinois at Urbana-Champaign, May 1997.

with association with other molecules. Suddenly it is possible to think about single molecules as mechanical objects, on which physics can be performed analogous to that done on macroscopic materials. Pipette aspiration allows equivalent mechanical manipulations of membranes and bilayer vesicles (see Box 5.6). The various moduli and deformabilities are just beginning to be codified and connected with the language of mechanics. The actual values of these material properties still often surprise us and require new thinking.

Force microscopes can be used to break bonds, stretch molecules, or observe spontaneous changes in macromolecular conformation. The breaking strengths of important bonds, such as those between an antigen and an antibody, have been observed using labeled polystyrene lattices. It has been found that the force needed to break a bond is a monotonically decreasing function of the time that the

BOX 5.6 The "Spring" of a Bubble to Tug Apart Large Molecules

Many biological processes depend on associations that require no chemical bonds between molecules. Important examples are the association between antigens and antibodies that allow cells to recognize alien matter or between integrins that hold cells together. Even without being pulled apart, these molecules will spontaneously come apart with time.

In a functioning cell, the precisely controlled duration of molecular association can be more important than the strength of association. Cells create characteristic times for many processes by controlling association times. The measurements illustrated here show how times of association between a single pair of large molecules can be measured and instructively varied.

The trick is to pull on the molecules with a spring whose rate of tugging (force/time) can be varied by 100,000,000 times (see Figure 5.6.1). Springs are made from lipid bilayer vesicles or from red cells—effectively bubbles or microscopic bags—to which associating molecules can be strongly attached. The bubble is held by being gently sucked into a pipette whose suction pressure can be varied to change tension in the vesicles.

To set up the measurement, a molecule on one bubble is brought near to another bubble that contains the mating molecule; the bubbles stick at the point of molecular contact. Then, by sucking on the pipette, stretching tension can be transmitted to the two molecules. The time to breakage of the antigen/antibody bond varies with varied rates of applied tension. From this measured time-to-breakage, it is possible to infer the kind of contact that the molecules made and, more important, to learn how the molecules create their important times of association.

In this particular example, binding is observed between biotin and avidin. Dissociation time ranges from 0.01 seconds to 100 seconds after the onset of a ramped force. These gentle-tug measurements show how association lifetimes can be exquisitely sensitive to small changes in applied structural forces. Physical forces can regulate cellular events.

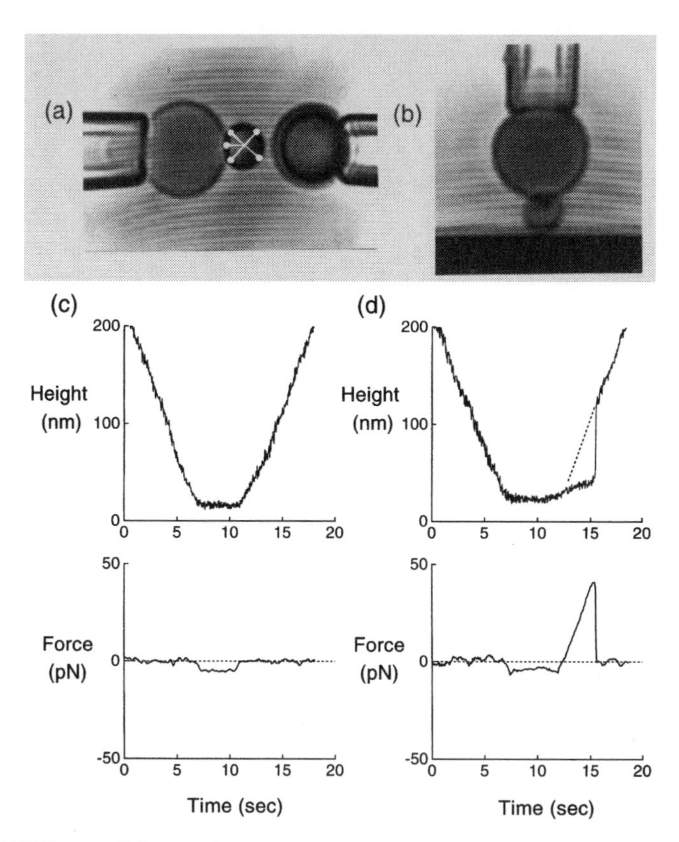

FIGURE 5.6.1 Using a bubble to tug apart large molecules. (Courtesy of Boston University and the University of British Columbia.)

force is applied. This connection between time and strength, understandable in terms of diffusion in the presence of a distance-dependent bonding potential, reveals the dynamics of molecular association in the context of biological control. One can expect many single-molecule systems to be observed and analyzed by probe microscopies to create a nanomechanics of molecular force and assembly. There is already a small literature on the spectroscopy of single proteins trapped in small spaces and illuminated by narrowly focused laser light.

Through the combined efforts of structural biologists, muscle biophysicists, and statistical physicists, force generation in muscle and in transport within cells is being seen as the combination of stochastic events and directed response. The conversion of chemical reaction to directed physical action has posed a fundamental problem because of the different symmetries of biochemical cause and physical effect. The puzzle is being solved through molecular tracking and stochastic models. We can now expect the design of new mechanical transduction systems.

Single ionic channels show electrically detectable transitions between "open" and "closed" configurations, whose probabilities are a function of solution conditions as well as the applied transmembrane electric fields. There is now a possibility of studying the dynamics of the channel molecules, as well as the solution components that affect them, simply by watching singular molecular events over very long times (see Figure 5.4). With channels, as with other proteins, it has been recognized for almost 30 years that occupation of the different conductance states follows simple Boltzmann statistics that allow us to relate the probabilities of different states with the changes in energy needed to achieve them. In fact, for a channel or other responsive molecule to sense changes in condition, it is necessary that there be small differences, comparable to the thermal energy kT, in the energies of differently functioning states.

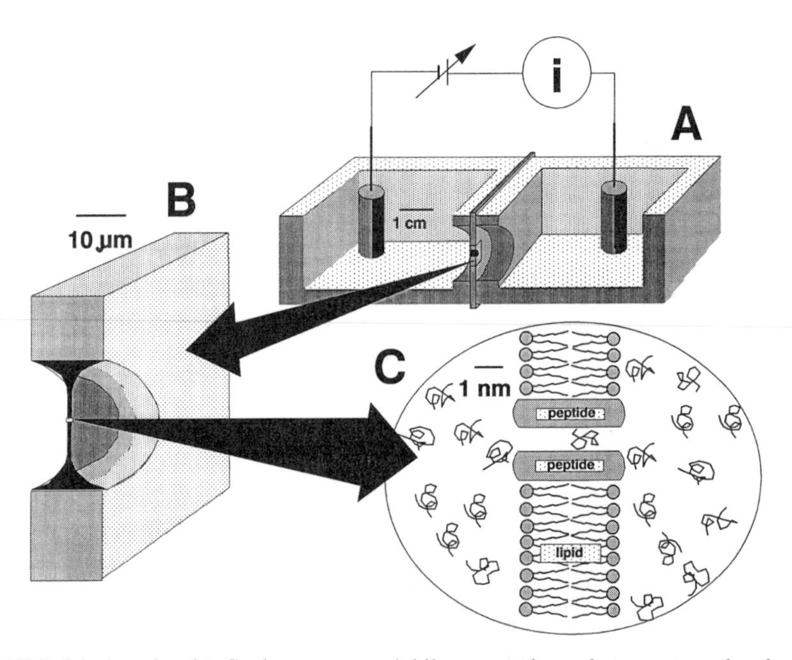

FIGURE 5.4 A molecular Coulter counter. A bilayer membrane between two chambers contains a single ionic channel of ~1 nm diameter. If a small molecule moves into an open channel, the event is seen as a reduction in electric current. The duration of this reduced current measures the residence time of the molecule diffusing into and out of the channel. [Reprinted with permission from S.M. Bezrukov et al., "Counting polymers moving through a single ion channel," *Nature* **370**, 279 (1994). Copyright © 1994 Nature.]

As a next step in time-resolved, small-current detection, it is possible to see changes in the "open" state when single molecules enter and exit the channel. In this way it is possible to measure the statistics and dynamics of flexible polymers as they move within the confines of a precisely defined single-molecule structure.

What does it mean to watch the changes in a single molecule? Can one use rigorous physics to describe these changes? Can this physics give us the energies and entropies that drive the system to different states? The answer seems to be yes. Statistics, statistical mechanics, and functional control present a perfect chance for physicists to bring their methods to help biologists. There can be ready extrapolation to create practical devices, such as detectors and computers, informed by biological designs, while single-molecule thinking inspires a new statistical mechanics of strongly coupled systems with enormous fluctuations.

It is known now that many biological systems, from single ionic channels to entire sensory systems, practice a kind of stochastic resonance. Physicists in the 1920s realized that adding some noise to a weak radio signal could improve signal detection. Today's physicists and biophysicists have succeeded in generalizing those ideas, raising the hope that they might be applied to helping the hearing-impaired.

Protein dynamics are a subject of intense interest. Binding of small molecules (see Figure 5.5) and changes in protein shape are essential properties that physicists are in an excellent position to explain. Timescales are especially worth recognizing here. Nerve signals, for example, occur on timescales of milliseconds, and different nerve channel proteins act at slightly different rates. Single molecules remain "open" or "closed" for milliseconds. These millisecond-characteristic times reveal the limits of computer simulation. Considering that these simulations can cover only nanoseconds, there is still a factor of 10^6 to cover to connect molecular dynamics computations with functionally interesting molecular times. The wit of physical theorists is the most promising way to compensate for this daunting factor in computation.

Molecular Association

There is justified pride in modern polymer synthesis, by which stretches of one or another kind of monomer allow polymers to associate in parts to multimolecular arrays of specific symmetry, packing, and material properties. Yet this kind of packing is rough compared to that of proteins or DNA, whose every monomer has functional consequence. One mutation in one amino acid of an antibody will qualitatively weaken its antigen-antibody binding strength and specificity. One change out of six nucleic acids will spoil that sequence for recognition by a protein that controls gene expression. Strength and specificity are what count when an antigen or a hormone binds to a cell-surface receptor at

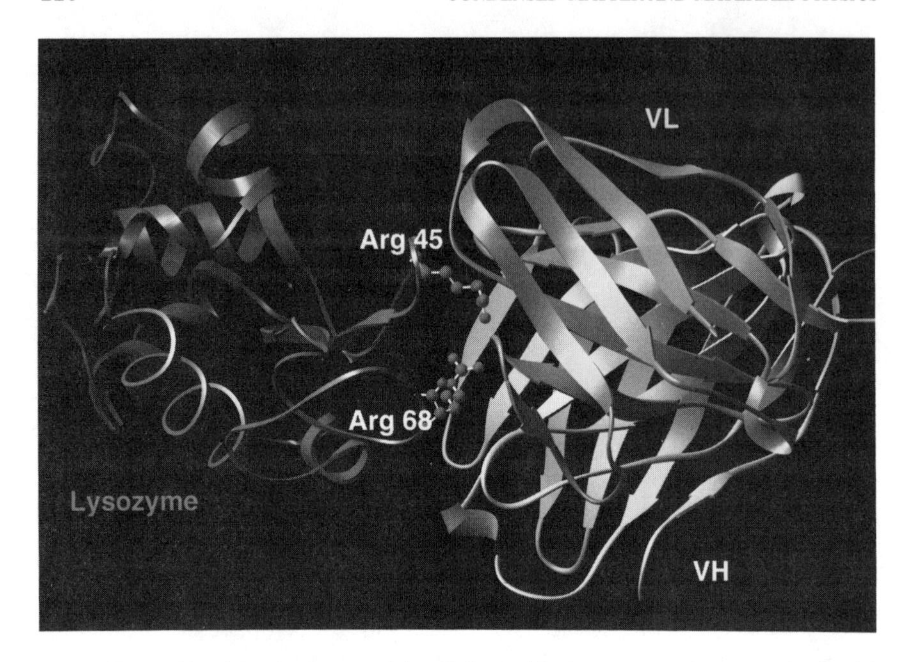

FIGURE 5.5 The right details: precision fit in the binding of an antigen to an antibody. The antigen, a lysozyme protein nestles into an antibody. A groove in the antibody matches itself tightly against a ridge in the lysozyme antigen. This ridge is formed by two arginines, at positions 45 and 68. Such is the precision of the match that mutation of Arg68 to a chemically similar Lysine reduces binding strength by a factor of 1000. [Reprinted by permission from S. Chacko, E. Silverton, L. Kam-Morgan, S. Smith-Gill, G.H. Cohen, and D.R. Davies, "Structure of an antibody-lysozome complex. Unexpected effects of a conservative mutation," *Journal of Molecular Biology* **245**, 261-274 (1995). Copyright © 1995 Academic Press.]

the end of a molecule that reaches through the cell membrane into the cell itself, where it can organize internal machinery just from the tension created by external binding. Even as x-ray diffraction reveals the intricacies of the essential contacts, we have no more than cartoon ideas of how energies are transmitted and applied. Physical opportunities abound.

Between the chemists' syntheses and nature's precise machining, there is the possibility to work both ways: to use new tricks in synthesis to create molecules with preferred properties, and to make changes in nature's design—intentional mutations of natural structures—to modify properties.

Great possibilities will be realized here when physicists trained to think about molecular organization are also trained in the much easier crafts of synthesis, mutation, modification, and manipulation. For example, biosensors are being designed with biological materials for contact with the species to be detected and electrodes with integrated circuitry for amplified response.

Consequences of the Human Genome Project and Other Genome Determinations

Probably the most difficult question to speculate about is the consequence to materials physics that will result from mapping of the human genome. It is hard even to describe the magnitude of the information being collected. It will be possible to synthesize and to mutate virtually every protein in our bodies or in the body of any known organism. There are about 100,000 identifiable human genes. The number of possible mutations is burdensome to behold. Given the 3 billion base pairs that compose our genome and the possibility of putting 4 different letter pairs at each of these 3 billion positions, the number of mutations easily exceeds anything one would try to change systematically and exhaustively.

In the context of this chapter and this survey, it is to the learning opportunities for materials research that these numbers direct our attention. Physicists can learn to produce any protein in desired quantities and to process and package it with increasing skill. Excellent demonstration of this comes from the synthesis of polypeptides of relatively simple sequence. The lure to create new industrial and medical materials compels us to think how these gene products can be designed and manipulated.

Simply knowing the amino acid sequence of a protein is yet not enough to predict its properties. Similar to the exasperation we suffer in studying collective quantum behavior (Chapter 3), difficulties in understanding proteins come from their ability to achieve the unpredictable properties that emerge because of their physical size.

As suggested at the beginning of this chapter, the nub of the matter is education. Ideally, physics students will soon be as comfortable with gene-processing procedures as they are with other ways to manipulate materials. New substances whose properties differ qualitatively from those usually considered by physicists, even in these advanced and exciting times, will also confront them.

It is probably pointless even to speculate about what we will be able to do with all these new techniques and manipulations. There is already a National Research Council study on new technology based in molecular biology.[4] Natural polymers such as silk and collagen will sooner or later be produced synthetically. It is possible to imagine using elastin, the animal's equivalent of rubber, as an industrial product. Talk of DNA computers and molecular motors is no longer strictly speculation or science fiction. Administration of gene therapy will come to be guided by the principles of organization learned from preparing condensed arrays of genetic material. Physicists who know how to make these materials will be in an excellent position to think about how they work and how to work with them using physical intuition.

---

[4] *Biomolecular Self-Assembling Materials: Scientific and Technological Frontiers*, National Academy Press, Washington, D.C. (1996).

Biological materials present a large opportunity for condensed-matter and materials physics research and application. At the same time, they present a large uncertainty in the time needed for that opportunity to be realized. It is difficult to predict the seminal moments in an untried marriage of new physics and new biology.

DIRECTIONS AND PRIORITIES

Despite the many different kinds of research described in this chapter, a few general themes emerge. From these themes we can identify needs and priorities for the support of research on soft materials.

Physicists have lost their pessimism about doing good physics on complex materials. Whether in physics departments or in other departments, they are studying biological materials for practical as well as fundamental research. Among many physicists there is huge optimism over the learning opportunities provided by biological systems, in spite of conflicting traditions of learning and reasoning. Although the necessary development of physics in biology can be expected to proceed naturally, the course and pace of this merger are uncertain.

The deformability and other macroscopic features of soft condensed matter are now routinely examined in terms of their microscopic structures and their large thermal fluctuations. We already think of the various forms of soft matter in terms of molecular arrangements, rather than their macroscopic properties. Biological processes are coming to be routinely discussed in atomic detail. With adequate support, structure determination and structural thinking can be expected to grow rapidly for at least another 10 years.

The ability to observe, to manipulate, and to characterize single molecules and to measure forces that create functional groupings of molecules is new and exciting. Mechanical properties and biological functions such as information transmission are increasingly discussed in terms of molecular structure. A science of single molecules or similarly small objects—as mechanical objects, as functioning machines, as interacting bodies—can be expected to have a large impact on biological science and on the development of artificial materials.

New tools of synthesis create new polymers and reproduce or modify natural biological polymers. These tools will become increasingly useful in the design of practical materials, medical, industrial, and domestic, and will enable systematic basic research to relate molecular cause and macroscopic effect.

Because of the complexity of the materials being examined, physicists are learning to manipulate a much larger range of material properties than had been thought possible before. Flow, deformability, microscopic patterning, strength, and durability are being evoked from substances previously not considered to have such possibilities. From better ways to design for particular properties, we may hope that industrial progress will grow less empirical and more logical.

Research on soft materials creates vast amounts of information. Whether in the possibilities built into the genetic code, the molecular details of protein structure, the many microscopic structures of liquid crystals, the nuances of medical scanning, the chemical possibilities of polymer synthesis, or the scrap heap of trial-and-error industrial innovation, the numbers that go into description are huge. The growing size, number, and kinds of data banks are teaching us new ways to organize and use information. We may expect physicists to grow increasingly comfortable working with such "rich" systems.

Priorities

Education

When physicists work with materials that were once the province of other fields, and when scientists in those fields use what physicists have learned, they discover that there are different ways of learning, thinking, and even speaking in the different fields. It is easy to say that education in physics, chemistry, and biology must be broad as well as deep. It is easy to argue that tomorrow's condensed-matter physicists should not fear to synthesize polymers or handle proteins or express genes. Although such skills are easily learned, there are many obstacles to such broad learning. Even if there were time in school, subject matter changes too fast to rely only on what is learned in school.

Several strategies can be tried:

- Interdisciplinary workshops;
- Summer schools with laboratories, for scientists at all career stages;
- New courses for biologists in elementary physics and for physicists in wet chemistry, biochemistry, and molecular biology;
- Introductory physics instruction that emphasizes soft systems; and
- Bilingual texts—e.g., in biology and physics—that teach the vocabulary and basic phenomena of particular systems. (This may be a good time for another review for physicists modeled on the landmark 1959 series in *Reviews of Modern Physics*.[5])

Basic Research

Industrial and medical results will follow naturally, as they have inevitably followed basic research in the past. Grant mechanisms can be established to encourage the necessary interdisciplinary work.

---

[5]Raymond E. Zirkle, "Biophysical science. A study program: General features of radiobiological actions," *Reviews of Modern Physics* **31**, 269 (1959).

- Special grant programs to compensate for the double jeopardy that goes with the present system when research is judged both as biology and as physics;
- Fellowships developed and expanded for physicists to work in biological laboratories—the NSF-NIH one-year visit program is one example; and
- Contact between university researchers and industrial scientists and between physicists, chemists, and biologists to foster collaborations, particularly with the chemical, medical, and pharmaceutical industries.

The residue from trial and error in industrial research is an abundant source of information for new physics. Biological systems are an inspiring source of solved problems for doing physics in a new place. We can work to create comfortable common ground for collaboration.

Undersupported research areas should be identified in which results will be needed. For example, polyelectrolytes and biological polymers will be increasingly used for products to displace environmentally unfriendly organic materials.

Research Facilities

For structure determination, neutron sources in particular are urgently needed. Synchrotron x-ray, ion beam, transmission electron microscope, and surface probe facilities are high on the list. Data processing is needed for the large amounts of information being generated and the large computations that will be undertaken.

Overall

Intellectually, industrially, and medically, soft-material research has a potential that justifies funding increases like those being given to research in biology and medicine.

6

New Tools for Research

Scientific progress is predicated on the observation of new phenomena, and there are two basic paradigms for making scientific observations. The first is the Galilean paradigm, which calls for building a better tool, such as a telescope, to investigate a familiar object, which in Galileo's case was Jupiter. The second could be called the Columbian paradigm, which calls for using existing technology, such as a small fleet of ships with the best available equipment, to investigate previously uncharted waters. Most of this report is dedicated to the Columbian paradigm and its technological consequences. This chapter summarizes where the Galilean paradigm has led us in the last 10 years and where building better tools might lead in the coming decade.

What we are really dealing with are new ways of seeing what has been there all along. The suite of small-to-large scale facilities that have enabled condensed-matter physicists to image atoms and electrons is as essential to the condensed-matter enterprise as the network of telescopes and detectors probing optical and cosmic ray spectra are fundamental to astronomy and cosmology. For more than a century, the condensed-matter suite has included small apparatus such as magnetometers and calorimeters. During the last few decades the suite has expanded to include synchrotrons and free-electron lasers, which produce highly coherent light of wavelengths from the far infrared to hard x-rays; nuclear reactors optimized for neutron yields; proton accelerators with targets for neutron and meson production; electron microscopes; and scanning-probe microscopes sensitive to everything from electron densities to magnetization at surfaces. Other exploratory tools include machines for subjecting matter to extreme conditions such as high magnetic and electric fields and pressures or ultralow temperatures.

Finally, in the past decade, direct computation or simulation has become an increasingly routine and reliable method for seeing and understanding condensed matter.

This chapter consists of sections devoted to each of the tools noted. Each section describes specific accomplishments these tools made possible in the last decade as well as opportunities and challenges for the future. Even though the sections deal with quite distinct facilities and techniques, there are certain overarching themes. An excellent example of an important scientific contribution over the last 10 years has been the effort to unravel the astonishing properties of the high-T_c cuprates and their siblings. It would be very difficult to imagine where our knowledge of the cuprates would be without the atomic coordinates given by neutron diffraction carried out at proton accelerators, the electronic bands given by photoemission at synchrotron sources, the defects found by electron microscopy, the magnetic order and fluctuations discovered using both reactor- and accelerator-based neutron sources, the charge transport measured in extreme pressures or magnetic fields, and the computer calculations of electronic energy levels. The experience with the cuprates shows that each of the facilities used is both unique and indispensable, and that their power is vastly amplified by combining data from the entire suite.

In addition to addressing specific scientific problems, another overarching theme includes the invigorating effects of new facilities, be they large national resources such as the Advanced Photon Source, the new hard x-ray synchrotron at Argonne National Laboratory; medium-scale installations such as the newly formed National High Field Magnet Laboratory operated in Florida and New Mexico; or electron microscopes and surface characterization equipment in central materials research facilities. The commercial availability of increasingly powerful workstations, electron microscopes, piezoelectric scanning-probe tools, and superconducting magnets have played an equally important but different role—namely, that of democratizing access to atomic resolution and high magnetic fields by giving individual investigators with small laboratories extraordinary capabilities formerly limited to those with access to large facilities.

A final thread linking the tools is a direct product of the information revolution seeded by condensed-matter physics and discussed at length elsewhere in the report—specifically, the proliferation of information the tools provide and the increasingly quantitative nature of the information. The most obvious manifestation is the trend away from simple black-and-white x-y plots and toward digital color images as experimental outcomes. Such images were exotic and laboriously produced 10 years ago. (The original scanning-tunneling microscopy images of silicon surfaces by Binnig and Rorer were actually photographs of cardboard models constructed from chart-recorder traces.) Today, color images are a routine feature of output from all of the techniques and facilities described below.

The future holds many opportunities and challenges including raising probe particle brilliance, improving instrumental resolution, extending spectral ranges, and diagnosing increasingly complex phenomena in areas from ceramic processing to biology. Less obvious but equally important is the need to continue to collect and take full advantage of the large and quantitative data sets that the tools of today and tomorrow promise. This implies a broad program including elements such as quoting results that had hitherto been considered qualitative in absolute units, modeling strong probe-sample interactions, and taking advantage of the most advanced data collection and display technologies available.

ATOMIC VISUALIZATION THROUGH MICROSCOPY

A quick glance at the illustrations in this report confirms that atomic visualization underpins much of condensed-matter and materials physics. Knowledge of the arrangements of atoms is a prerequisite for understanding and controlling the physical properties of solids. The techniques needed to visualize atoms in solids themselves challenge our scientific and engineering capabilities. Research in atomic visualization techniques has often lead to improved manufacturing technologies, for example, in semiconductor fabrication and quality control. Tools used for atomic visualization are small enough to fit into an average-sized laboratory and are inexpensive enough to fit into the budget of a small-instrumentation grant, but cooperative usage (as facilities) and especially cooperative instrumental development can be invaluable.

Our ability to see atomic arrangements and identify local electronic structure has progressed dramatically in the last decades. The Nobel Prize in Physics of 1986 recognized the development of the two most important techniques for this purpose—scanning-tunneling microscopy and transmission-electron microscopy (TEM) (see Table O.1). Since then there has been astounding progress. The tunneling microscope has given birth to a burgeoning industry of versatile "scanning-probe" microscopes that, while sharing many characteristics with the scanning-tunneling microscope, do not rely on vacuum tunneling for image formation. Whereas the tunneling microscope is sensitive to local electronic states, probe microscopies can examine chemical reactivity, magnetism, optical absorption, mechanical response, and a host of other properties of surfaces on a near-atomic scale. The United States is a leader in research with probe microscopes, and this is the only microscopy area in which we dominate commercially.

Probe microscopy is undoubtedly powerful, but it is to a large extent limited to surface imaging. There are interesting exceptions, such as ballistic emission electron microscopy (BEEM) in which fast electrons are injected into a layer and their propagation is influenced by interfacial structure. Other complementary surface microscopy techniques that have grown in the last decade include low-energy electron microscopy (LEEM) and near-field scanning optical microscopy (NSOM). TEM, however, remains the dominant tool used for the microstructural characterization of thin films and bulk materials because its images are not confined to the surface. In the transmission-electron microscope, a high-energy electron beam, guided by magnetic lenses, is scattered by a thin specimen. Diffraction makes it possible to study atomic structures inside solids and examine microstructure on scales from 0.1 nm to 100 μm. One example of the innovations achieved in the last decade with TEM is the discovery and structural solution of carbon nanotubes and nanoparticles. There has been significant progress in the last decade in the TEM field as well, for example, improved resolution (now at about 1 Å). Resolution is likely to be improved even further, using innovative aberration-correction techniques. Concomitant with improved spatial resolution in microscopy has been an improvement in efficiency and resolution in spectroscopy with electrons, which has enabled atomic-scale characterization of electronic structure. These techniques are complementary to, and synergistic with, improved neutron and x-ray tools described elsewhere.

Despite the undoubted value of improved resolution, a more important frontier in electron microscopy involves the ability to extract reliable quantitative information from images. An example is the use of fluctuation microscopy to go beyond the limits of diffraction in studying disordered materials. We anticipate much progress in the quantitative arena in the next decade. Although the proverb holds that a picture is worth a thousand words (no doubt true aesthetically), in science a few well-chosen words are sometimes worth a thousand pictures. This is because scientific questions involve precise answers, and pictures are by their nature imprecise. However, the theory of high-energy electron scattering is well developed, and continuing improvements in electron image detection and image analysis permit quantitative interpretation of images at the atomic level. We can expect that this capability will eventually reach a level at which nonexperts can use TEM as a quantitative structure analysis tool.

Similar progress can be expected in electron spectroscopy. Local spectroscopy allows not only atomic visualization, but also characterization of the electronic and chemical states of individual atoms or groups of atoms. Spectroscopy of surface atoms is the natural result of scanning-tunneling microscopy and can also be obtained (on groups of atoms) using TEM and surface-electron microscopy by electron energy-loss spectroscopy. Near-edge structure observed at characteristic x-ray energies can be used to determine band structure at buried interfaces, for example. Recent work has directly revealed the importance of metal-induced gap states in metal-ceramic bonding. One expects improvements both in the sensitivity of these techniques and in the quantitative modeling and data analysis needed to interpret their results. Ultimately, we need to obtain both atomic positional and chemical information for full structural characterization.

Although probe microscopes and some electron microscopes can flourish in the individual-investigator or small-facility setting, some instruments required for the future growth of atomic visualization will be of a scale such that they will need to be located in regional, if not national, centers. With computer network access, remote control of the instruments is likely to become widespread. So even though instruments may be located in only a few institutions, accessibility will be universal. It remains desirable to maintain centers of excellence where experts in the appropriate techniques can be available for consultation and collaboration. Also, instrument and technique development could be facilitated on the regional-center scale and should be encouraged because although it has historically been underemphasized, it is critical to scientific and technological success. In addition, centers facilitate education in instrumentation, so critical for industrial competitiveness.

Atomic Structure

Scanning-probe microscopes have made atomic resolution imaging of surfaces almost routine, with tremendous impact on surface science. We are finally beginning to understand the important subject of thin-film growth, one atom at a time, and can observe how atomic steps can prevent atom migration in one direction compared with another, leading to undesirable roughness in deposited films. Here, there is close interaction between experimental visualization and computer modeling. A particularly exciting development in scanning-probe microscopy has been the imaging of chemical and biochemical molecules and the possibility of monitoring chemical reactions. By choosing one molecule as the tip of the atomic-force microscope (AFM), the forces between molecules can be directly measured and chemical reactions sensed with unprecedented molecular sensitivity. This has already led to new insights into the rheology of macromolecules (see Chapter 5), and we can expect great advances in the near future, especially in the biological sciences. For example, the use of "smart" tips would allow recognition of molecules using specific receptors adhered to the tip.

The scanning-tunneling microscope (STM) views the local electronic structure, so careful image simulations must be made to deduce atomic structure. In general, for structural studies on surfaces, the best results have been obtained by a combination of direct STM imaging with diffraction—for example, by x-rays or electrons. The highest directly interpretable spatial resolution for atomic structure has been obtained with TEM (see Box 6.1); instruments capable of resolving 1 Å have recently been demonstrated. The committee notes that, partly because of the ~$50 million price tag for these instruments and partly because of the damage accompanying the high accelerating voltages required, no such instrument can be found in the United States. Researchers' hopes are pinned on lower accelerating voltage approaches to improved TEM resolution, such as holographic reconstruction, focus variation, incoherent Z-contrast, and aberration correction. However, it is troubling that work in these areas is predominantly located in Europe and Japan; a notable exception is work on incoherent Z-contrast imaging (see Box 6.1). A relatively recent study of trends in atomic resolution

BOX 6.1 Being Certain About Atom Positions at Interfaces

Identification of atomic structure at interfaces has been one of the important applications of high-resolution transmission electron microscopy. Interfaces control mechanical strength in ceramics, electrical transport in transistors, corrosion problems in aircraft, tunneling currents in superconductor junctions, and a myriad of other practical materials behavior. Yet, with rare exceptions, interfaces are not amenable to diffraction analysis because they are very thin and not usually uniform. Figure 6.1.1 shows an example of a high-resolution transmission-electron microscope image, using "z-contrast" of a grain boundary in MgO (courtesy of Oak Ridge National Laboratory), in which atomic columns at the boundary are revealed. Images like this are beginning to be analyzed in a quantitative manner, using accurate measurements of intensity, simulations of electron propagation, and computational modeling of atomic structure, to achieve unprecedented reliability in analysis of interfaces.

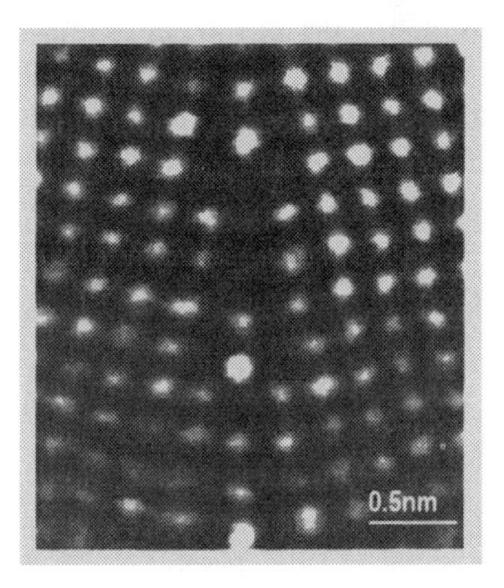

FIGURE 6.1.1 High-resolution transmission electron micrograph, using z-contrast, in MgO.

microscopy was published by the National Science Foundation.[1] Advances in electron microscopy enable advances in related industrial technologies, especially semiconductors; so the value of U.S. investment in this area extends far beyond atomic visualization.

A clear example of the value of improved resolution in TEM is tomography. Tomography has been widely used in biology to reconstruct objects at about 1-nm resolution. Only with a resolution of about 0.5 Å will it be possible to

[1] *National Science Foundation Panel Report on Atomic Resolution Microscopy: Atomic Imaging and Manipulation (AIM) for Advanced Materials*, U.S. Government Printing Office, Washington, D.C. (1993).

reconstruct objects in three dimensions at the atomic scale. This would be particularly exciting for amorphous and disordered materials; knowledge of their atomic structure is limited to statistical averages from diffraction. Instruments to enable this will require ~0.5 Å resolution combined with high specimen-tilt capability (>45°). Such will be possible either with very high voltages or with aberration correction.

Electronic Structure

For many research problems in condensed-matter and materials physics, it is important to visualize the electronic structure on a near-atomic scale. STM provides direct information about electronic states at surfaces but is often used for purely structural analysis and has had tremendous impact on surface science. Examples in the report include the germanium "huts" in Figure 2.13. In general, probe microscopy combined with electron microscopy has revolutionized our understanding of thin-film growth and epitaxy (see Chapter 2).

STM has been profitably used to examine surface electronic states and chemical reactions on the atomic level. Although detailed electronic structure calculations are needed to interpret STM images in terms of atomic positions, often the electronic structure information is directly useful. For example, Box 6.2 gives an example of direct STM imaging of the electronic states associated with individual dopant atoms in semiconductors.

Electron energy-loss spectroscopy in TEM provides an important method to obtain electronic structure from the interior of samples on a near-atomic level. Improvements in the sensitivity of detection, using more monochromatic field-emission electron sources and parallel detection, have led to important advances in the last decade. For example, dopant segregation at semiconductor grain boundaries has been identified.

Nanoproperties of Materials

One of the most significant developments of the last decade is the proliferation of scanning-probe techniques for measuring the nanoproperties of materials. Figure 6.1 shows a large variety of signals that are now detectable. Nanomechanical (force) measurements can be used to watch the behavior of individual dislocations; optical measurements can visualize single luminescent states; piezoelectric measurements can identify the effect of defects on ferroelectrics, which have potential for high-density nonvolatile memory; magnetic measurements can show the effect of single atoms on spin alignment in atomic layers; ballistic electron transport can identify the electronic states associated with isolated defects inside a film. We can expect these capabilities to revolutionize our ability to characterize the physical properties of nanoscale materials.

BOX 6.2 Single Impurity Atoms Imaged in Semiconductor Layers

One critical issue, as semiconductor devices are scaled down in size for higher density and speed, is the stochastic nature of the location of dopant atoms. These atoms, which lend electronic carriers to the active semiconductor layers, are typically present in densities of only about 1 in a million. Until recent years, it was an impossible dream to identify the exact location of these dopant atoms, but this has recently proved possible with scanning-tunneling microscopy. Figure 6.2.1 shows detection of the local electronic state generated by the impurity. When a semiconductor structure is cleaved in vacuum, the individual impurity atoms near the surface are clearly visible. The image (courtesy of Lawrence Berkeley Laboratory) shows the position of Si dopants in GaAs as bright spots. Also present in the image are Ga vacancies, which appear as dark spots.

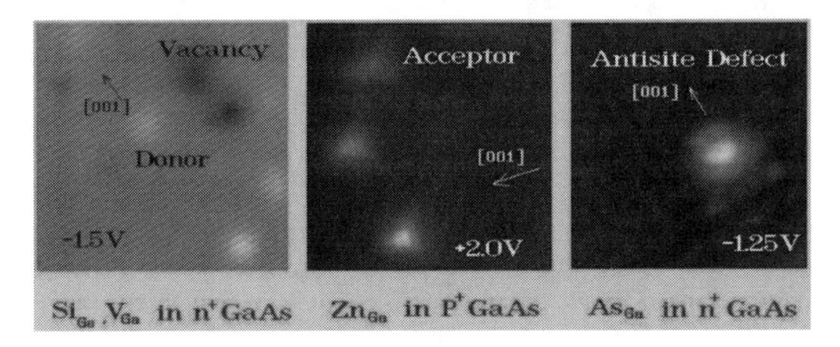

FIGURE 6.2.1 Local electronic states in GaAs generated by Si impurities.

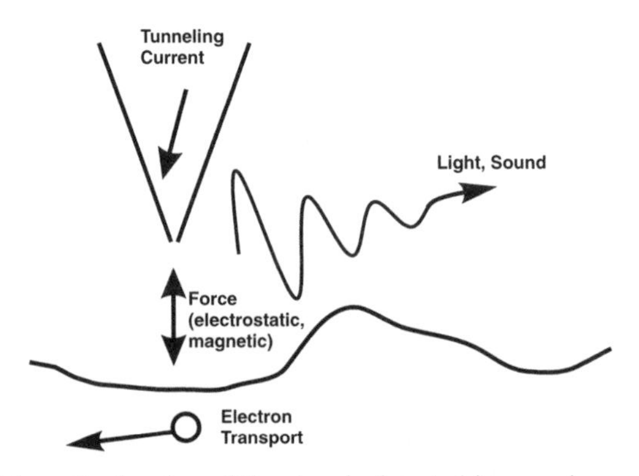

FIGURE 6.1 Schematic drawing of the signals detected in scanning-probe microscopy. (Courtesy of the University of Illinois at Urbana-Champaign.)

Similar developments have occurred in other imaging systems. A beautiful example of a technique known as "scanning electron microscopy with polarization analysis," which allows imaging of magnetic monolayers at surfaces in a modified scanning-electron microscope, is discussed in Chapter 1.

Atomic Manipulation

Whether intended or not, our atomic-scale characterization tools can change the structures they are examining. This can be used to our advantage in manipulating atoms on the atomic scale for making nanostructures. Figure 6.2 shows the classic example of a ring of iron atoms assembled by the tip of a scanning-tunneling microscope. The circular atomic corral shows the resonant quantum states expected from simple theory. The imagination boggles at the possibilities with related techniques. In principle, we can assemble arbitrary structures to test our understanding of the physics of nanostructures and perhaps make useful devices at unprecedented density. Two major issues will need to be addressed before these methods can reach their full potential. First, even when we place atoms where we choose, with few exceptions (such as the Fe atoms in Figure 6.2 at ultralow temperatures), they will not stay there. So, to assemble structures that

FIGURE 6.2 Atomic manipulation. The image shows the atomic scale capability for patterning that is possible with the scanning-probe microscope. Atoms of Fe (high peaks) were arranged in a circle on the surface of Cu and caused resonant electron states (the ripples) to appear in the Cu surface. The structure is dubbed the "quantum corral." Related structures might one day be useful for electronic devices, where as many devices as there are humans in the world could be assembled on an area the size of a pinhead (1 mm^2). (Courtesy IBM Research.)

retain their integrity, we need to understand the stability of materials on this scale. Second, the speed with which we can pattern structures with a single scanning probe is far too slow to allow practical device fabrication on the scale of modern semiconductor technology. Alternate methods involving massive arrays of tips, projection electron lithography, or other short-exposure techniques must be developed.

Conclusions

Atomic visualization is a crucial part of condensed-matter and materials physics. It is a thriving area in which advances usually driven by physics and engineering have wide impact on science and technology. Many manufacturing technologies depend on innovations enabled by atomic visualization equipment, so research in the field has important economic value. We expect continued developments, but attention must be paid to nurturing the development of appropriate instrumentation in close connection with scientific experiments. Depending on the nature of the visualization tool, the funding scope ranges from individual investigator to small groups, to national centers of excellence in instrumentation. From our success in probe microscopy, it appears we are stronger at the individual-investgator level but weaker at the medium- and larger-scale instrumentation development levels. A concern is that many new students are attracted by computer visualization rather than experimental visualization. The two methods are obviously complementary, and we are not yet near the point where we can rely only on computer experiments. Thus funding must be maintained at a level sufficient to create opportunities that will attract high-quality students into this field.

NEUTRON SCATTERING

The neutron is a particle with the mass of the proton, a magnetic moment because of its spin-1/2, and no electrical charge. It probes solids through the magnetic dipolar interaction with the electron spins and via the strong interaction with the atomic nuclei. These interactions are weak compared to those associated with light or electrons. They are also extremely well known, which makes it possible to use neutrons to identify spin and mass densities in solids with an accuracy that in many cases is greater than with any other particle or electromagnetic probe. The wavelengths of neutrons produced at their traditional source, nuclear research reactors with moderator blankets of light or heavy water held near room temperature, are on the order of inter-atomic spacings in ordinary solids. In addition, their energies are on the order of the energies of many of the most common collective excitations—such as lattice vibrations—in solids. To image spin and mass densities, condensed-matter physicists usually aim neutrons moving at a single velocity and in a single direction, that is, with well-specified momentum and energy, at a sample and then measure the energy and momentum distribution of the neutrons emerging from the sample. Such neutron-scattering experiments have been important for the development of condensed-matter physics over the last half century. Indeed, the impact of the technique has been such that C. Shull (Massachusetts Institute of Technology) and B. Brockhouse (McMaster University) were awarded the 1994 Nobel Prize in Physics for its development (see Table O.1). In previous decades, neutron scattering provided key evidence for many important phenomena ranging from antiferromagnetism, as originally posited by Neel, to unique quantum oscillations (called rotons) in superfluid helium. But what has happened in the last decade in the area of neutron scattering from solids and liquids, and what is its potential for the coming decade?

The Past Decade

Overview

Three major developments of the last decade are (1) the emergence of neutron scattering as an important probe for "soft" as well as "hard" condensed matter, (2) the coming of age of accelerator-based pulsed neutron sources, and (3) the revival of neutron reflectometry. The first development has expanded the user base for neutron scattering far beyond solid-state physicists and chemists, who had been essentially the only users of neutrons. The second development is associated with a method for producing neutrons not from a self-sustaining fission reaction, but from the spallation—or evaporation—that occurs when energetic protons strike a fixed target. As depicted in Figure 6.3, a spallation source consists of a proton accelerator that produces short bursts of protons with energies generally higher than 0.5 GeV, a target station containing a heavy metal target that emits neutrons in response to proton bombardment, and surrounding moderators that slow the neutrons to the velocities appropriate for experiments. Until the mid-1980s, the leading facility of this type was the Intense Pulsed Neutron Source (IPNS) at the Argonne National Laboratory. In the last decade, the clear leader by a very wide margin has been the ISIS facility in the United Kingdom. Successful developments, especially at ISIS, have given the neutron-scattering field growth prospects that it has not had since the original high-flux nuclear reactor core designs of the 1960s. This follows because pulsed sources are more naturally capable of taking advantage of the information and electronics revolutions and because the unit of cooling power required per unit of neutron flux is almost one order of magnitude less than for nuclear reactors.

The revival of neutron reflectometry seems at first glance less momentous than the emergence of neutron scattering as a soft condensed-matter probe or the emergence of accelerator-based pulsed neutron sources. However, as so much of modern condensed-matter physics and materials science revolves about surfaces and interfaces, neutron scattering could hardly be considered a vital technique

FIGURE 6.3 Drawing of the planned Spallation Neutron Source at Oak Ridge National Laboratory. The basic design features are similar to those for the Los Alamos Neutron Scattering Center (LANSCE) and the planned European Spallation Source (ESS). The linear accelerator takes protons to 1.33 GeV, while the accumulator ring groups them into 1 μsec bursts, occurring at a repetition rate of 60 Hz, which then impinge onto a liquid mercury target. The neutrons emanate in corresponding bursts from the target and feed scattering instruments with flight paths with lengths from 2 to 100 m. (Courtesy of Oak Ridge National Laboratory.)

without some clearly defined contribution in these areas. The revival of reflectometry has enabled neutrons, in spite of their weak coupling nature, to become a legitimate probe of surfaces and interfaces. Here we use long wavelengths and incident and reflected beams that nearly graze the sample, so that we are in the surface-sensitive regime near the condition of total external reflection.

Locating the Atoms

The major contributions of neutrons to condensed-matter and materials physics in the last decades come from using neutrons to answer the most fundamental question that always arises when new materials are discovered—"Where are the atoms?" Although this question is generally answered using x-ray diffraction, the unique properties of neutrons offer significant advantages in many important cases. However, since neutrons couple via the nuclear interaction to the atomic cores rather than via the electromagnetic interaction to the atomic electrons, neutrons can be equally sensitive to light (low-Z) and heavy (high-Z) atoms, whereas x-rays always couple much more strongly to the heavy elements. Neutrons are especially sensitive to the lightest and arguably most important element of all, hydrogen, and quite sensitive to its rival in importance, oxygen. In addition, it is possible to change atoms' visibility to neutrons, without appreciably changing the bonding or chemistry of a particular atom, by changing the isotope. Thus, particular sites in a material can be labeled for investigation of their microscopic coordinates and motion. Finally, the combination of various neutron sources, as well the ability to tailor wavelength distributions even at a single source, permits the examination of structures with characteristic length scales from angstroms to microns. The weak coupling nature of the probe means that even as the wavelengths of the neutrons used experimentally change over three orders of magnitude, the scattering cross sections do not and absorption and resolution corrections remain simply calculable.

One of the most lively areas in condensed-matter science over the last decade has been that of transition metal oxides, a field dramatically revived by the discovery of high-temperature superconductivity in oxides of copper. The materials are generally combinations of relatively heavy lanthanides, medium-weight transition metals, and light oxygen atoms. With this set of constituents, neutron scattering was ideally positioned to make an important contribution to the structure determination. The technique did not disappoint. First it has been demonstrated that the key structural elements common to all of the cuprate superconductors are nearly square planar arrays of copper and oxygen. The significance of this simple finding is impossible to overstate. That copper oxygen planes are the key feature of the high-temperature superconductors has been the starting point for essentially all of our thinking about high-temperature superconductivity as well as searches for materials with better superconducting properties. Beyond revealing the ubiquity of the copper oxygen planes, neutron diffraction has revealed how the planes appear singly, in pairs, or even as triplets, sometimes with and sometimes without copper oxide chains in intervening layers. The picture of the intervening layers as reservoirs that provide charges for the copper oxygen planes is largely the result of a combination of neutron diffraction and classical measurements of bulk electrical properties such as resistivity.

Extensive work has shown close correlations between structural details and superconducting properties. For example, in a mercury-based compound exhibiting an extraordinarily high T_c, which itself is very sensitive to pressure, neutron diffraction showed dramatic changes in the atomic coordinates with applied pressure (see Figure 6.4).

Even after 10 years of indispensable contributions to the understanding of high-T_c superconductivity, neutron diffraction retains its unique, driving role in this field. A recent illustration of this is the excitement generated by the discovery that certain materials very closely related to the high-temperature supercon-

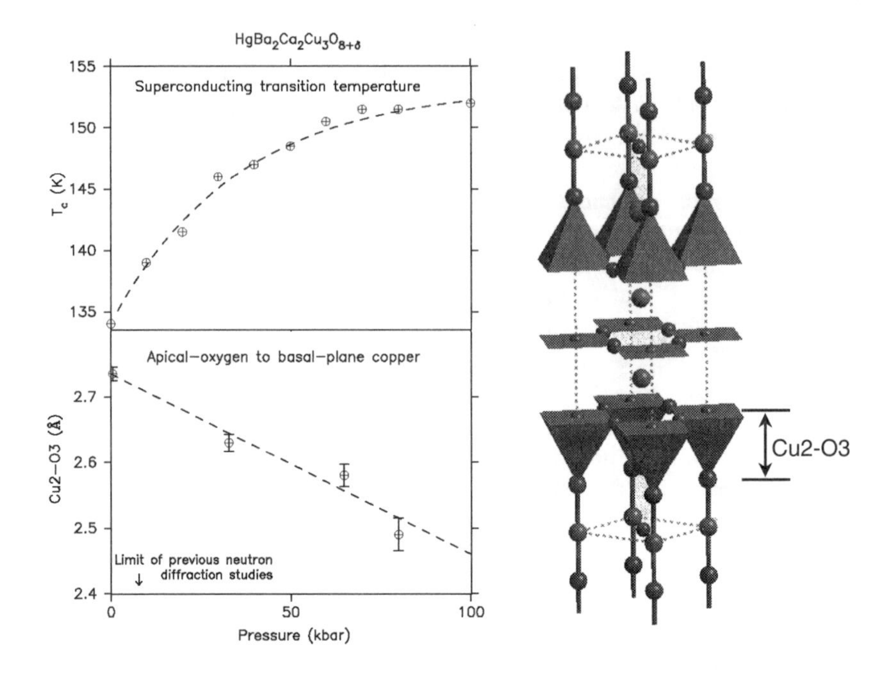

FIGURE 6.4 Mercury-based cuprates exhibit not only the highest transition temperatures T_c for superconductivity, but also extraordinarily pressure-dependent T_c's. High-resolution neutron diffraction at pulsed spallation sources has revealed the complex structures (at right) of these compounds. In addition, the penetrating power of the technique has been exploited to examine the pressure dependence of the structure. There is an astonishing 0.25 Å contraction of the marked copper-to-oxygen distance as pressure is applied to raise T_c from 138 to 160 K.

ductors undergo phase transitions to states with large-scale superstructures. Neutron scattering has provided atomic-scale information not only about the high-temperature superconductors, but also about many other transition metal oxides. Notable examples are the perovskite manganites whose "colossal" magnetoresistance—large changes in the electrical resistivity when external magnetic fields are imposed—has recently been rediscovered. Neutron diffraction has been used to identify the structural parameters most strongly correlated with the magnetoresistance.

Seeing the Spins

Once they know the locations of atoms in a particular solid or fluid, condensed-matter physicists generally would like to know what the electrons are doing. The electrons of particular interest are the outer electrons because they account for the chemical bonding and electrical and magnetic properties of a solid. The neutron couples to these electrons through the magnetic dipole interaction; because its energy is typically much too small to excite the electrons from the core where they form a closed shell with zero net orbital and spin-angular momenta, the cores are invisible. The outer electrons are easiest to see with neutrons when they live on a regular lattice and their spin orientations repeat periodically. In this case, they produce diffraction spots entirely analogous to those associated with atomic nuclei. As Shull showed in the 1950s, it is then possible to do magnetic crystallography to image the spin arrangements in virtually any magnet. In the last decade, magnetic crystallography with neutrons has continued to be among the most essential tools in condensed-matter physics. Again, high-temperature superconductivity has been an area of accomplishment. The important experiment was that which showed, shortly after the superconductivity's discovery, that the insulating and undoped parent compounds of the superconductors are actually very simple antiferromagnets. In the decade since this experiment, the superconductivity and magnetism of the cuprates have been inextricably intertwined. As for the neutron diffraction experiments that revealed the microscopic structures of the high-T_c compounds, the last decade's progress in high-temperature superconductivity would be unimaginable without the early magnetic diffraction data on the parent compounds.

Magnetic diffraction has played a similar role in other subfields that have been active in the last decade. For example, it established a definite link between the magnetism and exotic superconductivity of certain actinide and rare-earth intermetallics, also known as heavy fermion compounds. Also, a particularly important and elegant set of experiments explored the coupling between magnetic layers through intervening nonmagnetic layers in thin-film multilayer structures grown by molecular-beam epitaxy. The structures show great promise as "spin valves" for application to computer disk drive read heads. The optimization of their performance requires complete knowledge of the atomic and spin densities responsible for the desirable giant magnetoresisant behavior. Using polarized-neutron reflectivity, one can obtain a depth profile of the direction and magnitude of the magnetic moment in these materials with 2- to 3-Å resolution. Early polarized-neutron reflectivity studies confirmed that maximum giant magnetoresistance is correlated with an antiparallel alignment of the magnetic layers across the nonmagnetic interlayers. More recent experiments revealed the complex interplay between the magnetic structure and the physical characteristics.

Imaging Vortices in Superconductors

A seemingly different type of magnetic structure is that of mesoscopic field inhomogeneities. Mesoscopic inhomogeneities are seen by the neutrons in the same way in which they see the microscopic field inhomogeneities associated with the electron spins—namely, through the magnetic dipole coupling between the neutron spin and the magnetic fields. The relevant wave numbers and corresponding apparatus are different, but the concepts remain the same. The most famous mesoscopic field inhomogeneities in condensed-matter physics are those associated with type-II superconductors. Here, the superconductor accommodates an external field by admitting quantized vortices containing normal (metallic) state cores embedded in a superconducting matrix. The vortices typically arrange themselves to form a lattice with inter-vortex separations of order 100 to 1000 Å. One of the triumphs of neutron scattering in the 1960s was the verification of the vortex lattice picture for conventional, low-temperature type-II superconductors. Given this early success, it should come as no surprise that as unconventional superconductors such as actinide intermetallics and cuprates were discovered in the 1980s and 1990s, neutrons were used to image their vortex lattices. They provided key evidence for two of the most important new ideas about superconductivity. The first idea is that real solids could actually display superconductivity more akin to the superfluidity of helium-3 than to the superconductivity of ordinary solids like aluminum; the second is that collections of vortices can have intricate phase diagrams much like those of complicated organic molecules in solution.

Pictures of Soft Matter

The committee has focused so far on neutron scattering from intermetallic compounds and their oxides, which, although they are complex, are solids of long-standing interest to condensed-matter physicists. Indeed, manganites were among the first materials to be investigated by neutron diffraction shortly after the invention of the technique in the 1950s. The last decade has witnessed a huge growth in the use of neutrons to image structures formed at surfaces and interfaces, as well as the large-scale structures that emerge in materials with genuinely large molecular units, such as polymers and water-based biomolecules. The universe of such structures is actually much larger than that of traditional condensed-matter and materials physics and contains most of the matter essential for our lives. One particularly successful application of the neutron technique has been to diblock coploymers, which show a huge variety of disordered and mesoscopically ordered states. As in classic condensed-matter physics, the goal of the investigations is to relate the structures to properties, such as elasticity, which determine functionality. Neutron scattering has also measured the sizes and shapes of micelles that appear in microemulsions. Model systems mainly of interest to statistical physicists, as well as biologically interesting micelles such as ribosomes, have been examined. Another development of the last decade has been the application of neutron scattering to surfaces, interfaces, and membranes involving polymers and other large organic molecules. Knowledge of local structural features as well as interface profiles feeds into a vast array of scientific and technical fields from biology to integrated circuit packaging. Figure 6.5 shows the shape change undergone by diblock copolymers adsorbed on a glass substrate. As the conditions are changed, the copolymers undergo a transition from a mushroom- to a brush-like shape, which correlates with a change in the adhesive properties of the coated surface.

Dynamics

Nuclear and magnetic structure determinations represent the most common and widely understood application of neutron scattering. However, since the work of Brockhouse in the 1950s, the study of lattice vibrations and magnetic fluctuations has also had an impact on condensed-matter physics and materials science. As have neutron determinations of magnetic and nuclear structure,

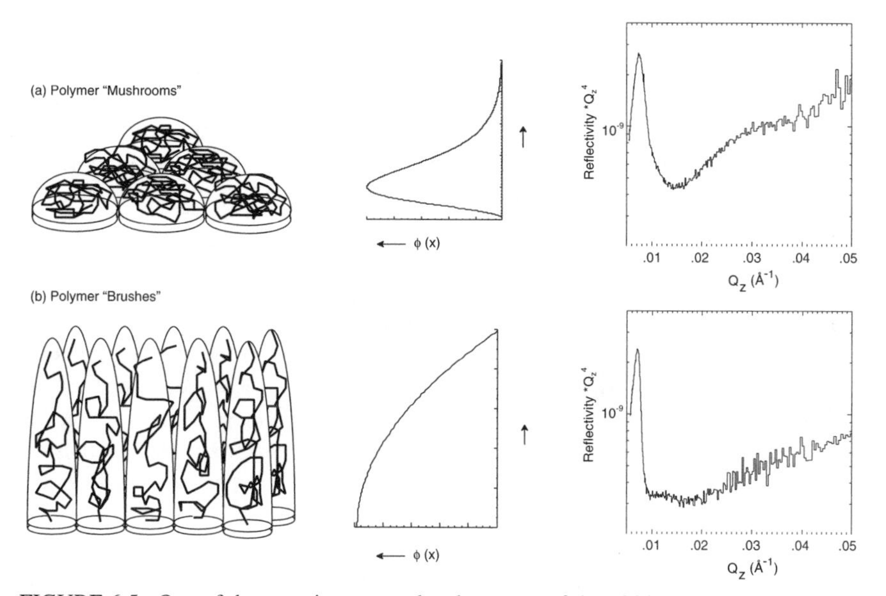

FIGURE 6.5 One of the most important developments of the 1990s has been the revival of neutron reflectometry. Formerly used as a tool for establishing absolute neutron-scattering cross sections, it has become a major technique for surface and interface science, with particularly significant accomplishments in the fields of soft matter and magnetoresistive films. The figure shows data for the "mushroom" to "brush" transition for polymers attached to a substrate. Raw data are at right, the directly deduced density profiles are in the middle, and the inferred morphology is shown at left. The radically different reflectivity profiles at right attest to the ability of the technique to discriminate between the different arrangements of the polymers at the surface. (Courtesy of Los Alamos National Laboratory.)

neutron scattering from excitations in solids has strongly influenced our thinking about transition metal oxides. For example, work on the cuprate superconductors includes measurements of the phonon density of states, which can be used as inputs into traditional calculations of the superconducting transition temperature. The apparent failure of such calculations remains an important motivation to search for a new theory for the superconductivity of the cuprates. Much more recent experiments provide similarly complete magnetic excitation spectra, which can now be used in analogous tests of "conventional" magnetic theories of high-temperature superconductivity.

An interesting development has been the use of neutrons to probe the electronic gap function and pair-breaking excitations in the superconducting state. Neutrons are unique for this application because they allow the only superconducting spectroscopy that is a true bulk probe capable of examining short-wavelength phenomena with high energy resolution.

The continuing work on the dynamics of ordered solids has coexisted with a rapidly growing enterprise concerned with the dynamics of fluids and soft matter. Important experiments include those that have verified one of the key concepts in polymer science—that polymers in a melt move in snake-like fashion within tubular structures formed by their neighbors. The experiments are noteworthy not only for their scientific impact, but also because they required the use of an instrument—the neutron spin echo spectrometer—that operates on a principle unknown to the founders of inelastic neutron scattering, Fermi and Brockhouse.

The Next Decade

Because of unique properties associated with their mass, charge, and spin, neutrons have a scientific future as bright as their past. Most likely, key accomplishments with neutrons will be as unexpected as were those of the last decade, when they were linked to a largely unexplored class of materials—the cuprates—that happened to display an extraordinary and unexpected property, high-T_c superconductivity. Thus, the cuprates offer for neutron scattering, as they do for condensed-matter physics as a whole, a lesson in humility to all who wish to plan future accomplishments. At the same time, the success of neutrons in meeting the challenges of high-temperature superconductivity was not entirely serendipitous. Indeed, the discovery and subsequent intensive study of the cuprates coincided with other developments:

1. The rapid development of accelerator-based pulsed neutron sources and instrumentation, whose operating paradigms are entirely different from those invented by Shull and Brockhouse for nuclear reactors;
2. Progress in electronics, data visualization, and computation driven by the microelectronics revolution;

3. Extension of the routinely examinable spectral range from its traditional 0.5- to 50-meV domain to its current 1-μeV to 1-eV band;

4. Development of increasingly efficient beam optics and scattered-neutron analysis and detection schemes; and

5. Extension of sample environment capabilities to lower temperatures, higher magnetic fields, and higher pressures.

These other developments not only coexisted with the great materials discoveries of the last decade, but were actually a prerequisite for the significant contributions ultimately made by neutron scattering to the elucidation of these discoveries. It is our judgment that further improvements in all five of the listed categories are inevitable in the next decade. The inevitability follows from the continued effects of accelerator-based pulsed neutron sources and instrumentation, and of advances driven by the microelectronics revolution on the entire field, that has been hampered by the limits imposed by the modest incident fluxes that even modern research reactors can provide. Advances in both accelerator-based pulsed neutron research and microelectronics have made it possible to multiplex many experiments on an enormous scale, for example, simultaneously collecting 10^6 usable pixels of information where the old reactor-based methods would yield a single pixel. Thus, the field of neutron scattering has changed qualitatively over the last decade, even though only one major new source (ISIS in the United Kingdom) has been completed. The figure of merit for many important experiments has been transformed from the reactor power to the information rate. In the coming decade, we expect the useable information rates as measured by the product of incident flux delivered by the beam optics and the number of independent pixels to grow in tandem with the microelectronics revolution (Figure 6.6). Beam optics are also on a growth curve driven by improvements in thin film-technology and x-ray and light optics, and so are also likely to improve. The continued growth in capabilities will make many new experiments possible, as well as allow old measurements to be performed with greater precision. The new experiments might include measurements of vortex lattice dynamics in type-II superconductors, investigations of the magnetic aspects of the quantum Hall effects, characterization of fluid flow in small capillaries, and studies of electromigration at silicon-metal interfaces. Of course, the most exciting experiments will be those dealing with phenomena we are unaware of today.

Neutron experiments have continued to be popular even in the absence of a new neutron source because of the neutron's uniqueness as a probe of condensed matter and because neutron experiments are so readily improved by ongoing advances in microelectronics and thin-film technology. However, merely transferring technology developed for other uses to its antiquated neutron-scattering centers will not allow the United States to recapture its lead in neutron science. There is no substitute for constructing a new high-power spallation source with many high-flux beam lines. In recognition of this, the government is supporting

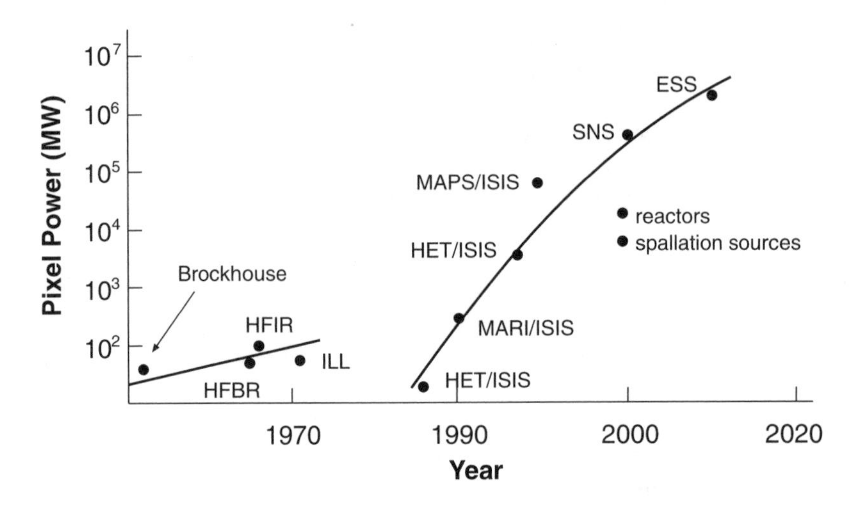

FIGURE 6.6 The information acquisition rate (left) for single-crystal inelastic experiments is the product of the flux at the sample (expressed in nuclear reactor equivalent MW units) and the number of useable pixels within which the scattered and incident neutrons fall. Brockhouse was co-recipient of the 1994 Nobel Prize for developing the single-pixel triple-axis spectometer (see Table O.1), which dominated inelastic neutron scattering until around a decade ago. The development of pulsed spallation sources and fast rotor chopper spectrometers has moved inelastic neutron scattering onto a growth curve (Moore's Law) driven largely by the electronic data-processing industry. The neutron sources identified are the HFBR (High Flux Beam Reactor, Brookhaven National Laboratory), HFIR (High Flux Isotope Reactor, Oak Ridge National Laboratory), and ILL (Institut Laue-Langevin, France) reactors and the ISIS (Rutherford-Appleton Laboratory, United Kingdom), SNS (proposed Spallation Neutron Source, Oak Ridge National Laboratory), and ESS (European Spallation Source, currently unsited) accelerator-based facilities. MAPS, HET, and MARI correspond to ISIS instruments at different stages of development. [*Physics World*, **33** (December 1997).]

the construction of precisely such a source, the Oak Ridge Spallation Neutron Source, whose completion will be the big event of the next decade for neutron science.

SYNCHROTRON RADIATION

In the past 30 years, the use of infrared, ultraviolet, and x-ray synchrotron radiation (SR) for condensed-matter and materials physics research, as well as research in the other natural sciences, engineering, and technology, has blossomed. The pace and scientific range of SR utilization has increased even more rapidly during the past decade because of source improvements, advanced instrumentation, and more beam time than was possible a decade ago. Further impetus was provided by the construction of new facilities with extreme performance. As a consequence of these developments, approximately 4,000 scientists from academia, industry, and government laboratories now use U.S. SR facilities.

In the 1960s and 1970s, research was initiated using SR produced by the bending magnets at storage rings designed for high-energy physics. As shown in Figure 6.7, such rings provided about four orders of magnitude greater brightness than the best in-laboratory sources. In addition, the radiation covered a very broad spectrum, in contrast to the line source x-ray tubes then available. These features made a number of previously unfeasible experiments possible.

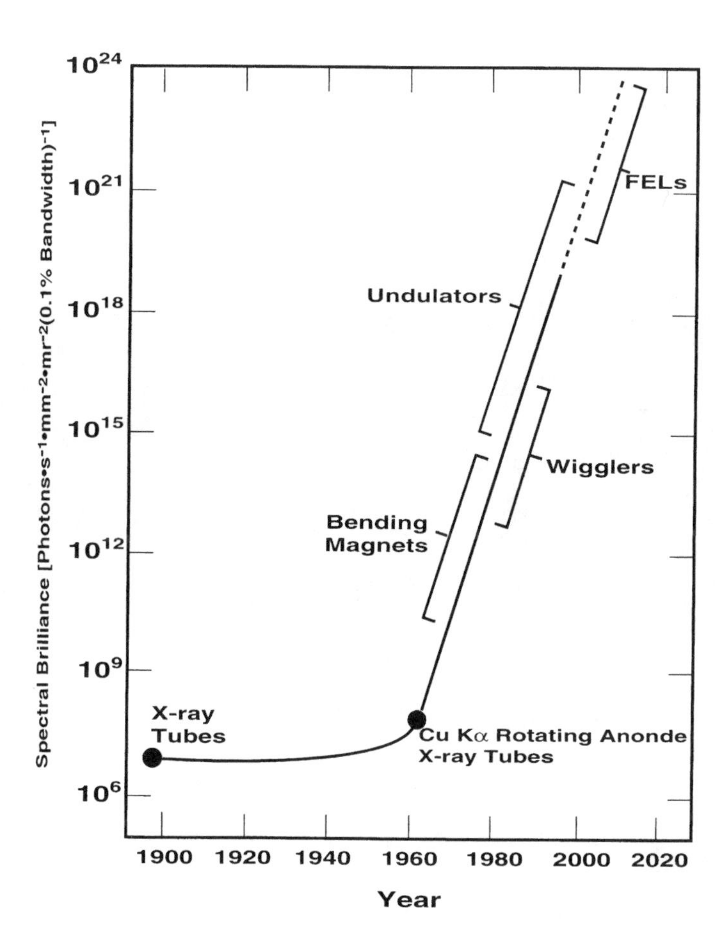

FIGURE 6.7 History of (8 keV) x-ray sources. Brilliance (or brightness) is defined as source intensity per illuminated solid angle. (Courtesy of Argonne National Laboratory.)

Because of the science and the large user communities resulting from the first-generation sources, they were joined by second-generation, high-brilliance rings designed specifically for SR research in the mid-1980s. The increased brightness and the greater availability of these sources, as well as the increased flux achieved by insertion devices at all sources, further expanded both the science and the user community. (Insertion devices, known as wigglers and undulators, are magnetic arrays that cause the charged particles to undergo quasi-sinusoidal paths, producing far brighter radiation than can be achieved with bending magnets at the same storage ring.)

During the past decade, third-generation rings [the Advanced Light Source (ALS) and Advanced Photon Source (APS; shown in Figure 6.8) in the United States, SPRING-8 in Japan, and the European Synchrotron Radiation Facility in France], with still higher brightness (by 4 to 5 orders) and many straight sections for insertion devices, have been constructed. At the same time, the first- and second-generation rings have been modified so that their performances have increased markedly. Such increases form the basis of revolutions in the research that utilizes SR—a process that is likely to continue well into the next century with new sources.

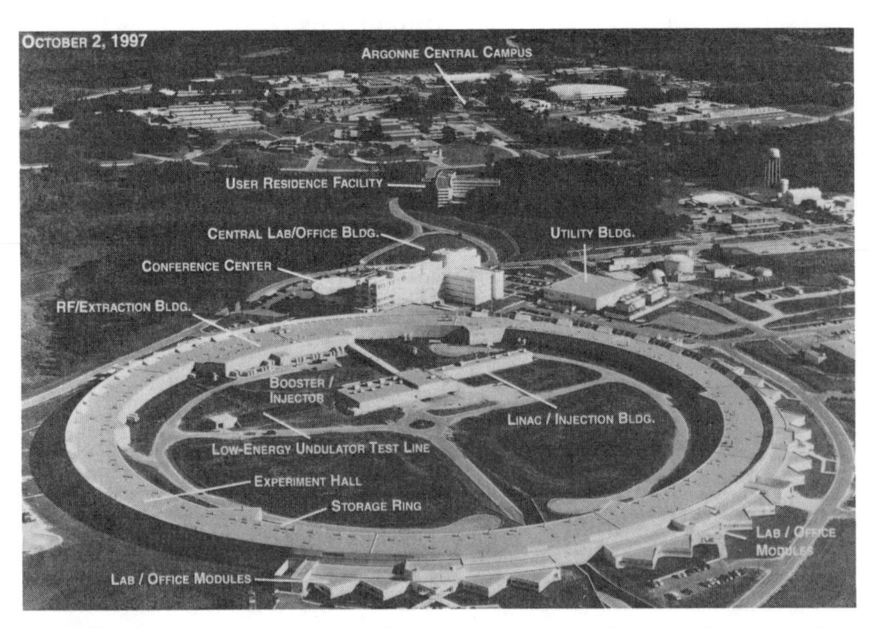

FIGURE 6.8 Overview of the recently completed Advanced Photon Source (APS) at Argonne National Laboratory. Electrons circulate in the storage ring and emit brilliant x-ray beams that are used to probe the structure of condensed matter at scales ranging from the atomic to the macroscopic. (Courtesy of Argonne National Laboratory.)

The Past Decade

Protein Crystallography

The goal of understanding life has evolved into a large interdisciplinary effort that integrates information extending from experimental results at the atomic and molecular levels to studies of organelle, cellular, and tissue organization and function. Atomic-level information will increasingly provide the means through which biological function, and malfunction that leads to disease, will be understood. Macromolecular crystallography has provided the vast majority of information about three-dimensional biological structure and will play an even greater role in the future. Information relating structure to function has also led to the development and successes of new approaches to drug discovery (often called structure-based drug design).

The unique properties of SR—namely, its tunability and high brilliance—have allowed it to play a seminal role in these advances. So important is SR to protein crystallography that 73 percent of new structures published in *Nature* and 60 percent of those published in *Science* in 1995 used synchrotron-based data, and this percentage continues to grow. Some of the most important results include the structure of the myosin head, which has led to a molecular-level interpretation of muscle contraction; the structure of cytochrome oxidase, which is the enzyme that carries out the final step in mammalian respiration; the structure of the enzyme nitrogenase responsible for production of most of the assimilable (fixed) nitrogen in our biosphere; the structure of the ribozyme, which is a catalytic form of RNA; numerous plant and animal virus structures (for an important example, see Figure 6.9), as well as studies of their interaction with potential antiviral drugs; and structures of a variety of enzymes, like topoisomerases, involved in DNA transformations and regulation.

Kinetic Studies of Structure

The five order of magnitude increase in photon flux provided by the first x-ray SR sources immediately enabled time-resolved diffraction studies. Structural biologists addressed the changes in muscle tissue as it contracts and expands and have come to a detailed understanding of the mechanism of force generation. Subsequently, time-resolved scattering and x-ray absorption spectroscopy (XAS) studies, as well as the related studies of systems in excited states, have blossomed, providing kinetic understanding of reactions and processes that cannot be obtained in other ways. Following are examples of other types of studies that have benefited from time-resolved scattering and XAS:

1. In situ studies of thin film growth by sputtering, organometallic vapor phase epitaxy, and molecular-beam epitaxy;

2. Both protein crystallographic and depolarization studies of atomic reorganization in photosensitive biological molecules subjected to light pulses (see Figure 5.3);

3. In situ small-angle x-ray scattering studies of the refinement of heavy crude oils; and

4. Structural studies of phases formed during various stages of welding.

The examples illustrate a portion of the range of time-resolved studies that have been performed thus far. With third-generation sources completed and fourth-generation sources being planned, applications to faster processes, shorter-lived states and more weakly scattering systems can be anticipated as we seek to understand aspects of reactions at the atomic and molecular levels.

Surfaces and Interfaces

Both fundamental and applied x-ray scattering studies of surfaces and interfaces have flourished over the past decade at all of the x-ray facilities. Among the

FIGURE 6.9 Structural information is central to the development of models and cures for disease, and today is largely established using methods and large facilities originally developed for the condensed-matter and materials physics community. The figure shows the exterior envelope protein (upper right) of the AIDS (acquired immunodeficiency syndrome) virus together with a neutralizing antibody (left) and the human CD4 receptor (lower right). The structures are deduced from x-ray diffraction data collected at the National Synchrotron Light Source (NSLS) at Brookhaven National Laboratory.

most exciting of the fundamental studies have been those of continuous phase transformations. SR studies of transformations between the different phases of monolayers absorbed on flat, well-ordered substrates, as well as reconstructed surfaces, have enabled significant tests of exact results from the theory of two-dimensional physics. Similarly, the step-bunching transition on single crystal surfaces, originally predicted theoretically, has been studied experimentally with SR. The surface-scattering methods have also been used for studies of liquid-surface and amorphous thin-film structures. A particularly interesting result is that near the surface of liquid metals, there is metal atom layering.

Surface sensitive SR techniques have been applied increasingly to significant technological problems. Several major projects have been aimed at understanding thin-film formation via vapor deposition and sputtering. Because the normal surface-sensitive techniques cannot be used in these "high-pressure" situations, synchrotron-based surface-scattering studies now account for a large portion of the existing in situ characterization of these processes. An important extension of the surface-scattering technique is grazing incidence fluorescence now being applied by several semiconductor manufacturing companies to micro-contamination analysis of Si wafers.

Also of technological importance are the surface-sensitive electron-yield techniques in the vacuum ultraviolet (VUV)/soft x-ray region, which measure bond lengths of adsorbate/surface bonds and orientations of molecules adsorbed on surfaces. These are now being used for practical applications such as determining the mechanisms governing the orientation of molecules of importance to liquid-crystal displays. In addition, x-ray magnetic circular dichroism is having a significant impact on the science of magnetic recording.

The number of applied problems involving surfaces, surface layers, and interfaces is enormous. We anticipate enormous growth in experiments related to corrosion, electrochemistry, tribology, environmental interfaces, and the like as more beam lines are commissioned around the world. Electrochemistry deserves special mention because already SR together with probe microscopies has transformed this field from one primarily dependent on electrochemical measurements and related modeling to the study of electrode processes at the molecular level.

Microspectroscopy

The availability of the third-generation sources has made possible higher resolution microspectroscopies. The higher brightness at ALS has enabled construction of an improved scanning transmission x-ray microscope (STXM) as well as a scanning photoelectron microscope (SPEM). The STXM is especially useful for micro-composition and orientation measurements in multicomponent polymers and organic systems. Spatially resolved x-ray photoelectron spectroscopy is now being applied to a range of materials issues, such as examining chemical structure of Ti-Al alloys reacted with graphite and chemical speciation on bond pads of integrated circuits in order to correlate chemical state with phenomena like adhesion and chemical residues in vias.

At APS, an x-ray microprobe with a FWHM focal spot size of 0.33 μm with a flux density exceeding 5×10^{10} photons/m^2 s (0.01 percent bandwidth) has been developed. Using a root specimen in its natural hydrated state, elemental sensitivity significantly better than 10 ppb and minimum elemental detection limit of 0.3 fg have been demonstrated. The x-ray microprobe is being used in variety of environmental and biological research projects. Figure 6.10 shows that it is also

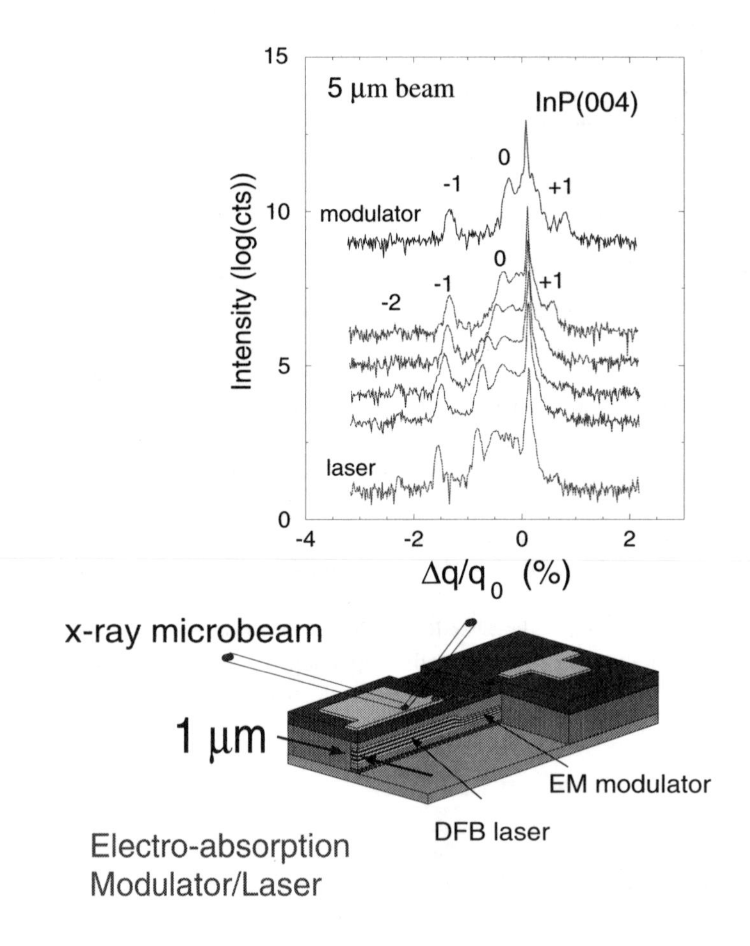

FIGURE 6.10 X-ray microbeam diffraction study at the NSLS on an electro-absorption modulator/laser device (lower figure). The upper figure shows the change in the strain distribution on going from the modulator region to the laser region of the device. (Courtesy of Bell Laboratories, Lucent Technologies.)

extremely useful for the characterization of technologically important man-made microstructures.

Insertion devices provide brilliance adequate for performing high-resolution inelastic scattering measurements of excitations in solids and liquids. X-rays have the special advantage that they are not subject to the kinematic restrictions that prevent neutrons from probing medium-energy excitations at long wavelengths. In addition, their coherence as they emerge from modern undulators is such that photon correlation spectroscopies, hitherto limited to visible wavelengths emitted by ordinary lasers, can be extended to examine slow dynamic processes at shorter distances than previously possible. Given these twin advantages, it should come as no surprise that x-rays have yielded some of the most exciting results in the physics of fluids and glasses, where they have been able to examine portions of phase space inaccessible to both neutron- and light-scattering techniques. Other progress has occurred in lower-resolution measurements to examine excitations in various simple metals as well as more complex oxides.

Photoemission Spectroscopy

Angle-resolved photoemission spectroscopy (ARPES) at SR sources has proven to be a unique tool when addressing the question of the electronic structure of solids, most notably the variation of the band energies with respect to momentum.

Over the last decade, such experiments have played an important role in advancing our understanding of high-temperature superconductors. The significant improvements in beam intensity and energy resolution obtained from undulators and new spectrometers have facilitated the discovery of a number of fascinating features in the electronic structure of the high-T_c superconductors. The most notable consequence is the beginning of a detailed view of how conventional band theory breaks down for these materials. In addition, ARPES has provided images of the unconventional superconducting gap functions of the cuprates.

Magnetic Scattering

Although the coupling of x-rays to magnetic moments is considerably smaller than that of neutrons or electrons, the extremely high-SR intensities have enabled qualitatively new kinds of experiments complementary to those performed with neutrons. In particular, the availability of radiation of tunable energy and polarization has led to spectroscopies that promise the separation of the orbital and spin magnetization densities in solids. SR has made very precise characterizations of the magnetic behavior of a variety of rare-earth, transition-metal, and actinide systems possible. The naturally high resolution not only allows the determination of the magnetic periodicities with unprecedented accuracy, but also allows magnetic correlations to be explored on micron length scales.

Infrared Investigations

One development that was not foreseen a decade ago was the rise of infrared techniques using SR. For example, the vacuum ultraviolet ring at the National Synchrotron Light Source (NSLS) at Brookhaven provides infrared light that is 10^3 times brighter than typical thermal sources and highly stable. Similarly, it produces more power than thermal sources in the far infrared and is also a pulse source suitable for time-resolved spectroscopy with subnanosecond resolution. This source has enabled infrared spectroscopy to be applied to problems such as the dynamics of adsorbates on metals and semiconductors, and photoconductivity.

X-Ray Absorption Spectroscopy

This is a very simple and powerful technique with applications too numerous to list. To give an idea of what is possible, the committee considers briefly its use in environmental science, a field driven by the tremendous need for new remediation and prevention technologies. Environmental science is therefore a growth area for the application of methods from condensed-matter and materials physics. The application of SR techniques, particularly XAS, to problems in environmental science has grown rapidly during the past decade. XAS is particularly useful because easily interpretable data on chemical states can be collected in environmentally relevant conditions (e.g., in the presence of water, at ambient pressures and temperatures, at dilute metal ion concentrations greater than 10 ppm). The resulting information is critically important for determining the toxicities, bioavailabilities, transport properties, and environmental fate of metal ions in soils and aquifers and for subsequently designing cost-effective and reliable remediation.

The Next Decade

The main focus of U.S. efforts will be to fully develop the third-generation sources that have only just come online. This means building functioning beam lines capable of what is currently routine at the European Synchrotron Radiation Facility (ESRF) in Grenoble (e.g., high-resolution inelastic scattering and photon correlation spectroscopies, high-pressure diffraction). However, past experience indicates that many of the new experimental techniques that will be developed at the third-generation SR sources have not yet emerged. It seems likely that the history of first- and second-generation sources, where some of the most fruitful techniques were advanced only after experience with the source and instrumentation had been obtained, will be repeated.

Nevertheless, SR scientists are already planning fourth-generation sources that will provide still greater brightness (about two orders of magnitude) and subpicosecond pulses at successively shorter wavelengths using single-pass free-electron lasers. The plans include a deep-ultraviolet source at Brookhaven National Laboratory, a soft x-ray source at Argonne National Laboratory, and an x-ray source providing radiation at wavelengths down to approximately 1.5 Å at the Stanford Linear Accelerator Center. The ultimate facility will be based on a superconducting linac of some 20-30 GeV feeding a farm of 50 or more 100-m-long undulators with radiation from the ultraviolet to the hard x-ray.

As indicated in Figure 6.7, the FELs promise about eight-orders-of-magnitude increases in peak brightness. They will provide diffraction-limited radiation at their operating wavelengths with sufficient numbers of photons so that data will generally be acquired with a single pulse. Thus, they are likely to usher in a new era of short wavelength coherent imaging and subpicosecond studies of electronic and atomic structure. Moreover the development of these sources will enhance the impact of SR on biology, soil science, agriculture, archeology, and other fields.

THE REINVENTION OF TRADITIONAL CONDENSED-MATTER EXPERIMENTS

At the same time that the capabilities of large-scale facilities have been dramatically expanding, there has been a quiet revolution in small-scale instrumentation. This revolution has had an enormous impact on both the efficiency and capabilities of single investigators and groups working on small-scale materials experiments in traditional laboratory settings. From spectrum analyzers, to top-loading dilution refrigerators, to personal computer (PC)-controlled parameter analyzers, the tools of the trade have evolved to the point where measurements that once took from weeks to entire graduate student careers, can now be routinely done in days, hours, or less. Laboratory instruments, computers, and software to enable quick and easy automation of most laboratory measurements are now available. Almost all commercial instruments come with IEEE GPIB interfaces. Inexpensive PCs are pervasive, and commercial software packages have been designed specifically for laboratory automation. The ongoing advances in microelectronics technology have spawned new generations of inexpensive yet extremely high-performance digital oscilloscopes, voltmeters, and all sorts of parameter analyzers. In parallel there have been continuous improvements in performance and reductions in the cost of systems such as dilution refrigerators, pulsed lasers, and superconducting magnets. There are also new types of commercial instruments that act as platforms for performing a number of complex experiments on a sample, all under computer control. For example, such probes had a huge impact on the development of the field of high-T_c superconductivity and are playing a very important role in the development and exploration of new magnetic materials.

Another very exciting area of "small-scale" instrumentation involves leveraging microfabrication technology, as driven by the microelectronics industry, to do or enable physics experiments. There are several different aspects of this. The first relates to the custom design of electronic circuitry specifically configured for some special laboratory instrumentation function. What used to require the effort of numerous people hand-wiring large numbers of components together to eventually produce a rack full of instrumentation can now frequently be reduced to an application-specific integrated circuit (ASIC) designed to the need, along with a few other high-function, but standard, integrated circuits.

A second aspect of leveraging microfabrication involves special-purpose technology developed to fulfill some engineering need, but using it for physics applications as well. A good example of this is low-T_c superconducting electronics. For the past several years high-quality foundry service has been available for producing prototype superconducting digital circuitry. This same foundry service has been used to fabricate on-chip experiments to study the physics of Josephson junctions, the behavior of arrays of superconducting devices, the performance of high-frequency mixers and antennae for radio astronomy, and so on. In addition, this technology can be used to fabricate all manner of integrated SQUIDs including, for example, magnetometers with small pickup-loop structures to be used in scanning SQUID microscopy applications. This is a fabrication service available to everyone at a very modest cost.

A third aspect of leveraging is related to microelectromechanical systems (MEMS). MEMS is a rapidly growing engineering field, closely linked to the microelectronics industry, that has developed a wide variety of devices such as micromotors, microactuators, and microflow-controllers. MEMS in the form of microcantilever structures are at the heart of many scanning-probe implementations. MEMS technology is beginning to provide some exciting opportunities to do physics in unconventional ways on very small quantities of matter. At present, cantilever structures in one form or another are the basis for many such experiments, but it is clear that MEMS can provide an ideal platform for a wide variety of physics experimentation. We can expect to see a rapid expansion of "laboratory-on-a-chip" concepts and implementations in the near future.

In their pursuit to gain control of entities of the very smallest dimensions, scientists are developing extremely sensitive sensors to detect and analyze very weak physical and chemical effects involving minute amounts of material. The basic sensing element used in one such study is a silicon microcantilever like that used in an atomic-force microscope. This microcantilever bends in reaction to the forces imposed on it by various phenomena under investigation. Several methods can be applied to detect the motion and deformation of the cantilever including optical and electrical techniques, the latter using piezoresistors.

Because of their very small size, such microcantilevers—if properly designed—feature high sensitivity and short response times.

There are two obvious prerequisites for experiments in condensed-matter and materials physics. The first is measuring equipment, the second is interesting specimens. As regards small laboratories, the key development is that interesting specimens are becoming more and more indistinguishable from the measuring equipment, in the shape of "experiments on chips." Certainly examples exist of outstanding physics being done with 20-year-old equipment on samples made in a great variety of traditional ways. However, the overall efficiency and productivity of the research will be significantly enhanced through ongoing upgrades of instrumentation and automation, both for measurement and fabrication. Incorporating upgrades in all the physics laboratories in the nation represents a substantial ongoing investment required to keep small-scale laboratory operations competitive.

The scientific and technological future of nanofabrication and nanoscale processing, which enables modern "experiments on chips," is bright and exciting. If, however, the momentum of this nanoscale revolution is to continue to grow and its full promise to be realized, steps must be taken to ensure that the broad research community has appropriate and effective access to the full arsenal of capabilities in this area, capabilities that must be at the very forefront of nanofabrication and nanoprocessing technology. Because of the complex and multifaceted nature of this technology, an essential component of this access must be via national user facilities, such as the current National Nanofabrication Users Network (NNUN) supported by the National Science Foundation. It is essential that each facility of this sort be adequately supported so that it can broadly provide world-class research capabilities to academic, industrial, and government researchers and thus continue to push the state of the art. It is also essential that each encompass a broad range of nanofabrication and processing capabilities so that users can take a nanoscale science or technology research project from concept through to a working device or functional structure. Further, it is essential that each facility be adequately staffed with highly competent professionals oriented toward introducing new users to the technology, educating the research community about nanofabrication, and facilitating users in successfully exploiting exciting new research opportunities.

MAN-MADE EXTREME CONDITIONS

The urge to discover new states of matter has been one of the deepest motivations in condensed-matter and materials physics. An important route to discovery has been the fabrication of new materials—new compounds, alloys, and combinations of metals, ceramics, and organic matter. Frequently it has been possible to design new materials for specifically desired properties. It seems likely that modern fabrication techniques will permit us to make a host of new objects and devices that are beyond our current imagination. Numerous advances in the fabrication of new materials are discussed in this report.

Another path to the discovery of new phenomena has been through subjecting matter to unusual or extreme conditions of low temperature, pressure, and magnetic field. Measurements under such unusual conditions have sometimes led to dramatic surprises, with important results that could not have been anticipated in advance. Here, the committee looks at a few of the results and describes the present state of the technology and future prospects. There are common themes for research under extreme conditions: (1) The limit in the equilibrium or static value of minimum temperature, maximum pressure, or maximum magnetic field is 10 or more times less than transient values that can be achieved. (2) The instrumentation required for preparing specimens and performing measurements has become increasingly sophisticated and frequently requires facilities available only at large laboratories. (3) The miniaturization of specimens and apparatus is becoming increasingly beneficial to each technology.

Matter at Very Low Temperatures

The classic unexpected result in condensed-matter physics was the discovery of superconductivity in 1911, just a few years after helium was first liquefied. The electric resistance of superconductors becomes zero. The low temperatures were critical for producing the phenomenon because thermal energy at higher temperatures disrupts the delicate interactions between electrons. At low-enough temperatures, the electrons form the paired state responsible for superconductivity. It took another 50 years before the underlying phenomenon was really understood. The discovery was made primarily because the experimenter wanted to see how the nature of conductors changed as they were cooled. Similarly, the superfluid states of the helium isotopes, ^{4}He and ^{3}He, were discovered because of developments in low-temperature technology and curiosity about how matter behaved at lower temperatures. When the thermal motion is reduced, delicate new phenomena appear.

The minimum temperature achieved under static conditions with the thermometer, coolant, and specimen under investigation is 1 mK. Temperatures down to 10 mK can be produced routinely with commercially available apparatus in which pumped ^{3}He is circulated through heat exchangers. Lower temperatures require a different principle: magnetic cooling. Nuclei of a metal (usually copper) are polarized at millikelvin temperatures in a magnetic field of 10 T. The copper and specimens attached to it are then thermally isolated and the magnetic field is removed. The polarized nuclear magnetic moments become very cold. Electrons in the copper transfer heat between the specimen and the cold nuclei.

An important challenge that has attracted quite a number of experimentalists in this field has been the attempt to cool dilute mixtures of ^{3}He in liquid ^{4}He to very low temperatures. The goal has been to discover a superfluid pairing transition similar to that which occurs in pure liquid ^{3}He. The paired state might be quite different from that in the pure liquid. No one has succeeded in cooling the dilute mixtures to temperatures less than 200 μK. Heat transfer between the metal coolant and the dilute mixture is quite difficult.

Significantly lower temperatures have been achieved for isolated systems not in thermal equilibrium with surrounding matter. One class of experiments has been the study of spontaneous nuclear magnetism in metals such as copper, silver, and platinum. At the end of the magnetic cooling process, after the large magnetic field used to polarize the nuclei is removed, the magnetic moments are quite cold. The thermal equilibrium times are quite long, sometimes more than 10^8 seconds. Through clever determinations of the spin-entropy, temperatures as low as a tens of picokelvin have been deduced. The method has been used to examine a variety of unusual states of magnetic order. As Figure 6.11 illustrates, some of the experiments have even been conducted in connection with neutron diffraction at reactor facilities. The neutrons were used to image the nuclear magnetic moments in the ordered state.

A spectacular example of the cooling of a metastable isolated system of matter has been the studies of Bose-Einstein condensation in gases of sodium, rubidium, and lithium. Modern optical techniques in conjunction with magnetic traps and radio frequency fields have been used to cool dilute gases of these atoms to sub-microkelvin temperatures. The hot atoms are kicked out of the magnetic trap by the radio frequency electromagnetic fields. At the very low temperatures, the atoms obeying Bose statistics can simultaneously occupy the same state—the condition for Bose-Einstein condensation. Quantum interference between clusters of atoms has been demonstrated. The effect is analogous to interference between two sources of coherent light.

Matter at Very High Pressures

For most of this century, progress in achieving high pressure in matter was achieved by building ever larger series of cascaded metal pistons to compress material. By 1980 a dead end had apparently been reached. Even with the strongest steel alloys, the limiting static pressure that could be produced on samples of milliliter volumes was in the range of several hundred kilobars. The invention of the diamond anvil pressure cell led to a major advance in high-pressure science. Small diamond crystals are formed into narrow tips to be pressed against each other. Typical contact regions have a diameter less than 10 μm. Forces of order of only a few Newtons can produce megabar pressures over such small areas. To achieve high pressures, the region around the narrow tips must be defect free. The smaller the region, the higher the probability that there are no defects.

The maximum static pressure achieved by the diamond anvil method is a

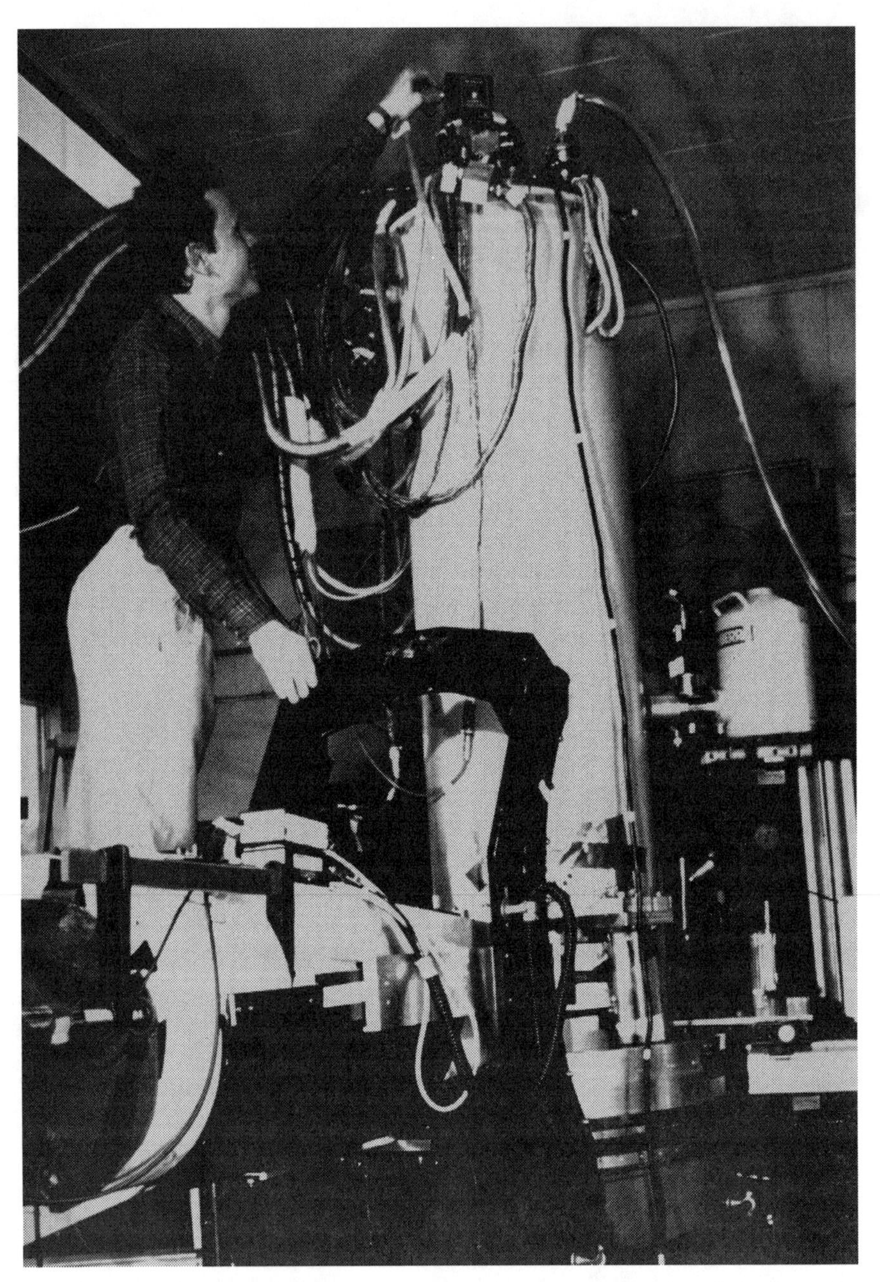

FIGURE 6.11 Photograph of nuclear demagnetization apparatus installed on a cold-neutron guide emanating from a nuclear reactor. The apparatus was used to discover the ordering of the nuclear spins in elemental copper at less than 58×10^{-9} K. (Courtesy RISØ National Laboratory, Roskilde, Denmark.)

little more than 1 Mbar. Quite a number of new dense phases of matter have been discovered. Of particular interest has been the transition of normally completely insulating materials such as solid xenon and sulfur in conducting metallic states. As the atoms are squeezed closer together, the outer electrons become free.

The small size of the specimens presents a special challenge in instrumentation. The entire experiment has to be built in micron-sized volumes. Nevertheless, nanotechnology has been used to apply electrical leads and even magnetic resonance coils to diamond anvil devices. The crystal structure of new high-pressure phases of matter has been determined with x-rays from synchrotron sources. The measurement of the pressure is also a difficult matter. Calculation of the force per unit area is frequently insufficient because the stresses are not uniformly distributed. Instead, a combination of calculated pressures and extrapolation of material properties such as fluorescence frequencies must be carefully compared in many experiments to establish a reliable pressure scale.

A special goal in high-pressure research in recent years has been the search to find the elusive metallic state of solid hydrogen. It would be an especially interesting discovery because hydrogen should be one of the easiest materials for which to calculate an equation of state with fundamental theory. The pressure predicted for the metallic transition in hydrogen is very close to the values currently being produced.

Beyond providing information about systems of fundamental interest to condensed-matter physicists, high-pressure research is essential for understanding the composition and properties of Earth's interior. Recent experiments have led to significant new findings on phase transformations associated with deep earthquakes, for example.

Further progress in achieving higher pressures will probably be achieved through use of stronger materials. For example, studies of tungsten and iron suggest that they become even stronger at megabar pressures.

Transient pressures greater than 2 Mbars have been obtained in shock waves. The maximum pressure lasts only a few nanoseconds. Nevertheless, most of the existing high-pressure and high-temperature data have been obtained with the use of gas guns, high explosives, and even nuclear detonations. The development of high-intensity lasers provides a potentially attractive complement to these methods, particularly for equation of state studies at high energy densities. By focusing a short-pulse, intense laser beam on a sample, a rapidly expanding plasma is created, which, in turn, drives a shock wave into the sample; laser-induced shock-wave experiments to obtain high-pressure data (in excess of a megabar) have been carried out for more than a decade. However, concerns have existed regarding the accuracy of the data owing to the lack of planarity of the shock front, preheating of the material ahead of the shock front, difficulty in determining the steadiness of the wave front because of the small sample size, and the absence of absolute pressure and volume data. Recent improvements in beam smoothing and other experimental developments have improved the quality of the propagating shock wave.

Matter in Large Magnetic Fields

Large magnetic fields represent yet another extreme condition to impose on matter. Frequently the large fields are imposed in conjunction with low temperatures. A large magnetic field can orient material, confine electrons in conductors to particular energy states and locations, and produce specially selected spin states of nuclei and electrons. Figure 6.12 illustrates how the range of accessible phenomena grows with the magnetic field strength, while Figure 6.13 shows the steady growth in man-made fields over the last century. Examples of dramatic discoveries in recent years are the integer and fractional quantum Hall effects. Both were discovered by accident—again—because the technology was available to produce the required extreme condition and experimentalists were interested in how matter changed under the new conditions. Both are discussed in other sections of this report.

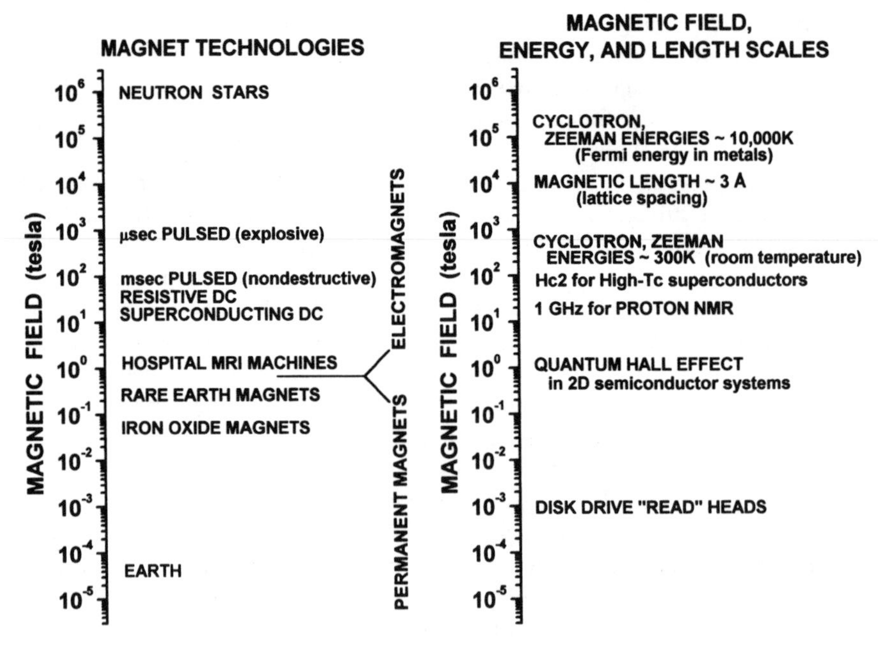

FIGURE 6.12 Higher fields are associated with higher energies, smaller length scales, and more extreme technologies and environments. (Courtesy of Bell Laboratories, Lucent Technologies.)

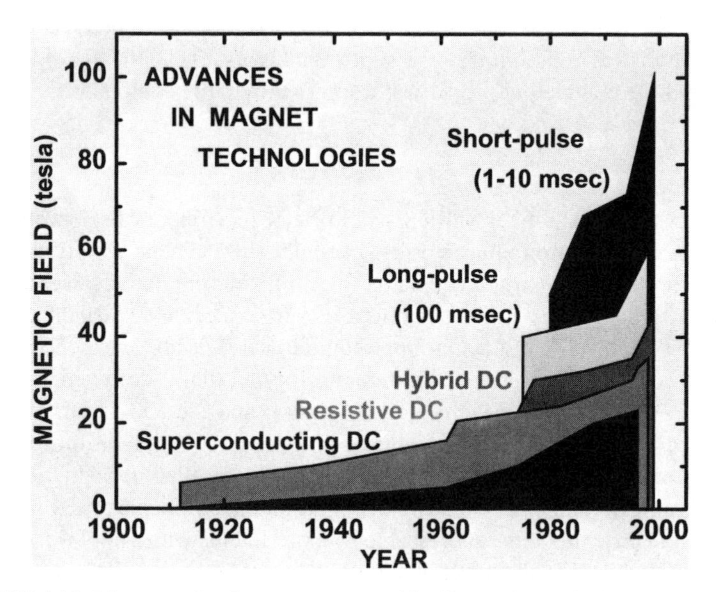

FIGURE 6.13 Magnet technology, as expressed in the maximum field reached in nondestructive experiments, has grown exponentially over the last century. We anticipate that the incorporation of high-T_c superconductors will assure continued growth. (Courtesy of Bell Laboratories, Lucent Technologies.)

The highest static magnetic field achieved is 37 T (80,000 times Earth's magnetic field strength). The large field is achieved with a combination of two concentric magnets—a water-cooled copper solenoid carrying a large current on the outside and a superconducting solenoid on the inside. Today's superconductors alone cannot carry enough current to produce a magnetic field greater than about 15 T. The outer copper (resistive) magnet produces the additional 22 T.

The ultimate design limit on the maximum strength of steady magnetic field is determined by the strength of the material that carries the current. The field produces a radial outward force proportional to the current passing through the solenoid wire. Using the strongest materials that exist, the maximum field is approximately 50 T.

Much larger fields can be achieved for shorter times. Large current pulses through wire coils have been used to produce fields greater than 75 T for tens of microseconds. These magnets can be repeatedly cycled with current pulses to obtain masses of data. The maximum fields, those in excess of 100 T, are produced by explosive technology. Typically, current-carrying cylinders are rapidly imploded and a very large magnetic field is produced as the tube collapses. The transient-field experiments share a challenge with the transient-pressure experiments—the data must be obtained in nanoseconds or less. The transient experiments have been important in gaining information about the high-field behavior of high-temperature superconductors, the optical properties of matter, and the conducting properties of unusual metallic compounds.

The Next Decade

Matter under extreme conditions is as obvious a frontier of science as high-energy physics or astronomy. It is also equally easy to state simply the future program for this field—to subject matter to ever higher pressures, lower temperatures, and higher magnetic fields and to use every conceivable visualization tool to see what happens. Spectacular opportunities will arise because of new infrastructure, such as the National Ignition Facility, actually designed for fusion research at the Lawrence Livermore Laboratory, and the 100 T pulsed magnet foreseen at Los Alamos. Other significant advances will arise from improved visualization capabilities, which will follow from installation of high-pressure cells and high-field magnets at advanced light and neutron sources.

The scientific problems addressable by experiments with samples in extreme environments will span the range of condensed-matter physics. Past performance suggests that such experiments will make important contributions to resolving the problems posed by the high-temperature superconductors as well as many other fascinating materials both known and unknown. We also look forward to breakthroughs in areas much further from the traditional core of condensed-matter physics. Recent examples include optical tweezers, whose development was closely interwined with the quest for ultra-cold laser-cooled matter, and magnetic resonance imaging.

COMPUTATIONAL MATERIALS PHYSICS

The modern high-speed computer is a remarkable device, made possible in part by fundamental discoveries and continuing advances in condensed-matter and materials physics. With clock speeds now routinely reaching as high as 500 MHz, small, mass-produced workstations have a computational power that would have been possible only with giant supercomputers (now viewed as dinosaurs) just a few years ago. Computers and their components are now so sophisticated that each succeeding generation cannot be built without making full use of the computational power of the existing generation. In addition to improving themselves, computers have become powerful tools in the study of a wide variety of condensed-matter phenomena and materials. Figure 6.14 shows the remarkable progress in high-end computing, together with its implications for one particular problem, the computation of turbulence in fluids.

The history of computing has been one of frantic attempts to find architectures and software that can deal with ever-changing hardware limitations. Less than 20 years ago, memory was very expensive (witness today's Year 2000

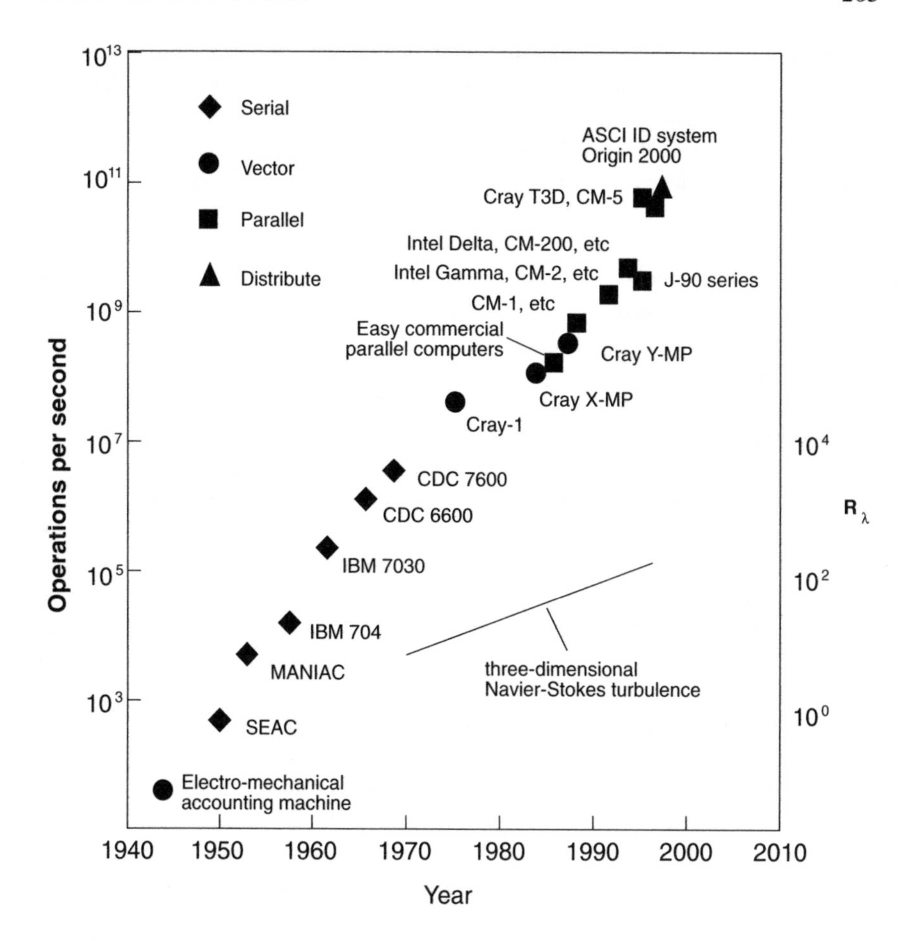

FIGURE 6.14. The plot shows the growth of the number of operations per second from 1940 to 2010 for the fastest available "supercomputers." Objects of different shapes are used to distinguish serial, vector, and parallel architectures. All processors until Cray-1 were single-processor machines. The line marked "three-dimensional Navier-Stokes turbulence" shows, in rough terms, the extent to which the increased computing power has been harnessed to obtain turbulent solutions by solving three-dimensional Navier-Stokes equations. Turbulence is used here as an example of one of the grand and difficult problems needing large computing power. The computing power limits the size of the spatial domain over which computations can be performed. The Reynolds number (marked on the right as Rλ) is an indicator of this size. (Courtesy of Los Alamos National Laboratory.)

problems) and microprocessors were much slower than discrete component designs. Today memory density has risen enormously and prices have fallen dramatically. Microprocessors have risen several orders of magnitude in speed and gone from 8- to 64-bit word lengths. After some tumultuous history involving exploration of different parallel architectures, shared-memory parallel systems combining many processors communicating via high-speed digital switches are now rapidly developing and have largely replaced pure vector processors. Clock speeds for microprocessors are now so high that memory access time is often far and away the greatest limitation on overall speed. One of the great software challenges now is to find algorithms that can take maximum advantage of parallel architectures consisting of many fast processors coupled together.

In addition to hardware advances, the last decade has seen some revolutionary advances in algorithms for the study of materials and quantum many-body systems. Improved algorithms are crucial to scientific computation because the combinatorial explosion of computational cost with increasing number of degrees of freedom can never be tamed by raw speed alone. (Consider the daunting fact that in a brute force diagonalization of the lowly Hubbard model, each site added multiplies the computational cost by a factor of approximately 64.)

In the last two decades computational condensed-matter and materials science has moved from the initial exploratory stages (in which numerical studies were often little more than curiosities) into the main stream of activity. In some areas today, such as the study of strongly correlated low-dimensional systems, numerical methods are among the most prominent and successful methods of attack. As new generations of students trained in this field have begun to populate the community, numerical approaches have become much more common. Nevertheless it is fair to say that computational physics is still in its infancy.

Pushing the frontiers of computational physics and materials science is important in its own right but also important because training students in this area provides industry and business with personnel who not only have expertise on the latest hardware architectures but also bring with them physicists' methods and points of view in analyzing and solving complex problems.

Progress in Algorithms

In spite of its great enthusiasm, the committee offers a warning before proceeding. Specifically, numerical methods have become more and more powerful over time, but they are not panaceas. Vast lists of numbers, no matter how accurate, do not necessarily lead to better or deeper understanding of the underlying physics. It is impossible to do computational physics without first being a good physicist. One needs a sense of the various scales relevant to the problem at hand, an understanding of the best available analytical and perturbative approaches to the problem, and a thorough understanding of how to formulate the interesting questions.

Electronic Structure Algorithms

The goal of electronic structure calculations is to compute from first principles or with approximate methods the quantum states of electrons in solids and large molecules. This information is then used to predict the mechanical, structural, thermal, and optical properties of the materials. The outstanding problems are ones of computational efficiency for large-scale calculations and convergence to the thermodynamic limit. Indeed, the calculations are so complex and time consuming that real-time dynamics can be followed only for pico- or nanoseconds.

Perhaps the single most dramatic development in the last decade has been the advent of the Car-Parrinello method, which has enormously enhanced the efficiency of electronic structure calculations. This method calls for adjusting the atomic positions and the electronic wave functions at the same time to optimize the Hohenburg-Kohn-Sham density functional. Additional efficiencies come from use of fast Fourier transform techniques to compute the action of the Hamiltonian on the wave functions without the necessity of computing the full Hamiltonian matrix.

Another area of intensive investigation has been the search for so-called "Order N" methods. The idea is to find approximation schemes in which the computational cost rises only as the first power of the number atoms or electrons, as opposed to some higher power (~3) as is typically the case. So far, this has been attempted only for tight-binding models involving spatially localized orbitals for the electrons. It is not yet clear that the problem will be solvable, but research in this direction is important if we are going to be able to do larger and more complex structures. Other techniques under investigation include adaptive coordinate, wavelet, and direct grid/finite element methods that are useful in situations in which the number of plane waves needed to represent atomic orbitals is very large.

The Kohn-Sham local-density functional approximates the many-body exchange-correlation corrections to the energy by a functional of the local density. It has been very successful and is finally winning support within the computational chemistry community. An important area of current research involves generalized gradient expansion corrections to the local-density approximation. In several examples, simple local-density approximations fail to give correct structures but appropriate gradient expansion functionals work. In general, however, it is often still difficult to obtain the chemical accuracy required.

Monte Carlo Methods

Fermion Monte Carlo techniques continue to be plagued by the "sign problem." Because of the sign reversals that occur in quantum wave-functions when two particles exchange places, not all time histories have positive weights in the

Feynman path integral. This means that the weights cannot be interpreted as probabilities that can be sampled by Monte Carlo methods. The fixed-node approximation attempts to get around this problem by specifying a particular nodal structure of the wave function. This has yielded very useful results in some cases in which the nodal structure is understood a priori. Some workers are now moving beyond small atoms and molecules to simple solids and have obtained good results for lattice constants, cohesive energies, and bulk moduli.

Fermion Monte Carlo path integral methods continue to be applied successfully to lattice models such as the Hubbard model, but again the sign problem is a serious limitation. For example, it is still difficult to go to low-enough temperatures to search for superconductivity, even in highly simplified models of high-T_c materials.

Bosons, which are much easier to treat numerically, also pose interesting problems. "Dirty boson" models have been used to describe helium films adsorbed on substrates and to treat the superconductor-insulator transition. With this model one makes the approximation that Cooper pairs are bosons and assumes (not necessarily justifiably) that there are no fermionic degrees of freedom at zero temperature.

Cluster Algorithms in Statistical Mechanics

One serious problem in the Monte Carlo simulation of statistical systems near critical points is the divergence of the characteristic timescales. The computer time needed to evolve the system to a new statistically independent state diverges as some power of the correlation length or system size, ${L_{MC}}^{d+z}$ where d is the dimensionality. Cluster algorithms have been extremely successful on certain classes of problems (such as the Ising and XY models and certain vertex models) and are able to reduce the dynamical exponent z_{MC} to nearly zero. This is accomplished by constructing clusters of spins, and for each cluster, choosing a random value of spin that is assigned to the individual spins it contains. Such a move cannot be implemented in an ordinary Metropolis algorithm because the Boltzmann factor would make the acceptance rate of the move essentially zero.

The trick is to have the probability that a cluster grows to a particular size and shape be precisely the Boltzmann factor for the energy cost of flipping the cluster. This is a very tiny probability, but it is canceled by the fact that there are a huge number of different possible clusters that could have resulted from the random-growth process.

This has been a very important advance. Unfortunately there are still many cases (such as frustrated spin systems) for which cluster methods cannot (as yet) be applied because of technical problems similar to the fermion minus-sign problem.

Density-Matrix Renormalization Group

A revolutionary development in computational techniques for quantum systems is the "density matrix renormalization group." The essential idea is to very efficiently determine which basis states are the most important to keep to be able to describe the quantum ground state. The procedure is the first one ever found that gives *exponentially* rapid convergence as the number of basis states is increased. It applies to essentially any one-dimensional model with short-range forces, even random systems without translation symmetry and fermion systems. Using this technique it is now easy to compute ground state energies and correlation functions to 10-digit accuracy on a desktop workstation. Ongoing work is extending the technique to excited states and to higher dimensions.

Computational Physics in a Teraflop World

In this section we contemplate questions of the future of computation and what can be (optimistically) done with the next factor of 1000 in computing power.

Glassy Systems, Disorder, and Slow Dynamics

At first sight, these problems do not seem well suited for more computer time. They are too hard. Experimentally, the phenomena are spread over fifteen decades in frequency, and even that dynamical range in the experiments is often not enough to reach firm conclusions. Current simulations span perhaps three decades: one might think that three more won't make an overwhelming improvement.

There are two reasons to be optimistic. First, in numerical simulations, it is straightforward to watch individual atoms/spins/automata relax. The last decade has seen tremendous progress in the visualization and study of spin glasses, charge density waves, glassy behavior in martensites, and "real" glasses. We are, however, barely into the scaling region for many of these simulations: even if the scaling region grows only logarithmically with the timescale, three more decades might make the patterns clear.

Second, there is every reason to believe that we can get around these slow timescales. There is no reason for our methods for relaxing glassy systems to be as inefficient as nature. Until now we have mainly developed techniques to mimic nature with as little wasted effort as possible. This was sensible for studying systems in which nature relaxes efficiently; when you are barely able to follow the system for a nanosecond, you study systems that relax rapidly. Now that we are turning to problems for which nature is slow (for example, glasses and phase transitions) we are making rapid strides in developing acceleration algorithms. In particular, because we are more likely to gain the next factor of 1000 in computing power by increasing the number of processors rather than through raw speed increases, we will naturally learn new algorithms for relaxation to exploit the extra processors.

Quantum Chemistry and Electronic Structure of Materials

With these problems we are not confused about the physical behavior. For these problems, the answers to the interesting questions inherently demand immense precision. Quantum chemistry is difficult not because the systems are complex and subtle, but because the standards are high. All of chemistry is controlled by reaction and binding energies that are tiny compared to total energies. Electronic-structure calculations for materials face exactly the same problem. We can now study only relatively simple molecules and crystal structures; with the next generations of machines and algorithms, this will change qualitatively.

Structured Systems: From Inorganic Industrial Materials to Proteins

These are systems for which there are huge ranges of length scales and timescales, which interact in nontrivial ways. We have to understand the physics and materials science on each scale and connect together the properties at different scales. The algorithms appropriate to the models at different scales can be quite different from each other.

The category of "industrial materials" includes ceramics, concrete, polycrystalline metals and alloys, and composites. Their important properties are normally almost completely removed from the world of perfect crystals and equilibrium systems often studied by mainstream physics. The wearing properties of steel, the resistance of concrete to cracks, the thermal and electrical properties of polycrystalline metals—all are dominated by the mesostructure, the detailed arrangement of domain walls, pebbles, and grains.

Three issues must be confronted to make progress. First, the materials are disordered. Second, they display history dependence; for example, the polycrystalline domains in metals are dependent in detail on how the metal was cast, rolled, and stamped during its manufacture. Third, the systems have a large range of scales. The dynamics of grain boundaries under external strain is determined by the dynamics of the individual line dislocations that make them up. The line dislocations interact logarithmically (in inscrutable ways), and one can only simulate them at the current level of knowledge. Their dynamics, in turn, is determined by atomic-scale motion; the diffusion of vacancies and the pinning to inhomogeneities (and to other line dislocations) are crucial to understanding their motions. It is this enormous range of scales that we can only hope to disentangle with large-scale simulations (see Figure 6.15).

Proteins and biomolecules provide similar problems. The molecular biologists separate their structures into primary, secondary, and tertiary precisely as a set of length scales on which the structure is organized. The functional behavior

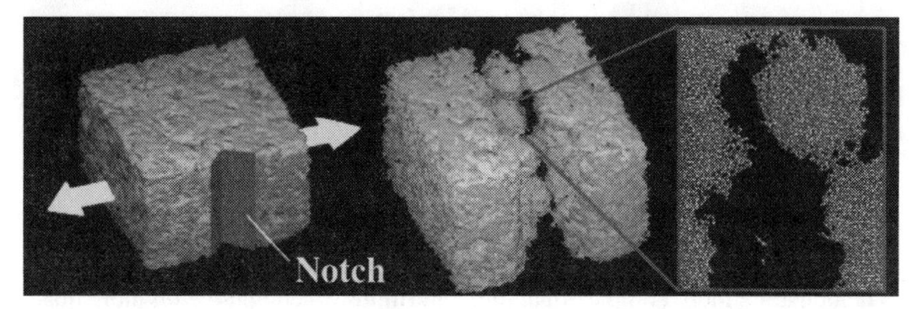

FIGURE 6.15 Million-atom molecular dynamics simulation of ductile behavior in nanophase silicon nitride, which is being explored for its extraordinary resistance to fracturing under strain: a 30 percent strain is required to completely fracture the nanophase system, while only 3 percent is required for single-crystal silicon nitride. Shown is the system before it fractures under an applied strain of 30 percent and a zoom-in to atomic scale visualizing that the crack front advances along disordered interfacial regions in the system. It is along the amorphous intercluster regions where the crack propagates by coalescence of the primary crack with voids and secondary cracks. (Courtesy of Louisiana State University.)

on the largest scales depends in detail on the dynamics and energetics not only down to the protein level, but even down to the way in which each protein is hydrated by its aqueous environment.

Quantum Computers

Theoretical analysis of the quantum computer, in which computation is performed by the coherent manipulation of a pure quantum state, has advanced extremely rapidly in recent years and indicates that such a device, if it could ever be constructed, could solve some classes of computational problems now considered intractable. A quantum computer is a quantum mechanical system able to evolve coherently in isolation from irreversible dephasing effects of the environment. The "program" is the Hamiltonian. The "input data" is the initial quantum state into which the system is prepared. The "output result" is the final, time-evolved state of the system. Because quantum mechanics allows a system to be in a linear superposition of a large number of different states at the same time, a quantum computer would be the ultimate "parallel" processor.

The basic requirement for quantum computation is the ability to isolate, control, and measure the time evolution of an individual quantum system, such as an atom. To achieve the goal of single-quantum sensitivity, condensed-matter experimentalists are pursuing studies of systems ranging from few-electron quantum dots to coherent squeezed photon states of lasers. When any of these reach the desired single-quantum limit, experiments to probe the action of a quantum gate could be immediately designed. Recent theory shows in principle how to form different types of gates and provides error-correcting codes to enhance robustness. At this point it is quite unclear if a practical system can be developed, but many clever ideas are being explored. Interesting physics is sure to result and there is at least a remote possibility of a tremendous and revolutionary technological payoff.

Several groups have reported an experimental realization of quantum computation by nuclear magnetic resonance (NMR) techniques. The race is now on to demonstrate more complex quantum algorithms, to compute with more quantum bits than the two bits of the first demonstration, and to verify error-correction techniques.

FUTURE DIRECTIONS AND RESEARCH PRIORITIES

Tools for visualizing atoms and electrons have been at the center of condensed-matter and materials physics since Bragg and von Laue first observed x-ray diffraction from crystals nearly 100 years ago. These tools will remain at the center of the field and many others, from catalysis to biochemistry. The last decade has seen great progress in research performed using apparatus of all scales.

In the area of medium-scale infrastructure, the three important developments have been widespread access to sophisticated electron microscopes and related equipment, the exploitation of the Cornell nanofabrication center, and the reinvigoration of U.S. high-field magnet research by the founding of the National High Field Magnet Laboratory. Access to equipment has fueled and will doubtless continue to fuel improved understanding and applications of bulk materials, surfaces, and interfaces. Beyond enabling U.S. academe to participate in and thereby greatly accelerate the development of mesoscale (between atomic and macrosopic scales) physics, the Cornell nanofabrication center has been an extraordinarily fertile training ground for the U.S. microelectronics industry. The National High Field Magnet Laboratory will provide access to a scientific frontier—a key site for discoveries and technological developments ranging from magnetic resonance imaging to the quantum Hall effect.

Turning finally to large-scale facilities of a type that can only exist at national laboratories, the major events have been the commissioning of third-generation synchrotrons at the Argonne and Lawrence Berkeley laboratories and the decision to recapitalize U.S. neutron science via construction of a pulsed spallation source at Oak Ridge. The synchrotrons will produce the x-rays and light necessary for the United States to compete in emerging areas such as time-resolved protein crystallography. Even though a U.S. scientist (Shull) shared the 1994 Nobel Prize for inventing neutron scattering in the 1950s (see Table O.1), the Europeans have since then established a clear lead. The Oak Ridge source will reestablish U.S. competitiveness in this area, which over the last decade has proven so vital for imaging atoms and spins in materials ranging from high-temperature superconductors to polymers.

In previous decades, key events in condensed-matter and materials physics have been the exploitation of inventions and investments in large facilities. The inventions and the facilities are devices with the special purpose of being tools for condensed-matter and materials physics. The last decade is unique in that the major event relating to such tools is actually not directly connected with inventions and facilities. Instead, it is the same phenomenon that has profoundly transformed nearly all other aspects of our society—namely, the information revolution. An obvious consequence of the information revolution for condensed-matter and materials physics is the recent progress in computational materials science. Less obvious but equally important is the ability to collect and manipulate progressively larger quantitative data sets and reliably execute increasingly complex experimental protocols. For example, in neutron scattering, data gathering rates and, more crucially, the meaningful information content, have risen in tandem with the exponential growth of information technology

What will happen in the next decade? Although we cannot predict inspired invention, we anticipate progress with ever-shrinking and more-brilliant probe beams and increasingly complete, sensitive, and quantitative data collection. One result will be the imaging and manipulation of steadily smaller atomic landscapes. Another will be the analysis and successful modeling of complex materials with interesting properties in fields from biology to superconductivity.

The promised performance improvements with applications throughout materials science will come about only if balanced development of both large-scale facilities and technology for small laboratories takes place. For example, determination of the crystal structures of complex ceramics and biological molecules is likely to remain the province of neutron and synchrotron x-ray diffraction, performed at large facilities, while defects at semiconductor surfaces will most likely remain a topic for electron and scanning-probe microscopy, carried out in individual investigators' laboratories and small facilities. Thus, the cases for large facilities and small-scale instruments are equally strong. Although the larger items such as the neutron and photon sources appear much more expensive than those that benefit a single investigator, recent European experience suggests that the costs per unit of output do not depend very strongly on the scale of the investment, provided of course that it is properly chosen, planned, and managed. Information technology is also blurring the difference between large and small facilities, as they all become nodes on the Internet. One important upshot will be that the siting of large facilities as well as the large-versus-small facility debates will largely cease to be of importance to scientists.

In addition to the construction of large facilities such as the SNS and APS, healthy research in instrumentation science is crucial to the development of improved tools for atomic visualization and manipulation. Although we have impressive success stories to point at, as in the dominance of the probe microscopy business, we strive for similar success in other areas of instrumentation that are important for both research and manufacturing. In the United States, scientific research and instrumentation have traditionally had an uncomfortable relationship. Although it is very important that instrumentation programs be science-driven and not isolated, sometimes long lead times and the need for expert research in the instrumentation itself (for example, in advanced lithography and electron, x-ray and neutron optics) require that special investment be allocated for instrumentation. The absence of such middle-scale investment as well as a perceived lack of intellectual respectability are key reasons why the nation is lagging in beam technology and science. A solution would be the development of centers of excellence in instrumentation research and education, the latter being an equally important role for this research. A model might be the National High Field Magnet Laboratory, which has recently revived magnet research in the United States. It is also clear that viable centers can exist in already strong centers of materials research.

The committee's list of priorities is designed to enable the United States to recapture its leadership in scientific tools for condensed-matter and materials physics and their exploitation. The goals to be achieved by the large neutron and synchrotron facilities are obvious—namely, to duplicate and then to exceed what the Europeans can do today. The recapitalization of the university laboratories will serve the similarly obvious purpose of maintaining the efficiency and quality of university research. The nanolithography investment will maintain user facilities in an area of extraordinary importance in materials research as well the U.S. economy. The medium-scale centers devoted to topics such as electron optics and high magnetic fields will serve not only to develop new technologies in the areas they are specifically devoted to, but also to establish a flourishing culture of scientific instrumentation within condensed-matter and materials physics. Finally, condensed-matter and materials physics needs to take advantage of all available information technology to continue to move toward its central goal of seeing all the atoms and electrons all of the time.

Outstanding Scientific Questions

- Can we manipulate single atoms fast enough to make devices?
- Can we use computation to predict superconductivity in complex materials?
- Can we make inelastic scattering using x-rays, neutrons, and electrons as important to materials science and biology as elastic scattering is today?
- Can we image and manipulate spins on the atomic scale?
- Can we develop a nondestructive subsurface probe with nanometer resolution in three dimensions?

Priorities

- Build the Spallation Neutron Source and upgrade existing neutron sources.
- Fully instrument and exploit the existing synchrotron light sources and do R&D on the x-ray laser.
- Build state-of-the-art nanofabrication facilities staffed to run user programs for the benefit of not only the host institutions but also universities, government laboratories, and businesses that do not have such facilities.
- Recapitalize university laboratories with state-of-the-art materials fabrication and characterization equipment.
- Build medium-scale centers devoted to single issues such as high magnetic fields or electron microscopy.
- Exploit the continuing explosion in information technology to visualize and simulate materials.

7

Changes in the R&D Landscape

Science has been a powerful agent of change throughout history. Science expressed through technology has redefined warfare, enabled economic growth, and extended the lifetimes and enhanced the well-being of billions of people. Over the past half-century, condensed-matter and materials physics has profoundly affected our lives, ushering in the information age and contributing to advances in communications, computing, medicine, transportation, energy, and defense. These advances have transformed the economy and dramatically altered our worldview. They have also changed the environment (the R&D landscape) in which science is performed.

FROM THE COLD WAR TO THE GLOBAL ECONOMY

Condensed-matter and materials physics is a young field. Although components of the field existed earlier, its modern development was enabled by the new discoveries in the 1930s of quantum mechanics and the wave nature of the electron. Its emergence as a discipline was heralded by the invention of the transistor in 1947. In the brief 50 years since then, an impressive understanding has been achieved of the structure and properties of materials on the atomic scale. Accompanying this understanding have been extraordinary technological developments, including the integrated circuit, optical fibers, solid-state lasers, and high-temperature superconductivity.

Powerful forces have driven the development of condensed-matter and materials physics. In the beginning, the desire was to replace inconvenient and bulky vacuum tubes with solid-state devices. Then came World War II with radar and defense technology. The arms race, Sputnik, the energy crisis, and the information revolution stimulated continued growth in the field over the subsequent decades. For most of this period, there was sustained growth in the federal investment in science, including condensed-matter and materials physics. This federal role in fundamental research, originally articulated by Vannevar Bush at the end of World War II in *Science: The Endless Frontier*,[1] was substantially justified on the basis of national defense.

In the late 1980s, the end of the Cold War, the emergence of the global economy, and the growing federal deficit combined to shake the foundations of the national R&D enterprise. In the absence of a major military threat, investments in the defense establishment were reduced, including support for R&D. Overall federal R&D investments, which peaked at $80 billion (in 1997 dollars) in 1987, declined 20 percent in the following decade (see Figure 7.1) as priorities shifted away from defense, and the desire to reduce the deficit applied increased pressure to the discretionary part of the federal budget. Federally supported basic research, performed mostly at universities, fared much better, increasing by 30 percent between 1985 and 1995 (see Table 7.1). This increase was dominated by increased investment in the life sciences; only modest gains were recorded for physics. At the same time, competition in the global economy (which itself was enabled by communications advances rooted largely in condensed-matter and materials physics) forced industry to sharpen the focus of its R&D investments. Industrial R&D turned away from long-term physical sciences and toward projects with more immediate economic return, reducing fundamental research investments that have been essential to the development of new technologies.

A DECADE OF CHANGE

The transition to the global economy represents a significant opportunity for condensed-matter and materials physics. Competitiveness in a fast-moving economy is critically dependent on advances in materials for a broad range of applications from information technology to transportation to health care. Condensed-matter and materials physics has responded effectively over the past decade, supporting continued innovation in electronic and optical materials, while developing new thrusts in complex fluids, macromolecular systems, and biological systems (collectively known as "soft materials"), and nonequilibrium processes. At the same time, science has become increasingly international, and U.S. leadership in many areas of science and technology, including condensed-matter and materials physics, is being challenged. Continued progress in condensed-matter and materials physics is critical to sustained economic competi-

[1]Vannevar Bush, *Science the Endless Frontier: A Report to the President*, U.S. Government Printing Office, Washington, D.C. (1945), reprinted by the National Science Foundation, Washington, D.C. (1960).

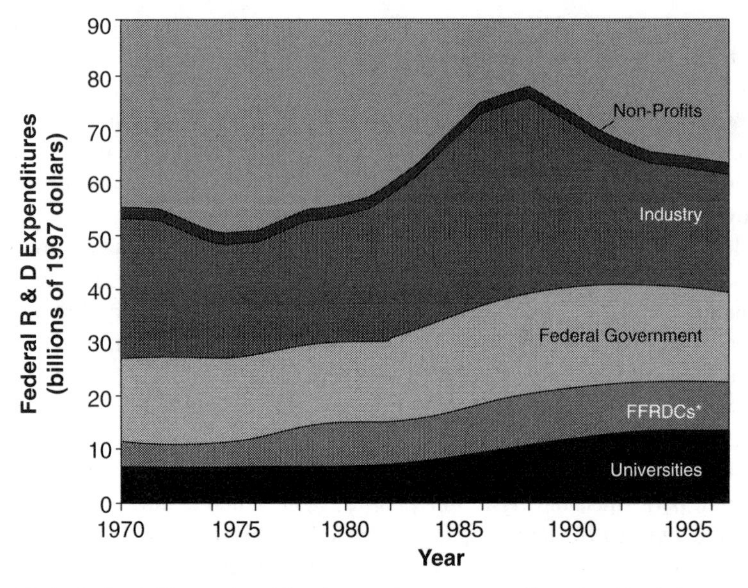

*Federally Funded Research and Development Centers operated by universities, industry, and non-profits.

Source: National Science Board, *Science and Engineering Indicators—1998*, National Science Foundation (NSB98–1).

FIGURE 7.1 Federal investments for R&D by performer, 1970-1997.

TABLE 7.1 Trends in Federal Investments in Basic Research by Discipline

Research Area	Expenditures (billions of 1995 dollars)			% Change
	1985	1990	1995	(1985-1995)
Life Sciences	5.20	5.98	6.94	33.5
Physical Sciences	2.50	3.08	2.91	16.4
(Physics)	(1.32)	(1.71)	(1.50)	(13.6)
Environmental Sciences	0.96	1.46	1.54	60.4
Mathematical & Computer Sciences	0.36	0.47	0.58	61.1
Social Sciences	0.19	0.17	0.21	10.5
Engineering	1.22	1.27	1.29	5.7
(Materials)	(0.30)	(0.30)	(0.28)	(–6.7)
Other	0.14	0.35	0.53	78.6
Total Basic Research	10.57	12.78	14.00	32.5

SOURCE: National Science Board, *Science and Engineering Indicators—1996* (Table 4-22), National Science Foundation, Washington, D.C.

tiveness in a broad array of current and emerging technologies. Successfully navigating the changes in the R&D landscape while strengthening condensed-matter and materials physics is a strategic imperative for the field and for the nation.

Within the major industrial laboratories, investments in fundamental physical sciences have been reduced, and much of the remaining effort has been focused on nearer-term projects. The situation has been exacerbated by the increasing tendency to conduct manufacturing offshore to reduce costs. This change is particularly alarming because the special environment of these laboratories sparked many of the important discoveries that led to the major technological advances of the twentieth century (see Box 7.1). This environment—which balanced enlightened management with opportunities for independent inquiry, assembled a critical mass of researchers from a diversity of disciplines, and encouraged "unfettered" research within a framework of strategic intent—was unprecedented in its scientific and technological impact. It was also dependent on monopoly or pseudomonopoly positions, since the profits associated with many of the scientific advances accrued to other companies. There was a high probability that scientific results would fall outside the commercial interests of the parent company, and in many cases the technological value of the new discoveries was not apparent for many years, allowing the information to diffuse throughout the community. This system was inefficient for the parent company, and unsustainable in the current world of global competition and corporate raiding, but very productive for the economy overall. Recreating this environment within the present R&D system represents a major challenge and opportunity.

Within the universities, there has been a steady decline during the past decade in the number of new students entering the physical sciences (see Table 7.2). In fact, undergraduate enrollment in physics (including condensed-matter and materials physics) is at its lowest level in 30 years. This appears to be a response to an apparent oversupply of new Ph.D. physicists in the early 1990s, in combination with a slow job market for physical scientists in fundamental research during this period. The only scientific discipline to increase undergraduate enrollments during the period was biology, which is consistent with the increasing federal investment in biomedical research. The declining numbers of physical science undergraduates raise serious questions about the availability of future human capital in the technologies that drive large sectors of the economy. For example, shortages in many areas of software research and semiconductor processing are already apparent. Physicists, who often possess skills that are attractive to high-technology industry, provide some of this human capital by pursuing careers in industry. In addition, the economy relies heavily on foreign students who remain in the United States to work after completing their Ph.D. Currently half of the graduate students at U.S. universities in the physical sciences are foreign nationals. There is evidence that many of these students are choosing to return to their homelands as global opportunities in science and technology improve. Unless

BOX 7.1 The Legacy of the Industrial Laboratories

For condensed-matter and materials physics related to applications, the tweni-eth century has been the century of the large industrial laboratory. A handful of corporate research laboratories—Bell Laboratories (see Figure 7.1.1), IBM Research, DuPont, and others—have dominated the scene with developments such as the transistor, the solid-state laser, optical fiber, synthetic polymers, high-temperature superconductivity, scanning-tunneling microscopy, and electron diffraction. These organizations operated at the frontier of science in a strategic context, developing broad new understanding to advance both science and technology and making profound contributions with impacts far beyond their corporate borders.

The corporate research laboratories emerged to exploit the promise of the physical sciences for the development of revolutionary new technologies and products. Condensed-matter and materials physics, invigorated by the new quantum mechanics, was on the verge of an intellectual explosion. Many corporations recognized the importance of being part of that explosion. Some became leaders by

FIGURE 7.1.1 Bell Laboratories at the time of the invention of the transistor. (Courtesy of Bell Laboratories, Lucent Technologies.)

committing to fundamental research in condensed-matter and materials physics as a path to new technologies.

A number of forces combined to promote the rise of the corporate laboratories. First, the time was right. The development of quantum mechanics, advances in the understanding of electrons in solids, and new tools such as x-ray diffraction, electron microscopy, and neutron scattering provided fertile ground for research. At the same time, it was apparent that advances in condensed-matter and materials physics were key to the new materials and devices that would drive modern technology. In many cases the military was willing to pay initial development costs, and the success of the Manhattan Project generated optimism about the power of physics and corporate-scale research. Finally, many of these laboratories enjoyed monopoly or pseudomonopoly positions because of regulatory policy or market dominance. As a result, these institutions took on many of the characteristics of "national laboratories."

The corporate research model was very successful in developing new science and technology. Corporations brought together scientists from different disciplines, provided freedom within the context of strategic intent, and had the resources and vertical integration to support large-scale, long-term R&D. Technical management, drawn from the research ranks, was empowered to make financial decisions and to move quickly without formal peer review. The resulting research environment was extremely productive—but the economic benefits did not always accrue to the parent corporation. The unpredictability of research results and applications, the diffusion of knowledge, and the ability to bring new technologies to market were all factors in spreading the economic impact of corporate research. This widespread impact was good for the economy (AT&T maintained an open license for the transistor) and for the development of condensed-matter and materials physics.

As we enter the twenty-first century, condensed-matter and materials physics has become much too large to be dominated by a few corporate research laboratories. Furthermore, corporate research has become more focused, and the extent of corporate participation in long-term condensed-matter and materials physics research in the future cannot be predicted. In addition, many industries are emphasizing software and systems research over hardware, and there is a trend toward more research being done in small companies. As a result, the special research environment that led to many of the fundamental condensed-matter and materials physics discoveries of the twentieth century no longer exists. Today, this special environment can best be emulated by government laboratories and universities working together with industry to create distributed, multidisciplinary networks in condensed-matter and materials physics. Within these networks, industry must continue to play a significant role in fundamental research in order to provide the vision needed to connect the research to technological applications.

Computers, new synchrotrons and neutron sources, and other instrumentation advances place us on the verge of another revolution in condensed-matter and materials physics. Cooperation among universities, government laboratories, and industry is essential to maintaining U.S. leadership in this revolution. This will require mechanisms, including intellectual property provisions, that encourage this cooperation.

TABLE 7.2 Bachelor's Degrees Awarded in Selected Disciplines in the United States, 1985-1995

Field	Number of Bachelor's Degrees 1985	1990	1995	Change (1985-1995)	1997	Change (1985-1997)
Biological Sciences	39,405	38,040	56,890	(+44%)		
Physics *	5,013	4,950	4,263	(–15%)	3,826	(–24%)
Chemistry	10,701	8,289	10,016	(–6%)		
Mathematics	15,389	14,674	13,851	(–10%)		
Engineering	77,572	64,725	63,371	(–18%)		
Geosciences	7,001	2,256	3,820	(–45%)		
Computer Sciences	39,121	27,695	24,769	(–37%)		
Materials/Metallurgy Engineering	1,276	1,166	1,046	(–18%)		
All Bachelor's	990,877	1,062,151	1,174,436	(+19%)		

*American Institute of Physics (AIP), *Enrollments and Degrees Report,* AIP, New York (April 1997).
SOURCE: National Science Foundation (NSF), *Science and Engineering Degrees: 1966-1995,* NSF 97-335, NSF, Washington, D.C. (1996).

there is an increase in the attractiveness of the physical sciences to U.S. undergraduate and graduate students, the nation risks a future with insufficient human resources to maintain leadership in science and technology. Intervention at the secondary school level is an essential component of any effort to stimulate interest in physical science careers.

Federal investments in condensed-matter and materials physics have increased slightly in the past decade (see Figure 7.2), despite the general downturn in federal support for R&D. However, these increases have been more than offset by the operating costs of major new research facilities that have come online during this period, notably synchrotrons at the laboratories of the U.S. Department of Energy (DOE). These facilities, which include neutron sources and microcharacterization centers as well as synchrotrons, serve more than 5,000 users per year, more than half of whom come from disciplines other than condensed-matter and materials physics (see Figures 7.3 and 7.4). Setting aside this stewardship responsibility for national facilities, federal investments in condensed-matter and materials physics research have actually declined more than 10 percent since 1985. Although budget statistics are not available, head counts in physical research departments at major industrial laboratories in physics-related industries have declined by a factor of two during the same period, reducing further the nation's research effort in condensed-matter and materials physics.

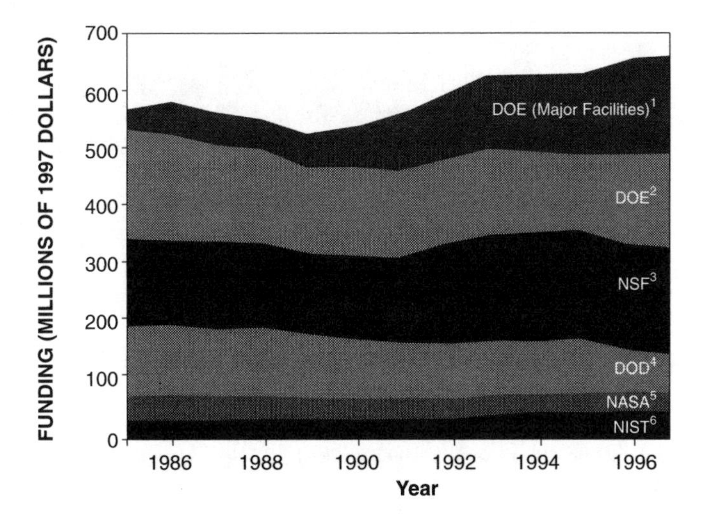

FIGURE 7.2 Trends in federal investment in condensed-matter and materials physics, 1985-1997.

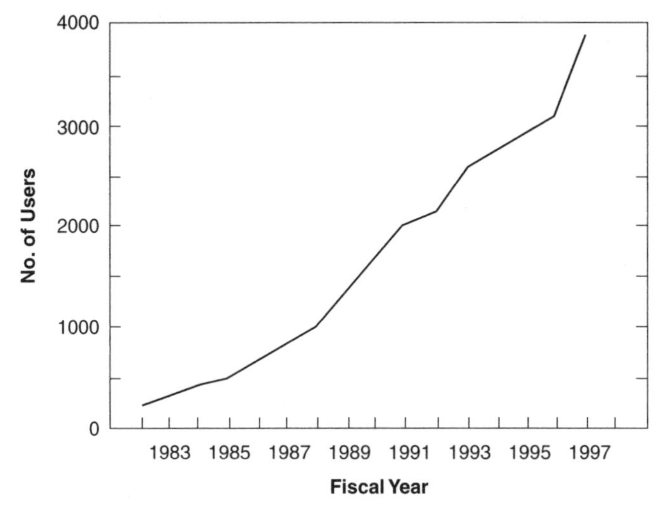

FIGURE 7.3 Growth in the number of users at Department of Energy synchrotron facilities, 1982-1997.

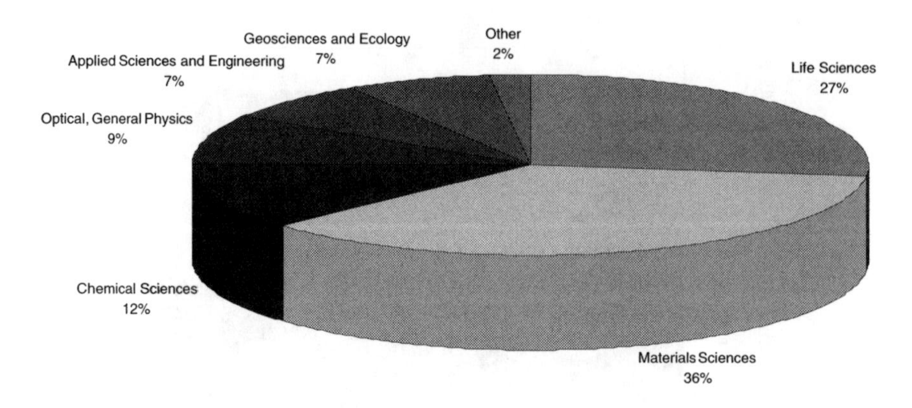

FIGURE 7.4 Use of national synchrotron facilities by scientific discipline shows that more than half of the 4000 users in 1997 worked in fields other than condensed-matter and materials physics.

CONDENSED-MATTER AND MATERIALS PHYSICS TODAY

The evolution in the practice of condensed-matter and materials physics in response to these external forces and to developments within the field itself has been dramatic. Partnerships across disciplines and among performers have proven to be essential to continued progress in the field. Powerful new research facilities have provided individual investigators with unprecedented access to the world of atoms and electrons in materials. These facilities, developed and supported by the condensed-matter and materials physics community, now provide unique research capabilities to thousands of researchers from a wide range of scientific disciplines. Finally, many of the institutions that practice condensed-matter and materials physics have undergone fundamental change in response to changing priorities and economic realities.

Condensed-matter and materials physics is inherently interdisciplinary, with advances increasingly occurring at interfaces with chemistry, materials science, atomic and molecular physics, engineering, biology, and other disciplines (see Box 7.2). Major professional societies, including the Materials Research Society and the Divisions of Materials Physics and High Polymer Physics of the American Physical Society, have positioned themselves to foster and serve this interdisciplinary materials community. Interdisciplinary research represents a significant challenge to the disciplinary boundaries inherent in university departments, funding agencies, and the peer-review process. Bridging these barriers is an important priority for the future of condensed-matter and materials physics. In particular, mechanisms must be found to ensure that compelling interdisciplinary

BOX 7.2 Meeting the Interdisciplinary Challenge

Condensed-matter and materials physics is inherently interdisciplinary. Linkages to chemistry, materials science, atomic and molecular physics, and engineering have been essential to progress in the field. New linkages to biology are critical to the future. Forefront research transcends boundaries within condensed-matter and materials physics as well and often depends on an integration of theory and experiment with the synthesis of special research samples and a variety of advanced characterization techniques. Advances in condensed-matter and materials physics cluster at the interfaces between established disciplines, interfaces that face institutional, funding, and disciplinary barriers.

Traditional scientific disciplines maintain and nurture the foundations of knowledge. This is important to the scientific enterprise, but it also presents the potential for barriers to interdisciplinary research as scientific disciplines evolve, developing language and culture not readily understood by the practitioners of other disciplines. For example, biology is at a point where the atomic view of physicists is having enormous impact, but physicists and biologists have difficulty communicating. This structural defect in the disciplinary organizational scheme must be addressed through education and budgetary incentives.

Institutions also present barriers to interdisciplinary research. University departments are structured around scientific disciplines. As a result, new faculty who wish to pursue research at the boundaries between disciplines can have difficulty finding a home (and earning tenure). Furthermore, the individual-investigator mode does not foster multidisciplinary teaming. The emergence of the multidisciplinary Materials Research Science and Engineering Centers, Science and Technology Centers, and Engineering Research Centers, sponsored by the National Science Foundation, is a response to this issue.

Finally, the funding process can present barriers to interdisciplinary research. Projects that fall outside traditional disciplines can easily be overlooked by a peer-review community structured around those disciplines. The peer-review process must include explicit capabilities for handling interdisciplinary proposals.

Universities can meet the interdisciplinary challenge through joint appointments for faculty, by encouraging multidisciplinary centers, and by recognizing the value of interdisciplinary research in tenure decisions. Government laboratories, which have an easier time putting together multidisciplinary teams, should be encouraged to involve universities in those teams. Funding agencies have the most leverage, as seen in the success of agency-sponsored multidisciplinary centers. Proposals that assemble diverse teams, including physicists and biologists for example, to tackle high-profile multidisciplinary problems should be encouraged. Institutions that ignore the interdisciplinary challenge risk abandoning the scientific frontier.

research proposals are not lost in the competition with other proposals that more neatly fit the boundaries of established disciplines.

Partnerships across disciplines and among universities, government laboratories, and industry are becoming increasingly important in bringing together the resources and diverse skills needed to continue advancing knowledge in condensed-matter and materials physics. For many leading-edge research projects, it is neither practical nor cost-effective to assemble the required capital and intellectual resources at a single location. Teams form and dissolve as research directions change, and the diversity of institutions and performers ensures that a wide range of projects and approaches can be accommodated. This is a fundamental strength of the U.S. R&D system. Modern communications, an outgrowth of condensed-matter and materials physics, is essential to these partnerships.

Another significant change in the practice of condensed-matter and materials physics has been the emergence of major national research facilities. These facilities, which include synchrotrons, neutron sources, and microcharacterization centers, have had an extraordinary impact on the ability of researchers to investigate ever-smaller, lower cross-section, more dilute, and more complex systems. Accordingly, there has been a spectacular increase in the use of these facilities. These powerful tools transcend condensed-matter and materials physics to serve large user communities from other disciplines, including biology, which now consumes more than 25 percent of the beam time at national synchrotron facilities. As a result, condensed-matter and materials physics is having a significant impact on many fields with which it had little connection just a decade ago.

Institutional change is never comfortable, and it is a continuing challenge to U.S. space science. Research organizations are being expected to improve organizational effectiveness and resource utilization, create new partnerships, and serve customers better. Customers, ranging from corporate manufacturing arms to sponsors to facility users, are increasingly involved and demanding. All sectors of condensed-matter and materials physics underwent profound change in recent years. Industrial laboratories were downsized and redirected. Government laboratories struggled with substantial reductions in resources, increased regulation, and mission and operational reform. Research universities came under increasing pressure to reduce overhead, cut costs, and become more responsive to the public and to industry. All of these changes have potentially positive outcomes, and condensed-matter and materials physics is particularly well positioned to contribute effectively in this new environment. However, great care will be required to navigate these changes while preserving the research infrastructure of the nation for the long term.

MEASURING PERFORMANCE AND ECONOMIC IMPACTS

The Government Performance and Results Act (GPRA), passed in 1993, provides a timetable for agencies to develop strategic plans and criteria for measuring their performance against established goals. These plans and performance measures, which were intended to be in place by 1997, will form the basis for evaluating the effectiveness of agency programs and developing budgets. This represents a major challenge for fundamental science, in which the important impacts tend to be unpredictable, dilute across the spectrum of research activities, and frequently are separated by decades from the initial research results.

GPRA establishes a framework in which agencies provide *inputs* in order to produce *outputs*, which have intended *outcomes* for society and the economy. Inputs might be person-years and equipment-years of effort, for example, while outputs are the direct results of an agency's inputs, and outcomes are the broader impacts that result. For research, these concepts are summarized in Table 7.3. Within condensed-matter and materials physics, the discovery of high-temperature superconductivity represents an output, while the commercialization of superconducting technology would be an outcome. Agencies are required to develop performance criteria for both outputs and outcomes. This requirement is a substantial challenge for agencies involved in fundamental research, for which the outputs

TABLE 7.3 Inputs, Outputs, and Outcomes of R&D

		Performance Indicators	
Category	Concepts	Proxies	Correlates
Inputs	Person-years Equipment-years	Expenditures	
Outputs	Ideas, discoveries Inventions	Papers, prizes Patents, invention-disclosures	
	Human capital Technology transfer	Degrees awarded CRADAs, licenses	 Cost-shared dollars
Outcomes or Impacts	Broad advance of human knowledge	Papers, citations, expert evaluations	
	New products	Patents, citations	Licenses, license royalties, product announcements, new product sales
	Productivity improvements	Measured productivity growth	
	Income growth	Benefit/cost ratio or rate of return	New firms, induced investment
	Excitement about science		*Science News* articles
	Health, environment, etc.	New drug applications	Emissions levels
	Cooperation and knowledge flow	CRADAs	

SOURCE: Adam B. Jaffee, "Measurement issues" in L.M. Branscomb and J. Keller, eds., *Investing in Innovation,* MIT Press, Cambridge, Mass. (1997).

(ideas and discoveries) are difficult to measure, and the outcomes (the advance of knowledge or the introduction of new products), are difficult to quantify or relate to specific programs. Consequently, proxy indicators related to the desired outputs or outcomes are developed for research activities. These proxies might include papers, prizes, and patents to take the place of ideas and discoveries in measuring research outputs, and citations and productivity growth to take the place of ad-

TABLE 7.4 Economic Growth Rates Attributable to R&D Investments

Author(s) and Year of Study	Rate of Return[a] (Percent)
Firm-level Studies	
Link (1983)	3
Bernstein-Nadiri (1989b)	7
Schankerman-Nadiri (1986)	13
Lichtenberg-Siegel (1991)	13
Bernstein-Nadiri (1989a)	15
Clark-Griliches (1984)	19
Griliches-Malresse (1983)	19
Jaffe (1986)	25
Griliches (1980)	27
Mansfield (1980)	28
Griliches-Malresse (1984)	30
Griliches-Malresse (1986)	33
Griliches (1986)	36
Schankerman (1981)	49
Minasian (1969)	54
Industry-level Studies	
Terleckyj (1980)	NS[b]
Griliches-Lichtenberg (1984a)	4
Patel-Soete (1988)[c]	6
Mohnen-Nadiri-Prucha (1986)	11
Terleckyj (1974)	15
Wolff-Nadiri (1987)	15
Sveikauskas (1981)	16
Bernstein-Nadiri (1988)	19
Link (1978)	19
Griliches (1980)	21
Bernstein-Nadiri (1991)	22
Scherer (1982, 1984)	36

[a] For studies for which Nadiri (1993) reports a range of possible returns, the midpoint of that range is provided in this table.

[b] Not significantly different from zero in a statistical sense. This result, however, may be a reflection of limitations in the quantity of data used in the study.

[c] Economy-level study (all industries grouped together).

SOURCE: M.I. Nadiri, "Innovations and Technological Spillovers," Working Paper No. 4423, National Bureau of Economic Research, Cambridge, Mass. (1993).

vances in knowledge or the introduction of new products in measuring outcomes. Although inherently imperfect, the use of multiple proxy indicators in combination with peer review probably represents the most likely (and most reasonable) performance measurement approach for fundamental research under GPRA.

This discussion of performance measures raises a key issue: the rate of economic return on R&D investments. Specific data are not available for condensed-matter and materials physics, although powerful evidence abounds, such as the economic impact of the transistor, magnetic materials, fiber optics, and the solid-state laser. Numerous studies of economic return have been performed for the broader R&D enterprise. These studies, many of which are listed in Table 7.4, indicate average rates of return of 15 to 20 percent per year. This is an extraordinary indication of the value of research. Unfortunately, these data do not provide information on the impact of adjustments (either up or down) in the level of R&D investment and are therefore not useful in determining absolute funding levels.

8

The Next Decade

Condensed-matter and materials physics lies at the heart of many of the scientific and technological challenges of our time. Progress in condensed-matter and materials physics drives our fundamental understanding of the materials and phenomena that enable technological advances; and condensed-matter and materials physics is entering a new era driven by new capabilities in synchrotron and neutron research, atomic-scale visualization, nanofabrication, and computing. These capabilities provide opportunities to examine the behavior of materials at levels of complexity and with degrees of microscopic control that are unprecedented. The new era promises to revolutionize our understanding of materials, expanding our knowledge beyond the physics of idealized systems to touch the real materials that enrich our lives. Fundamental understanding of electronic and optical phenomena, complex assemblies of atoms and multicomponent materials, nonequilibrium phenomena, and biological phenomena will fuel advances in technologies ranging from microelectronics to structural materials to medicine.

The stage is set. The new era holds the promise of revolutionary developments in condensed-matter and materials physics that will contribute to economic growth, national security, and the quality of life. Success will require investing in human capital and research infrastructure, establishing partnerships across disciplines and institutions, integrating research and education, and maintaining excellence with relevance.

MAKING THE RIGHT INVESTMENTS

Progress in condensed-matter and materials physics has been enabled by sustained investments in long-term research by federal agencies and at large industrial laboratories. In recent years, in response to new competitive environments, industry has shifted away from long-term physical sciences research and toward nearer-term research and development. At the same time, the government's discretionary expenditures (which include R&D investments) have been constrained by efforts to balance the federal budget amidst growing entitlement outlays. Additional pressure on condensed-matter and materials physics funding comes from the field's responsibility to develop and operate large national facilities for materials research, such as synchrotrons and neutron sources, that are heavily utilized by a growing community of users from many scientific and engineering disciplines. As a result, although the resources available to condensed-matter and materials physics are substantial, there are severe constraints in comparison to the overall need to maintain the nation at the forefront of fundamental research in this technologically critical area.

As a fraction of gross domestic product, federal investment in R&D has dropped by about half over the past 30 years. This trend of declining investment threatens U.S. leadership in science, including condensed-matter and materials physics. At the same time, it is estimated that half of the economic growth in the last half century has come from technological innovation that requires leadership in science. The President's budget request for FY 1999 reflects these concerns, placing increased priority on science and technology and showing strong gains for many federal research agencies. In addition, the bipartisan Fritz-Rockefeller bill (S. 2217) calls for a doubling of federal investment in civilian research over the next 12 years. This bill, known as the Federal Research Investment Act, is supported by a coalition of more than 100 science, engineering, and technology organizations. A parallel effort to increase support for defense R&D is also under way.

Human Capital

Many economists attribute current economic growth to investments in human capital, the capacity to generate new ideas that organize and rearrange existing resources to achieve productivity gains. Examples range from new ways of processing steel and polymers, to the soaring performance of electronic and optical systems, to the growth in software and computer applications. These advances share common characteristics of innovation and integration of knowledge—the economics of ideas. Human capital, enabled by investments in educational and research institutions, drives economic growth by providing the new ideas that allow escape from a traditional economic future limited by scarcity of resources and the law of diminishing returns.

Unlike physical resources, which are limited in a finite world, the potential of human capital is nearly limitless. But it is not free. A commitment to education, to research, and to the free exchange of information and ideas is essential. In the modern global economy, world leadership is impossible without leadership in human capital.

In condensed-matter and materials physics, human capital is a product of the education system and the collective learning of universities, industry, and government laboratories. It is nourished by sustained investments in fundamental research and by maintaining close interactions among condensed-matter and materials physics performers, with other scientific disciplines, and with industry. The reservoirs of human capital include school teachers, the professoriate as both educators and researchers, and researchers and policy makers in government and industry. These reservoirs also include institutions—research universities and government and industrial laboratories—that provide the environment and infrastructure for generating and preserving knowledge and from which new ideas can emerge.

Human capital is probably the single most important investment for science and technology. Human capital in condensed-matter and materials physics occupies a special place in the national economy, underpinning many of the technological advances that drive economic growth. The U.S. system of graduate education, research universities, government and industrial laboratories, and national facilities for condensed-matter and materials physics is the envy of the world. Maintaining this leadership requires continued commitment to strengthening these institutions. In addition, condensed-matter and materials physicists must play a crucial role in engaging undergraduates in research and improving their understanding of science and technology. These investments are needed to develop the human capital essential for leadership in condensed-matter and materials physics and related technologies.

Facilities and Infrastructure

Condensed-matter and materials physics encompasses a broad array of institutions and research modes, ranging from individual investigators to multidisciplinary teams, from bench science to large national facilities, and from fundamental to applications-oriented research. This diversity is representative of the diversity of the field and is essential to its success. Maintaining an appropriate balance among performers, institutions, and research modes is a continuing challenge for condensed-matter and materials physics. There are no clear boundaries. For example, large facilities are used primarily by individual investigators (often from fields other than condensed-matter and materials physics), and applications-oriented research often leads to breakthroughs in fundamental science. Priorities for infrastructure investments in facilities, laboratories, and institutions must be assessed in the context of this diversity and interdependence.

Infrastructure

Laboratories, instrumentation, and facilities for performing state-of-the-art condensed-matter and materials physics are becoming increasingly expensive to develop and operate. At the same time, more universities are competing effectively for federal research dollars. It is becoming increasingly apparent that the needed infrastructure cannot be duplicated at even a few dozen universities, let alone the more than 180 institutions nationwide that grant physics Ph.D.s. It is estimated that nearly half of university laboratories in the physical sciences require refurbishment in order to be used effectively. Government institutions, including the DOE laboratories, are also burdened with an aging infrastructure.

At the same time, there has been a significant increase in the availability of modern research infrastructure at major national and regional research facilities and centers. These facilities provide needed infrastructure on a shared basis. In addition, there is substantial research infrastructure at government laboratories (beyond the major facilities) that is contributing to alleviating this problem. The number of guest researchers from universities and industry at DOE national laboratories has skyrocketed in the past 15 years, and more could be accommodated with modest investments. An integrated solution, combining revitalization of university laboratories with modernization and increased community utilization of government laboratories, seems to provide the most cost-effective option to serve the infrastructure needs of the condensed-matter and materials physics community (see Box 8.1).

Materials Microcharacterization and Processing Facilities

The dozens of materials microcharacterization and processing centers distributed among universities and government laboratories provide access to electron microscopes, accelerators, and other microanalytical and processing equipment that is beyond the means of individual investigators. They also provide expertise in the operation of these facilities that greatly enhances their accessibility. As a result, state-of-the-art microcharacterization and processing capabilities are available to virtually all researchers on a shared basis. In addition, like larger facilities, these centers establish an environment where cross-disciplinary research is naturally encouraged.

BOX 8.1 Recommendations for Materials Research Infrastructure and Microcharacterization and Processing Facilities

- Increased investment in modernization of the condensed-matter and materials physics research infrastructure at universities and government laboratories.
- Increased investment in state-of-the-art instrumentation and fabrication capabilities, including centers for instrumentation R&D, nanofabrication, and materials synthesis and processing.

Microcharacterization facilities also provide a resource for advancing microcharacterization science and developing new and better instrumentation. The United States has lagged in this area except in scanning-tunneling microscopy. Modest investments in research at these facilities could contribute to impressive performance gains in electron optics, visualization, virtual operation, and other improvements. Such investments are needed to ensure that the existing infrastructure is used effectively and that U.S. scientists have access to the best technology.

The committee recognizes the essential role of regional and national centers for atomic-scale visualization, nanofabrication, materials synthesis and processing, high-field magnetism, and other specialized capabilities to support leading-edge condensed-matter and materials physics research. There should be explicit recognition of the importance of these centers and the need to strengthen their role in instrumentation development. In addition, attention should be given to the education of the next generation of instrument scientists (see Box 8.1).

Neutron and Synchrotron Facilities

In recent years, federal expenditures for the operation of large materials research facilities, such as neutron-scattering and synchrotron radiation sources, have received considerable attention because of the magnitude of these expenditures in comparison to the core research budgets of the agencies that fund them. In FY 1998, the estimated U.S. Department of Energy (DOE) Basic Energy Sciences (BES) operating budgets for these facilities exceeded $253 million. The bulk of this amount, $235 million, was provided by the materials sciences and chemical sciences programs of BES. This represented approximately 38 percent of BES's total budget authority. Over the past decade, facility costs have almost doubled (in inflation-adjusted dollars), while the core research budgets of the BES materials sciences and chemical sciences programs have remained essentially constant.

At the same time, there is increasing recognition that much of the research performed at these facilities (particularly at synchrotron sources) is in scientific and technological areas other than materials and chemical sciences. There have, therefore, been proposals that federal programs more closely linked to these other research areas should provide significant facility operating funds.

Two national synchrotron radiation facilities have been constructed recently: the Advanced Light Source (ALS) at Lawrence Berkeley National Laboratory, for ultraviolet and soft x-ray research, and the Advanced Photon Source (APS) at Argonne National Laboratory, for x-ray research. In addition, the National Synchrotron Light Source (NSLS) at Brookhaven National Laboratory, the Stanford Synchrotron Radiation Laboratory (SSRL) at the Stanford Linear Accelerator Center, the Cornell High Energy Synchrotron Source (CHESS) at Cornell University, the Synchrotron Radiation Center (SRC) at the University of Wisconsin, and the Synchrotron Ultraviolet Radiation Facility II (SURF II) at the National Institute of Standards and Technology remain highly active and productive.

In contrast, construction of the Advanced Neutron Source, which was to have been a reactor source at Oak Ridge National Laboratory, was canceled in 1995. In addition, the High Flux Beam Reactor at Brookhaven National Laboratory is currently not operating, and there is opposition to restarting it. On a positive note, the neutron-scattering facilities at the National Institute of Standards and Technology have been recently upgraded and the High Flux Isotope Reactor (HFIR) at Oak Ridge is being upgraded. Nevertheless, the neutron-scattering field now depends on an array of facilities that is even smaller than what was already found inadequate by national review committees in the 1980s and early 1990s.

As a consequence of these concerns, the DOE's Basic Energy Sciences Advisory Committee (BESAC) recently established reviews of its existing and proposed neutron and synchrotron radiation facilities. BESAC considered the neutron situation at a meeting in Washington, D.C., on February 5-6, 1996. Drawing on reports from several national panels, BESAC made the following recommendations for neutron-scattering facilities:

- Construct a 1 MW, upgradable short-pulse spallation neutron source, now known as the Spallation Neutron Source (SNS).
- Upgrade existing neutron scattering facilities at the High Flux Beam Reactor at Brookhaven National Laboratory, the High Flux Isotope Reactor at Oak Ridge National Laboratory, and the Los Alamos Neutron Scattering Center at Los Alamos National Laboratory.

BESAC further emphasized that the proposed construction and upgrades, while critically important to the future of neutron-scattering science in the United States, must not come at the expense of other BES research activities and must take explicit recognition of the additional operating and instrument development needs involved.

Considerable progress has been make toward implementing these recommendations. The conceptual design for SNS has been completed, and construction funding for the project has been included in the FY 1999 budget. SNS will be constructed at Oak Ridge National Laboratory by a consortium of five DOE national laboratories. In addition, the proposed upgrades at the High Flux Isotope Reactor and the Los Alamos Neutron Scattering Center are under way.

The committee recommends priority construction of SNS as well as upgrades to existing neutron-scattering facilities, provided that these projects do not come at the expense of the core research programs of BES (see Box 8.2).

To address issues related to the operation and scientific roles of synchrotron facilities, BESAC established the Panel on Synchrotron Radiation Sources and Sciences in 1997. This panel was charged to assess the scientific importance of

BOX 8.2 Recommendations for Major Materials Research Facilities

- Prompt construction of the Spallation Neutron Source as well as upgrades to existing neutron scattering facilities.
- Increased funding for operations and upgrades at synchrotron facilities including research and development on fourth-generation sources.
- Consideration of the broad utilization of synchrotron and neutron-scattering facilities across scientific disciplines and sectors when establishing budgets for the agencies that operate these facilities.

synchrotron radiation over the next decade, determine the size and nature of the user community both globally and by facility, and assess the operation of the facilities including their plans and vision for the future. The panel was also asked to make detailed recommendations under various budget scenarios and to consider the consequences of closing one or more of the BES synchrotron facilities.

In its report to BESAC, the panel concluded unanimously that ". . . shutdown of any one of the four DOE/BES synchrotron light sources over the next decade would do significant harm to the nation's science research capabilities and would considerably weaken our international competitive position in this field." The panel recommended the following actions (in priority order):

1. Continue operation of the three hard x-ray sources (APS, NSLS, and SSRL) for their large user communities, with a modest investment for general user support and for R&D on a fourth-generation x-ray source. (Recommended expenditures at both NSLS and SSRL were $3 million per year above the FY 1998 DOE-requested levels.)
2. Develop new beam lines at APS and modernize existing facility beam lines at NSLS. (Recommended expenditures were $8 million per year at APS and $3 million per year at NSLS.)
3. Fund ALS at the FY 1998 DOE-requested level of $35 million.

The panel also recommended funding proposed upgrades to the NSLS and SSRL facilities at an estimated cost of $27 million per year over 3 years. These upgrades should be carried out under a special initiative separate from the normal budgeting process. For example, BES might seek partnerships with other divisions within DOE and with other agencies such as the National Institutes of Health (NIH) or could request a budget add-on. This recommendation was intermediate in priority between the second and third priorities above.

The committee recommends support for operations and upgrades at existing synchrotron facilities (including modest investments for user support), as well as

R&D on a fourth-generation x-ray source. Synchrotron and neutron-scattering facilities are used extensively by researchers in various scientific disciplines, but the operating funds for the facilities are drawn from the agencies and programs that developed the facilities. The committee recommends that this fact be taken into account when the budgets of agencies operating these facilities are formulated (see Box 8.2).

In light of the extensive use of the synchrotron radiation facilities by fields other than materials and chemical sciences, the panel considered funding models that included contributions from other agencies. Their conclusion was that this is not practical. Rather, the panel endorsed BES's stewardship of synchrotron radiation sources, and urged BES to build on the broad impact of these facilities, especially in fields related to health and the environment, to increase its own base budget. The panel did, however, recommend diversification of the funding sources for special initiatives such as the proposed SSRL and NSLS upgrades.

There are synchrotron and neutron facilities supported by agencies other than DOE, including the National Science Foundation (NSF) and the U.S. Department of Commerce (DOC). These facilities are used by scientists supported by these and other federal and private agencies and institutions. Given the considerable cost of operating and improving these facilities, it is important that there be coordinated, interagency (including at least DOE, NSF, DOC, and NIH) consideration of, and planning for, neutron and synchrotron radiation facilities. The respective roles of these agencies in funding the construction, instrumentation, upgrading, and operation of these facilities should be delineated. However, each facility should have one agency that provides support for basic operations, and the broad utilization of synchrotron and neutron-scattering facilities across scientific disciplines and sectors should be considered in establishing the budgets for the agencies that operate these facilities. A committee of the National Research Council is considering interagency issues related to national facilities.

REDEFINING ROLES AND RELATIONSHIPS

The national R&D enterprise includes the funding agencies as well as essential components in industry, universities, and government laboratories. Industry is by far the largest performer of R&D in the United States, with expenditures of $133 billion in FY 1997, compared with $72 billion for the remaining sectors. Industrial R&D, however, is heavily weighted toward near-term development; industry provided only one-fourth of the $31 billion of U.S. investment in basic research in FY 1997. Universities and government laboratories are the largest performers of basic or fundamental research. Universities play a unique role in education, while government laboratories provide the infrastructure for multidisciplinary research and large facilities. Recently, many states have become significantly involved in the R&D enterprise, providing funds to stimulate the research competitiveness of their states. The states, responding to correlations between R&D activity and regional economic development, are becoming important resources for research support.

Within condensed-matter and materials physics, there are many approaches to the conduct of research, ranging from individual investigators to large multidisciplinary teams and from bench-scale experiments to studies at major national facilities. There is also a diversity of federal sponsors for condensed-matter and materials research, led by DOE, NSF, and the defense agencies. No single approach can span the diversity of research problems, and an effective national research program requires balance among a variety of performers and approaches. Achieving this balance requires an appreciation of the R&D roles of industry, universities, and government laboratories and of how to establish relationships among performers that encourage research synergy, funding leverage, and scientific productivity. The diversity of performers, institutions, and funding sources is a fundamental strength of condensed-matter and materials physics, essential to progress in a field that embraces both fundamental and applications-oriented research and spans both small and big science.

Role of Research Universities

Research universities are the bedrock of the U.S. R&D system. They are embedded in our communities with a holistic mission in knowledge creation, integration, and transfer. The desired outcomes are increased human capital (particularly in the form of trained students), opportunities for an improved quality of life (created by the advancement of knowledge), and an enlightened general public. Long the envy of the world, U.S. research universities face serious challenges over the next decade. Curricula will have to be overhauled to respond to the needs of industry in the global marketplace. Costs, which have escalated faster than the inflation rate for more than 2 decades, will have to be contained. Outreach to communities and businesses will have to be improved to create a public that understands and supports research and can compete in a technological economy. Underrepresented groups must be attracted to research careers in order to ensure an adequate supply of future talent. These challenges will have to be met in an environment of increasing research infrastructure costs, while adjusting to the impact of new information technology that will make distance learning and remote participation in research a reality.

Within condensed-matter and materials physics, universities face the daunting challenge of determining how to support and distribute new R&D infrastructure. It will simply not be possible to duplicate the infrastructure now available at major research universities across the university system. A system of teaming will have to be established, among universities and between universities and government laboratories, to ensure broad access to the best research facilities.

The time it takes to obtain a physics Ph.D. is approaching 7 years. This is costly and undesirable, particularly when industry, the permanent employer of the majority of physics Ph.D. recipients, places a higher premium on flexibility and just-in-time learning than on in-depth knowledge within a narrow field. Graduate programs in applied physics and engineering physics appear to have bridged this gap successfully at several universities. Perhaps the time has come to redefine the physics Ph.D. or develop a professional degree for the industrial physicist.

The steady decline in the number of undergraduate physics majors over the past 2 decades represents a major challenge to the field. Continued reliance on foreign students to fill graduate physics programs and provide human capital to U.S. industry is unwise in a global economy when offshore educational and employment opportunities can be expected to improve. The survival of the field and its continued impact on the U.S. economy depend on making physics relevant to U.S. students. Diversity presents the major opportunity here. Although the enrollment of women in physics has doubled over the past 2 decades, women only accounted for 12 percent of physics doctorates granted in 1997. African-American and Hispanic enrollments have not changed in recent years and only represented 1 to 2 percent of new physics doctorates in 1997. National demographic trends dictate that continued leadership in physics will require participation by these underrepresented groups.

Research universities serve a variety of customers, including students, industry, research sponsors, and the general public. Continued success over the next decade will require increased attention to the special needs of these customers, from outreach programs to engage students and the public, to exchange programs with industry to promote better understanding of market drivers. Within condensed-matter and materials physics, particular attention must be given to designing a curriculum that communicates the excitement and impact of physics to beginning undergraduates and that is more responsive to the needs of industry in graduate programs.

Role of Government Laboratories

Government laboratories, particularly the large multiprogram laboratories, represent a national R&D asset of enormous capability. These laboratories have the infrastructure and human capital to address large-scale problems of national importance that transcend traditional disciplinary boundaries and require access to special facilities. Within condensed-matter and materials physics, government laboratories conduct multidisciplinary research related to national missions in energy, defense, commerce, and space. These laboratories also develop and operate the nation's most powerful research tools for materials research, including synchrotrons, neutron sources, and microcharacterization facilities. Such facilities are an essential part of the R&D fabric of the nation, serving thousands of scientists from universities and industry.

A particular strength of the laboratories is the performance of long-term, large-scale, multidisciplinary research in an applications context. Such research requires a critical mass of resources to integrate across disciplines and apply a variety of tools to address a problem. In the past, this approach led to the development in large industrial laboratories of the transistor, synthetic polymers, and the solid-state laser. Today, with industry focusing on global competitiveness and nearer-term development, government laboratories represent the principal national resource for research on this scale. Realizing the full potential of this resource requires a continuing commitment to long-term, multidisciplinary research and development at the laboratories and effective research integration with universities and industry. Facilitating this integration and the related research and development partnerships is an important role of the government laboratories. In addition, the connection of program offices to research should be strengthened through exchanges at all levels between the agencies and the research community.

The government laboratories also represent a powerful resource for research infrastructure and integration. The success of the major materials-research facilities at these laboratories suggests that broader use of the entire infrastructure by universities and industry should be encouraged. This is already happening; the number of guest scientists performing research on site has more than doubled at many laboratories over the past decade. The government laboratories can also assist by providing access to state-of-the-art infrastructure for thesis research and by facilitating the formation of teams involving shared resources and effort with universities and industry to address appropriate research topics. Better utilization of the government laboratories can significantly reduce the infrastructure problems currently being encountered in other sectors of the R&D establishment.

Interactions with Industry

Condensed-matter and materials physics occupies a special position in science: fundamental research at the technological frontier. It is one of those rare fields for which the distance between basic research and technological development is small and the concept of "strategic intent" is applicable to research. As a result, condensed-matter and materials physics finds itself closer to industry than any other subfield of physics, and industry has a tradition of involvement in condensed-matter and materials physics as a practitioner, partner in research, and employer of condensed-matter and materials physicists. Interactions with industry are vital to the development of the field and its impact on the economy.

A primary interaction is between industry and universities. Here, industry is looking for talent to promote growth and innovation—talent that is the product of graduate schools. Curricula in physics (including condensed-matter and materials physics) have evolved little over the past decade. At the same time, industry has undergone extensive change, and the pace of the worldwide technology enterprise has accelerated greatly. Industrial input is needed to help physics education adapt and become more flexible so that it can better serve the future needs of industry and the nation.

R&D interactions with industry involve both universities and government laboratories. For large companies with in-house R&D capabilities, access to unique skills or facilities at universities or government laboratories drives the interaction. These interactions often involve a financial commitment by the company to the partner organization. For smaller companies, many that have no R&D capabilities of their own, interactions with universities and government laboratories may be the only way to assemble the necessary R&D resources to address a technical barrier. The success of cooperative research interactions depends critically on pursuing projects that contribute to the core missions of all involved organizations. An urgent need is the development of workable intellectual property arrangements, particularly between industry and universities (see Box 8.3).

Industry interactions with universities and government laboratories help provide a strategic context for condensed-matter and materials physics research. This is extremely important in a field that has such a direct impact on the economy and for which there are insufficient resources to explore every opportunity. Choices have to made, and interactions with industry provide useful input as to what may be important. As a first step, physics departments should become more involved in the industrial liaison programs at their universities, and government laboratories should engage in cost-shared research in their competency areas with industry to provide a window on technology. These interactions should not drive condensed-matter and materials physics research at universities and government laboratories, but they can provide a context for appreciating the broader implications of the research.

The Importance of Partnerships

Condensed-matter and materials physics is an ecosystem that involves a wide variety of performers, institutions, and research styles. The vitality of this ecosystem depends on establishing productive relationships among the participants. Partnerships are important in all branches of science, as noted in the 1996 report *Endless Frontier, Limited Resources* by the Council on Competitiveness. The central finding of that report was that "R&D partnerships hold the key to meeting the [R&D] challenge that our nation now faces." This point is especially important in condensed-matter and materials physics, where a range of performers and approaches is often required in order to span the expertise and capabilities required to make progress.

Within condensed-matter and materials physics, there is a tradition of partnerships among universities, government laboratories, and industry. These partnerships include informal collaborations, the use of unique facilities, personnel exchanges, consulting, and subcontracts and other formal relationships including

BOX 8.3 The Intellectual Property Bottleneck

A major impediment to collaboration between scientists from different institutions arises from policies and expectations concerning intellectual property (IP). When investigators from different institutions wish to work together, they must typically enter into a formal written agreement that addresses, among other things, inventions, patents, copyrights, and trade secrets. Such agreements are often prepared by attorneys representing their respective institutions. Negotiating these agreements has always been problematic. With intensified scrutiny of the economic rewards of research, these negotiations have become increasingly time consuming and frustrating.

No two institutions view IP in the same way, and views seem to evolve over time. The patent concerns of very large companies center on freedom of action. As both sources and users of IP, large companies are frequently cross-licensed with major competitors in common fields of endeavor. This helps to prevent a large company from being denied access to important patents. The strength of a large company's patent portfolio plays a major role in establishing the terms and conditions of these cross-licensing agreements and also can provide additional income through direct licensing to other companies. Extensive cross-licensing of patents implies a lack of exclusivity, although know-how and trade secrets may still be handled in an exclusive fashion.

Small companies typically view things very differently. One or two key patents can be the basis for the company's existence. Exclusivity can mean the difference between success and failure. Universities are sources of IP but are generally not users in the commercial sense. Sale or licensing of IP, exclusively or nonexclusively, can provide income to the institution, income to the inventors, and evidence of the institution's value to society. Government laboratories are similar to universities, although legislation authorizing cooperative research and development agreements (CRADAs) for jointly sponsored research between government laboratories and industries has helped to facilitate IP negotiations. In today's environment of intense economic pressure, it is not surprising that agreement on terms and conditions for joint work with IP-generating potential is often difficult to achieve, particularly when the agreements are being negotiated by individuals other than those who want to work together.

With sufficient perseverance IP agreements can usually be put in place, although often with considerable delay and expense. Too often the enthusiasm for the interaction or the timeliness of the work expires before an agreement is struck. In addition, existing agreements tend not to be precedent-setting, and negotiations between the same institutions frequently begin anew, often with different sets of attorneys, when a new project is proposed. A particular irony is that the likelihood that valuable patents will be generated decreases as the proposed work becomes more fundamental in nature, yet the fervor of the negotiations seems to endure undiminished.

Although difficult to quantify, it is clear that the present system is time consuming, inefficient, expensive, and a major obstacle to the investment of industry in university research. Industry has established its IP practices over decades, and these practices appear to work smoothly in the industrial sector. Universities and government laboratories, on the other hand, have IP practices that reflect different priorities and are still evolving. CRADAs are an important first step, but improved industry-government-university cooperation in research depends critically on achieving mutual understanding and convergence on IP issues among the sectors.

cooperative research and development agreements (CRADAs). CRADA partnerships, although not without controversy, have been enormously successful in bringing together government laboratories and industry. Box 8.4 summarizes the important characteristics of successful R&D partnerships. Partnerships within and among funding agencies are also becoming increasingly common as traditional barriers yield to the advantages of leverage and working together. The states are also playing an important role, providing key support to facilitate partnerships that have an impact on regional economies.

Recreating the fertile research environment of the major industrial laboratories of the past 50 years is a high priority for condensed-matter and materials physics. That environment, which has significantly diminished in U.S. industry, was extremely productive in both science and technology. In effect, these laboratories functioned as national laboratories, before divestiture and global economic forces required them to adopt a nearer-term, more focused approach to R&D. The extraordinary success of these laboratories resulted from their ability to integrate long-term fundamental research, cross-disciplinary teams that included experimentalists and theorists, materials synthesis and processing, and a strategic intent. The elements of this fertile ground still exist in condensed-matter and materials physics in the form of potential partnerships among universities, government laboratories, and industry. Federal R&D agencies should encourage partnerships that recreate

BOX 8.4 Recommendations for R&D Partnerships

The committee encourages R&D partnerships among universities, government laboratories, industry, and government agencies in order to

- Optimize the use of infrastructure and facilities,
- Enable cross-disciplinary research,
- Improve university and government laboratory appreciation of industry priorities and needs,
- Share the risks and returns of long-term research, and
- Assemble teams that can emulate the fertile research environment of the large industrial research laboratories of the past half century.

These partnerships should be fostered by

- Making resources available through special programs that encourage partnerships,
- Developing effective protocols for intellectual property issues in cooperative research,
- Encouraging university and government laboratory internships and sabbaticals in industry, and
- Requiring partners to have a stake in the partnership (e.g., for universities and government laboratories, the partnership should add value to core missions).

this environment in appropriate subfields of condensed-matter and materials physics. This will require the development of management systems and intellectual property practices appropriate for such multisector initiatives.

INTEGRATING RESEARCH AND EDUCATION

Support for fundamental research in an education-rich environment characterizes the U.S. research university and distinguishes it from universities in many countries. The U.S. research university has indeed been, over the past 50 years, the envy of the world. During the Cold War, much of the research activity in these institutions was concentrated in physical sciences and engineering. As a branch of physics with intimate links to engineering, condensed-matter and materials physics played a key role during this period in contributing to the research strength, national defense, and economic health of the country. Its strong quantum mechanical foundations, coupled with intimate links to the world of technology, have been key features. Soon after the end of World War II, many of the U.S.'s leading universities created applied physics departments and programs that emphasized this link between physics and technology.

Today, with strong currents of change in the external environment, discussed elsewhere in this report, we must reexamine the role in society of physics in general and condensed-matter and materials physics in particular. To succeed in coming decades, we must continue to pioneer new, often interdisciplinary, research directions that address societal needs. In parallel, we are challenged to do a better job of educating our students in a time of diminishing resources. To do both well, we need to be more effective in integrating the teaching and research components of academe's mission, as is increasingly recognized nationally.

The National Science Foundation (NSF) highlighted, in its 1995 document *NSF in a Changing World*, the importance of integrating education and research. Along with the development of intellectual capital, physical infrastructure, and promotion of partnerships, the integration of research and education forms one of the four core strategies of NSF. A decade ago, NSF pioneered the creation of Science and Technology Centers (STCs) and Engineering Research Centers (ERCs), which focus both on interdisciplinary research and on the integration of research and education. Many other excellent examples of undergraduate participation in research exist both in independent study, such as the NSF Research Experience for Undergraduates (REU) Program, and in classroom-based activities. In addition, there is a national call to better integrate research and education in graduate programs. To encourage this integration, NSF is supporting a number of summer workshops for beginning science and engineering faculty and graduate students planning faculty careers.

The experience gained from these efforts, as well as the intrinsic nature of condensed-matter and materials physics research, should allow physics, applied physics, and materials-oriented faculty to make key contributions to campus-wide efforts to integrate research and education. To make this possible, we must work proactively on many fronts. Universities and departments must be at the forefront of this effort and can greatly increase the attraction of physics in basic ways.

1. Bring the excitement of research and discovery into education at an earlier stage. The intimate relationship between technology and daily life provides us with many opportunities to show this relationship to our students: quantum mechanics in the real world, "seeing atoms" with tunneling microscopes, superconductivity, magnetism, and so on.

2. Take a more holistic approach to education, combining depth with breadth. The importance of interdisciplinary education and research are particular strengths of condensed-matter and materials physics. New linkages need to be forged with other sciences, applied sciences, and engineering. Team-teaching that both highlights the fundamentals and illustrates concepts from different fields can broaden horizons for both students and faculty.

3. Departments should consider new professional degree programs that link undergraduate physics education with, for example, engineering-oriented disciplines. Professional master's programs in engineering physics areas such as instrumentation science, materials synthesis, and biotechnology, for example, would be of particular importance to the condensed-matter and materials physics community. Such programs would enhance the value of these degrees and be particularly suited to training of the industrial physicist.

4. Joint academic appointments across departments, to break down disciplinary barriers, need to be encouraged. Campuses should experiment with the creation of "virtual departments," which would aid intellectual restructuring to better achieve their research and education missions in changing times.

5. Most Ph.D. dissertations in condensed-matter and materials physics are experimental. Many of today's graduate students are very strong in computer skills but have little hands-on experimental experience. This critical imbalance in experimental skills can be corrected by requiring undergraduates to have research experiences in faculty laboratories or summer internships in industry or at national laboratories.

6. Applied physics departments and programs can serve as a critical link to industrial liaison programs, which generally are strong in colleges of engineering. The inclusion of an appropriate subset of physics and condensed-matter and materials physics faculty and students would help to provide a critical link with our technological future.

A RESEARCH STRATEGY FOR CONDENSED-MATTER AND MATERIALS PHYSICS

Managing scientific research is a delicate matter. If one invested only in winning projects in the right fields, the impact would presumably be enormous.

In reality, science proceeds on a broad front, with many advances dependent on progress in other branches of science and technology, and breakthroughs coming at unexpected times with unanticipated benefits. One need only consider the discovery of high-temperature superconductivity to appreciate this unpredictability. Nevertheless, there are important choices to be made. The desired output of federal investments in science and technology is the creation of new knowledge and discoveries; the desired outcome is improved economic growth, national security, and quality of life. Although specific scientific breakthroughs cannot be planned, the environment in which science is performed can be optimized to encourage successful outputs and outcomes.

Discovery

Encouraging discovery is critical to the strategic success of condensed-matter and materials physics. Incremental progress is not sufficient to maintain leadership in science or technology. Although discovery cannot be predicted, it often occurs when researchers explore the boundaries between fields and when advances in instrumentation make possible new measurements. Both can be encouraged within the federal R&D system. Funding must be made available for research at the interfaces between disciplines. For example, the new field of molecular mechanics falls between structural biology and macromolecular physics. New mechanisms must be developed to encourage and evaluate interdisciplinary proposals, which are often lost in a peer-review process organized according to traditional disciplines. A multiplicity of funding sources is also essential to ensure that bold, new ideas are given an opportunity to succeed. Increased flexibility for agency program managers to take risks in funding decisions should also be encouraged. New facilities and instrumentation create new opportunities in condensed-matter and materials physics, and continued support for facilities and for broad access to them must be emphasized. Finally, the strategic context of the research should be understood, particularly in condensed-matter and materials physics, where the coupling to technology is so strong. The strategic context of a research area encompasses the related technological issues and opportunities. A broad appreciation of strategic context is important both in planning research and in recognizing significant potential research developments. This appreciation is most effectively acquired through interactions with industry through research partnerships, personnel exchanges, and consulting arrangements.

Scientific Themes

Chapters 1 through 6 of this report identify the key scientific questions that are expected to drive the subfields of condensed-matter and materials physics for the next decade. Specific areas of emphasis for future condensed-matter and materials physics research are suggested. In this section, the committee addresses a broader question: Where is the field headed? In particular, what strategic themes are expected to unite the field and catalyze scientific and technical progress (see Box 8.5). Maintaining scientific excellence, a long-term perspective, and a world-class environment for research are essential. The research environment has been the subject of much of this chapter, and recommendations have been given for investing in facilities and infrastructure, encouraging partnerships across disciplines and institutions, integrating research and education, and encouraging discovery. We turn now to strategic scientific themes for condensed-matter and materials physics.

The committee identified 10 scientific themes that span and underpin the subfield-specific scientific priorities of condensed-matter and materials physics as described in the body of this report. These themes, which are listed in Box 8.5, represent high-level strategic priorities for condensed-matter and materials physics research over the next decade.

- The quantum mechanics of large, interacting systems focuses on the emergent phenomena that result when large collections of atoms are brought together to form a material. Important examples include Bose-Einstein condensation, high-temperature superconductivity, and colossal magnetoresistance. These emergent phenomena are bringing quantum mechanics into the world of our experience.
- The realm of reduced dimensionality includes thin films, surfaces and interfaces, artificially structured materials, polymer chains, and nanoclusters. An improved understanding of thin-film growth, self-assembly, and materials properties at reduced dimensions is essential to technological advances ranging from

BOX 8.5 Strategic Scientific Themes in Condensed-Matter and Materials Physics

- The quantum mechanics of large, interacting systems
- The structure and properties of materials at reduced dimensionality
- Materials with increasing complexity in composition, structure, and function
- Nonequilibrium processes and the relationship between molecular and mesoscopic properties
- Soft condensed matter and the physics of large molecules, including biological structures
- Controlling electrons and photons in solids on the atomic scale
- Understanding magnetism and superconductivity
- Properties of materials under extreme conditions
- Materials synthesis, processing, and nanofabrication
- Moving from empiricism toward predictability in the simulation of materials properties and processes

displays to catalysis. Reduced dimensionality also provides opportunities to extend understanding of fundamental phenomena including phase transitions, magnetism, morphological development and strain, and novel quantum effects.

- Continued progress in condensed-matter and materials physics depends on the ability to understand materials at increasing levels of compositional, structural, and functional complexity. Advances in atomic-scale visualization, synchrotron and neutron sources, and computational capabilities are providing opportunities to extend fundamental understanding beyond model systems to the structure and properties of real materials. Examples include highly correlated systems, multicomponent magnets and superconductors, and polymer blends.
- Nonequilibrium processes include phenomena such as friction, fracture, microstructural evolution, and pattern formation. These phenomena occur away from mechanical and thermal equilbrium and, in many cases, are controlled by processes that develop both at the atomic and mesoscopic scales. The ability to bridge length scales and to understand complex patterns in fluids and solidification is essential to continued progress.
- There has been a spectacular increase in soft condensed-matter research over the past decade. This field emphasizes the softness and fluidity of materials—the physics of large molecules. Its importance derives from the pervasive use of polymers, complex fluids, and macromolecular systems in medicine, industry, and consumer products. Research in soft condensed matter has strong connections to biology, especially through fundamental understanding derived from synchrotron and neutron sources and from investigations of molecular mechanics and energy flow.
- The fundamental understanding of electrons and photons in solids underpins the Information Age. Driven by the need for faster, cheaper, more compact information-processing and communication technologies, the limits of electronic and photonic phenomena are relentlessly challenged. The future of these technologies depends on the ability to control electrons and photons in solids at the near-atomic level.
- Magnetism and superconductivity are interrelated phenomena of enormous fundamental and technological importance. Although much progress has been made, the basic understanding of magnetism is incomplete, and there is no agreement as to why high-temperature superconductivity occurs. Research is needed in many areas including low-dimensional magnetism, nanoscale magnetism, and high-temperature superconducting phenomena.
- Research on the structure and properties of materials under extreme conditions of pressure, temperature, and magnetic field continues to provide one of the most powerful means to test theories and explore novel phenomena in condensed-matter and materials physics.
- Materials synthesis is an area of extreme importance to condensed-matter and materials physics. In many areas of condensed-matter and materials physics research, the availability of research samples of sufficient quality and size is the

limiting step to continued progress. The United States has lagged in the development of materials-synthesis and processing capabilities despite strong recommendations from the National Research Council report, *Materials Science and Engineering for the 1990s: Maintaining Competitiveness in the Age of Materials.*[1] Access to facilities for nanofabrication and crystal growth is needed, as well as increased emphasis on processing research.

- The increasing power of computers foreshadows a shift from empriricism toward increased predictability in materials development. Although in its infancy, this shift presents significant challenges to and opportunities for materials theory and computional physics. The prospects for accelerating progress in condensed-matter and materials physics through simulation of complex systems are truly revolutionary.

Excellence with Relevance

Condensed-matter and materials physics is science at the technological frontier. The fundamental understanding of materials and materials phenomena is central to continuing advances in almost all areas of modern technology. Enormous societal benefits have been derived from condensed-matter and materials physics research. The continued impact of condensed-matter and materials physics depends on maintaining leadership across the broad spectrum of condensed-matter and materials physics research activities. It requires strategic investments in research, facilities and infrastructure, and human capital. It requires a research system that encourages discovery and partnerships. It requires an integration of contributions from a diversity of research approaches, institutions, and disciplines.

The recommendations of this report are intended to encourage continued excellence with relevance in condensed-matter and materials physics. Guidance is provided on strategic priorities for scientific themes and the research environment. Urgent facilities and infrastructure needs are addressed. The importance of partnerships involving universities, government laboratories, and industry and spanning disciplines and agencies is emphasized. These partnerships are essential to leverage resources, enable cross-disciplinary research, and provide a strategic context for condensed-matter and materials physics research. The integration of research and education is also discussed along with recommendations for improving condensed-matter and materials physics education.

Condensed-matter and materials physics lies at the heart of revolutionary advances in broad areas of science and technology. The next decade promises exciting new discoveries and powerful technology impacts as new capabilities in synchrotron and neutron research, atomic-scale visualization, nanofabrication,

[1] *Materials Science and Engineering for the 1990s: Maintaining Competitiveness in the Age of Materials,* National Academy Press, Washington, D.C. (1989).

computing, and many other areas probe the secrets of materials and materials-related phenomena. This is a new era, as vast new arenas ranging from subtle quantum phenomena, to macromolecular science, to the realm of complex materials become increasingly accessible to fundamental study. It is a time of exceptional opportunity to perform pioneering research at the technological frontier—a frontier enabled by advances in condensed-matter and materials physics.